Foundations of Engineering Mechanics

Jacek J. Skrzypek • Artur Ganczarski

Modeling of Material Damage and Failure of Structures

Springer-Verlag Berlin Heidelberg GmbH

Jacek J. Skrzypek • Artur Ganczarski

Modeling of Material Damage and Failure of Structures

Theory and Applications

With a Foreword by
Professor Holm Altenbach

With 157 Figures

Springer

Series Editors:
Prof. Dr V. I. Babitsky, DSc
Loughborough University
Department of Mechanical Engineering
SE11 3TU Loughborough, Leicestershire
UK

Prof. Dr J. Wittenburg
Universität Karlsruhe (TH)
Institut für Mechanik
Kaiserstraße 12
D-76128 Karlsruhe/Germany

Authors:
Professor Dr. Jacek J. Skrzypek,
Dr. Artur Ganczarski
Politechnika Krakowska
Cracow University of Technology
Institute of Mechanics and Machine Design
Division of Deformable Solids
Jana Pawła II 37
31-864 Cracow
Poland

ISBN 978-3-642-08353-2

Cataloging-in-Publication Data applied for

Die Deutsche Bibliothek – CIP Einheitsaufnahme

Skrzypek, Jacek:
Modeling of material damage and failure of structures: theory and applications / Jacek Skrzypek.

(Foundations of engineering mechanics)
ISBN 978-3-642-08353-2 ISBN 978-3-540-69637-7 (eBook)
DOI 10.1007/978-3-540-69637-7

Cover design: de´blik, Berlin
Typesetting: Camera ready by authors

SPIN: 10650938 62/3020 - 5 4 3 2 1 0 - Printed on acid-free paper

Foreword

Since the pioneering papers of L.M. Kachanov (1958) and Yu. N. Rabotnov (1959) the Continuum Damage Mechanics (CDM) is a successfully developing branch of solid mechanics, which interlinks the experiences in continuum mechanics, fracture mechanics, materials science, physics of solids, etc. Surprisingly rapid development in this field during next decades has shown how important the problem is and how difficult is the proper irreversible thermodynamics based modeling of the material damage response. A lot of scientific papers, monographs and textbooks were published by leader researchers working in the field: J.L.Chaboche, D. Krajcinovic, J. Lemaitre, S.Murakami, C.L.Chow, G.Z.Voyiadjis, to mention only some names. Several international and national scientific societies have stimulated and sponsored different conferences, symposia and workshops in this field. In addition, there were organized numerous research programmes on damage mechanics and related problems. But the conclusion that the Continuum Damage Mechanics reaches a saturated level is not justified and there is a necessity till now for writing new books in this field in order to summarise the established results and develop the new directions such as damage anisotropy, local or non-local approach of damage and fracture, unilateral damage response, probabilistic approach of damage, etc.

Nevertheless, there are several reasons for publishing the present work. The very first is that the authors (J.J. Skrzypek and A. Ganczarski) have been taking part in the scientific discussion concluding the own knowledge influenced by the best traditions in Continuum Damage Mechanics. So the reader gets a good and understandable introduction to some sub-branches of CDM (Part I). This introduction corresponds with some extensions missed in other monographs and textbooks. Examples of such extensions are the thermo-damage coupling, the orthotropic damage accumulation in case of variable principal directions of the stress and damage tensors, local approach of fracture. In general, the authors discuss mostly used CDM-theories for different material behavior models such as creep damage, elastic brittle damage, etc. Chapter 2 deals with coupled isotropic damage and creep-plastic behavior. The introduced models correspond to irreversible thermodynamics and the reader can find many models proposed by other authors. Anisotropic three-dimensional theories of damage evolution are presented in Chapter 3 and 4. Till now, there is no unique approach to formulate a theory: different damage variables (scalar, vectorial, tensorial) and equivalent effective variables are proposed and discussed. In the Chapter 5 the non-classical coupled thermo-damage and damage-fracture approaches are developed.

The second reason for publishing this new book on CDM-problems is the excellent structuring of the contents. The discussions on the use of the CDM are connected not only with the theoretical foundation and their experimental proof, but also with the introducing of CDM models in structural mechanics analysis. While the understanding of the specific problems of damage and failure analysis in connection with the analysis of complex structures is cumbersome the authors include many examples of simple structural elements such as axisymmetric structures, etc. (Part II of the book). They present a great number of original results. It is very easy for the readers to obtain the influence of different effects (heat transfer in damaged solids, damage accumulation under shear condition, etc.) on the mechanical behavior of the structural elements. The analysis and design of damaged structures, with help of the CDM, demands the use of numerical methods. For this purpose a short introduction to finite element or finite difference based methods is given. Axisymmetric problems are discussed in Chapter 6 whereas the Chapters 7–10 deal with creep-damage accumulation in presence of shear deformations. The corresponding models are applied to plate problems (membrane state, bending state, Reissner's theory, von Kármán's theory etc.).

It should be underlined that the authors have included a Part III which directs the attention of the readers to the problem of optimal design in connection with CDM problems. This part of the book extends many years of research tradition of the authors' department influenced by the former head of the department M. Życzkowski. The general structural optimization problem under damage conditions is presented in Chapter 11. Effective optimization procedures under creep damage conditions are discussed in Chapter 12 and 13. The corresponding examples (axisymmetric disks and plates) allow, e.g., the optimization of the lifetime under some constraints (constant volume, stability conditions, geometric constraints, etc.).

Writing the foreword, general characteristics of the book should be given. However, it is very difficult to say whether the present book is a monograph or a textbook because it contains the elements of both. But I think that it can be recommended for a first reading and for detailed studying the CDM to any scientist, engineer or graduate student dealing with modern problems of structural safety and integrity.

Professor Holm Altenbach
Lehrstuhl für Technische Mechanik
Martin-Luther-Universität Halle-Wittenberg
Merseburg, September 11, 1998

ABSTRACT Continuum Damage Mechanics is a quickly developing branch of Solid Mechanics. The book provides, in a systematic and concise way, a broad spectrum of one-dimensional and three-dimensional constitutive and evolution models of isotropic and anisotropic damage theories, as well as damage coupled constitutive equations of elastic or inelastic time-dependent solids in the presence of damage. The effective numerical procedure and computer applications, mainly based on FDM and FEM computer codes, are developed and adopted to simple structural members under thermo-mechanical loadings.
Part I of the book provides in a systematic fashion a survey of coupled damage-constitutive theories of engineering materials. Influence of isotropic damage evolution on the constitutive equations of creep-plastic solids is discussed in Chapter 2, both from the phenomenological point of view (single or two state variable mechanism-based coupled damage-creep-plasticity models) and unified irreversible thermodynamic formulation (the Lemaitre and Chaboche kinetic law of damage evolution, the Mou and Han unified model of ductile isotropic damage, the Saanouni, Forster and Ben-Hatira model of coupled isotropic damage-thermo-elastic-creep-plastic solids). Anisotropic three-dimensional theriores of damage evolution are presented in Chapter 3 and Chapter 4. Damage variables, scalar, vectorial and tensorial, are reviewed and stress, strain and energy based equivalence principles are compared (Lemaitre and Chaboche, Simo and Ju, Taher et al., Cordebois and Sidoroff, Chow and Lu) in order to introduce a concept of the effective variables as well as the fourth-rank damage effect tensors in terms of the second rank damage tensors. Particular attention is paid to the orthotropic creep-damage accumulation models of crystalline materials in case of non-proportional loadings (Skrzypek and Ganczarski) as well as the elastic-brittle orthotropic damage model (Litewka and Hult, Kuna-Ciskał). An objective damage rate measure is introduced for the case when effect of shear deformation results in the damage and stress tensors which are not coaxial in their principal axes. Unified irreversible thermodynamics-based theory of anisotropic damage-elastic-brittle rock-like materials is also presented and discussed when applied to High Strength Concrete (Murakami and Kamiya, Chaboche). Matrix representation of fourth-rank coupled damage-elasticity tensors is reviewed (Chow and Lu, Chaboche, Litewka and Hult, Murakami and Kamiya). Nonclassical coupled thermo-damage and damage-fracture CDM-based approaches are presented in Chapter 5. Coupled constitutive thermo-creep-damage equations are developed when partly (scalar) or fully (tensorial) coupled creep-damage models are applied combined with the damage affected heat flux eqaution where heat conductivity in the damaged material follows the anisotropic damage evolution to yield a nonstationary temperature field (Ganczarski and Skrzypek). Local approach to fracture by the use of CDM together with FEM is presented as well. A discussion of convergence and possible regularization methods proposed in cases of crack growth in the presence of the continuum creep-isotropic damage (Murakami and Liu) and the elastic-brittle-anisotropic damage (Skrzypek and Kuna-Ciskał) are also enclosed.
In Part II of the book a computer analysis of damage accumulation and failure or fracture mechanisms of simple engineering structures is illus-

trated. Effective FDM and FEM based numerical procedures and computer methods are adopted and developed when classical and nonclassical creep-damage, thermo-creep-damage, elastic-brittle damage, damage-fracture problems result from either stationary or nonstationary, proportinal or nonproportional mechanical and thermal fields. Axisymmetric creep-damage problems are discussed in Chapter 6, axisymmetric coupled thermo-creep damage applications are presented in Chapter 7, whereas creep-damage accumulation in presence of shear deformation and corresponding failure mechanisms are illustrated in Chapter 8. CDM-based creep-damage theory of axisymmetric Love–Kirchhoff's plates (coupled Kármán equations extended to elastic-creep-damage range) is developed for both uniform and substitutive sandwich cross-sections of plates of variable thickness and FDM approach is developed for computer analysis. Discussion of boundary excitations of the membrane or bending type, applied as initial prestressing, is enclosed in Chapter 9. Examples of two-dimensional coupled creep-brittle anisotropic damage applications are shown in Chapter 10. In the case of Reissner's plate precritical 2D damage field growth is analysed numerically by FDM when the coupled orthotropic creep-damage model is used. In the case of 2D plate structure loaded by in-plane forces both the precritical (continuum damage accumulation) and postcritical (crack propagation in presence of damage) analysis are done numerically by ABAQUS FEM code when extended Litewka's model of elastic-brittle damaged orthotropic material is implemented.

Examples of optimal design of structures made of time-dependent materials in presence of damage fields are discussed in Part III. General formulation of structural optimization problems under damage conditions (local or global optimality criteria, constraints, control variables) is presented in Chapter 11. Effective optimization procedures and their applications to optimal design of axisymmetric disks and plates in creep-damage conditions are discussed in Chapter 12 and Chapter 13. Structures of uniform creep strength are, in general, not optimal with respect to the lifetime prediction t_I or t_{II} for crack initiation or complete structure failure, when both the thickness and the initial prestressing of the membrane or the bending type are considered as the design variables. Important improvement of structure when two-step optimization procedure is used (first, uniform creep strength and, second, maximum lifetime under constant volume or minimum volume under constant lifetime) may be recommended for practical applications.

Table of Contents

Part I

Modeling of continuum damage in structural materials

1

Continuum damage mechanics: basic concepts

1.1 Material damage

1.1.1 Concept of a quasicontinuum approximation of the damaged material

Failure of most structural members on the macroscale follows the irreversible heterogeneous microprocesses of time and environment dependent deterioration of materials. The existence of distributed microscopic voids, cavities, or cracks of the size of crystal grains is referred as material damage, whereas the process of void nucleation, growth, and coalescence, which initiates the macrocracks and causes progressive material degradation through strength and stiffness reduction, is called damage evolution (cf. Murakami, 1987; Chaboche, 1988).

With respect to their scale, the damage models may be referred to the atomic scale (molecular dynamics), the microscale (micromechanics), and the macroscale (continuum mechanics) (cf. Woo and Li, 1993; Krajcinovic, 1995).

On the atomic scale, material structure is not continuous at all, but is represented by a configuration of atoms in the order of a crystal lattice or molecular chains bonded by the interatomic forces. The state of material damage on this level is determined by the configuration of atomic bonds, the breaking and re-establishing of which constitute the damage evolution. On the microscale, material structure is piecewise discontinuous and heterogeneous. The state of damage in a volume of material can be determined by the number of microcracks or microvoids and their size and configuration.

On the macroscale, a concept of "quasicontinuum" is introduced where the discontinuous and heterogeneous solid, suffering from damage evolution, is approximated by the ideal pseudo-undamaged continuum by the use of the couples of effective state variables, e.g., $(\tilde{\boldsymbol{\epsilon}}, \tilde{\boldsymbol{\sigma}})$, $\left(\tilde{r}, \tilde{R}\right)$, $\left(\tilde{\boldsymbol{\alpha}}, \tilde{\mathbf{X}}\right)$, in the state and dissipation potential instead of the classical state variables, e.g., $(\boldsymbol{\varepsilon}, \boldsymbol{\sigma})$, (r, R), $(\boldsymbol{\alpha}, \mathbf{X})$, representing strain and stress tensors, state variables of isotropic hardening and tensorial variables of kinematic hardening, for the idealized (pseudo-undamaged) and the true (damaged) solid, respectively. The definition of the effective state variables can be based on the so-called equivalence principles, among which strain equivalence (Lemaitre, 1971; Lemaitre and Chaboche, 1978), stress equivalence (Simo and Ju, 1987), elastic energy equivalence (Cordebois and Sidoroff, 1979; Sidoroff, 1981), and total energy equivalence (Chow and Lu, 1992), are most known (cf. Sect. 4.3). In other words, the effective state variables, associated with the pseudo-undamaged state, are defined in such a way that the strains, the stresses, the elastic energy, or the total energy in the true (damaged) and the (undamaged) states are the same.

In all cases of various equivalence hypotheses, in an idealized quasicontinuum, the true distribution of the interatomic bonds, dislocations and vacancies (atomic scale), or individual microvoids and microcracks (microscale), is smeared out and homogenized by a selection of the properly defined internal variables that characterize the damage state and are called the damage variables. Among them, the scalar variables ω, D, or ψ (damage or continuity parameters; Kachanov, 1958), the vector variables ω_α, D_α, or Ψ_α (Davison and Stevens, 1973), the second-rank tensor variables $\boldsymbol{\Omega}$, $\mathbf{D}$ or $\boldsymbol{\Psi}$ (Rabotnov, 1969; Vakulenko and Kachanov, 1971; Murakami and Ohno, 1981), or the fourth-rank tensor variables, $\widetilde{\mathbf{D}}$ (Chaboche, 1982; Krajcinovic, 1989), are frequently used (cf. Sect. 4.2). This approach is known as continuum damage mechanics (CDM), as initiated by Hult (1979) and developed by Chaboche (1981), Krajcinovic (1984), Murakami (1987), and others.

Finally, the effective stiffness or compliance of a damaged solid may also be defined in terms of the actual damage state represented by the properly selected damage variables (scalars, vectors, tensors). This fully coupled approach, where the damage evolution affects both the viscoelastic strain (Leckie and Hayhurst, 1974, and others) and also the elastic properties of the material (Chaboche, 1977, 1978; Cordebois and Sidoroff, 1979; Lemaitre, 1984; Litewka, 1985; Simo and Ju, 1987; Murakami and Kamiya, 1997, etc.), eventually leads, in a general sense, to the concept of fourth-rank elasticity tensors for damaged material (stiffness $\widetilde{\boldsymbol{\Lambda}}$, or compliance $\widetilde{\boldsymbol{\Lambda}}^{-1}$ tensors), where the damage induced material anisotropy is characterized by the properly defined fourth-rank damage effect tensor $\mathbf{M}(\mathbf{D})$ (Voyiadjis and Kattan, 1992; Chaboche, 1993; Chen and Chow, 1995; Voyiadjis and Park, 1996, 1998, to mention only some representatives of this approach).

1.1.2 Concept of representative volume element

The proper transition between the microscale and the effective material properties on the macroscale requires an adequate definition of the representative volume element (RVE), which maps a finite volume of linear size λ in true discontinuous and heterogeneous solids suffering from damage at a material point of the equivalent idealized quasicontinuous, pseudo-undamaged solids (cf. Murakami, 1987; Nemat-Nasser and Hori, 1993; Krajcinovic, 1995). The RVE of linear size λ must be large enough to include a sufficient number of microvoids and microcracks, but, at the same time, it must be small enough for the stress and strain state to be considered as homogeneous, or with a small inhomogeneity allowed. In other words, "a volume is a RVE if the average effective stiffnesses determined from two sets of tests during which the volume is subjected: a) to the uniform displacement, and b) to the uniform tractions over its external surfaces, are equal" (Krajcinovic, 1995). Hence, the material is on the scale $\lambda \geq \lambda_{\mathrm{RVE}}$ statistically homogeneous (if λ_{RVE} exists), and the finite element size must be at least as large as the RVE. A minimum size of the RVE depends on the microstructural nonhomogeneity of the material considered and, loosely speaking, the following characteristic magnitudes might be suggested (cf. Lemaitre, 1992):

i.	metals and ceramics	$(0.1\ \mathrm{mm})^3$
ii.	polymers and composites	$(1\ \mathrm{mm})^3$
iii.	wood	$(10\ \mathrm{mm})^3$
iv.	concrete	$(100\ \mathrm{mm})^3$

The fundamental assumption for the CDM method is that the influence of spatial correlation between defects on the effective properties of the continua is of second-order magnitude and, hence, the exact microvoid configuration within the RVE can be disregarded. The effect of all other voids within the RVE is measured only through the change of effective properties (effective stiffness). This local continuum theory (LCT) is objective as long as the damaged macrostructure can be divided into a number of subsystems, each of the size of the RVE, to allow for homogenization. If, on the other hand, the direct interaction between the microvoids and microcracks is essential with regard to their growth, coalescence and stability, the so-called non-local theory must involve the distance between the neighbouring defects as the scale parameter, and the local approach is no longer sufficient (Woo and Li, 1993).

1.2 Definitions of material damage of crystalline materials on the atomic scale

On the atomic scale, physically-based (atomic) material damage definitions, considering the inter-atomic energy and the actual configuration of the atomic bonds, are used. Two scalar measures of damage are discussed by Woo and Li (1993); the first of them is based on the reduction of the interatomic energy, and the second on the number of broken bonds:

$$D_1 = \frac{\sum_{i=1}^{n} \left[\mathbf{b}^i \left(\Delta G_0^s\right) - \mathbf{b}_0^i \left(\Delta G^s\right)\right] \mathbf{N}}{\sum_{i=1}^{n} \mathbf{b}^i \left(\Delta G_0^s\right) \mathbf{N}} = 1 - \frac{\sum_{i=1}^{n} b^i \left(\Delta G^s\right) \mathbf{L}_b^i \mathbf{N}}{\sum_{i=1}^{n} b^i \left(\Delta G_0^s\right) \mathbf{L}_b^i \mathbf{N}}, \quad (1.1)$$

or

$$D_2 = \frac{\sum_{i=1}^{n} s\left[b^i\left(\Delta G^s\right)\right] \mathbf{L}_b^i \mathbf{N}}{\sum_{i=1}^{n} \Delta G_0^s \mathbf{L}_b^i \mathbf{N}}. \tag{1.2}$$

In the above definitions $\mathbf{b}^i\left(\Delta G^s\right)$ denotes the i-th bond force between two atoms of the direction vector $\mathbf{L}_b^i$ of the intensity b^i in terms of the single interatomic energy ΔG^s (for broken i-th bond $b^i(0) = 0$); ΔG_0^s denotes the interatomic energy of the perfect (undamaged) materials. The influence of the orientation of bonds is included through the scalar product of the unit direction vector $\mathbf{L}_b^i$ of i-th bond and the unit normal vector to the plane considered $\mathbf{N}$. Symbol s denotes a selector factor introduced in order to distinguish the broken bonds, $s\left(b^i = 0\right) = 1$, from the active ones ($s = 0$), and n denotes the total number of bonds through the plane considered. Let the characteristic area of the single i-th interatomic bond of the direction vector $\mathbf{L}_b^i$ be ϑ^i, then the projective area of ϑ^i onto the plane of unit normal vector $\mathbf{N}$ be $A^i = \vartheta^i \mathbf{L}_b^i \mathbf{N}$ (Fig. 1.1).

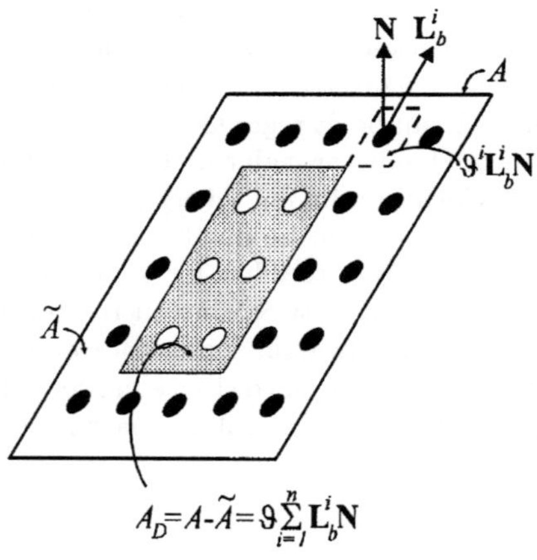

Fig. 1.1. The characteristic area of interatomic bonds of the direction vectors $\mathbf{L}_b^i$ through the surface area A of the normal vector $\mathbf{N}$: • active bands, ○ broken bonds (cf. Woo and Li, 1993)

If the average characteristic area ϑ is used for the regular n atoms configuration, the total cross-sectional area associated with n atomic bonds crossing the plane is $A = \vartheta \sum_{i=1}^{n} \mathbf{L}_b^i \mathbf{N}$. Applying the above for n atomic bonds crossing through the area A, the definition (1.2) may be rewritten in the following fashion:

$$D_2 = \frac{\vartheta \sum_{i=1}^{n} s\left[b^i\left(\Delta G^s\right)\right] \mathbf{L}_b^i \mathbf{N}}{A}. \tag{1.3}$$

Note that D_2 varies from 0 (all bonds are active) to 1 (all bonds are broken). For the partial damage the effective area $\widetilde{A}$ is defined as

$$\widetilde{A} = A - \vartheta \sum_{i=1}^{n} s \left[b^i \left(\Delta G^s \right) \right] \mathbf{L}_b^i \mathbf{N} \tag{1.4}$$

which, finally, reduce the above definition to the scalar damage parameter $\omega = 1 - \psi$, proposed earlier by Kachanov (1958), on the macroscale:

$$D_2 = \omega = 1 - \frac{\widetilde{A}}{A}, \qquad \psi = 1 - \omega, \qquad \omega \in [0, \omega_{\text{crit}}]. \tag{1.5}$$

The critical damage state ω_{crit} in Kachanov's sense is $D_{\text{crit}} = \omega_{\text{crit}} = 1$, but in general it is an additional material constant that characterizes the fracture resistance of the specific solids (macrocrack initiation), and for the majority of metals the following holds: $0.2 < D_{\text{crit}} < 0.8$ (cf. Chaboche, 1988).

1.3 Definitions of material damage on the microscale

1.3.1 One-dimensional surface damage parameter

To characterize a gradual deterioration process of a microstructure, via microcrack and microvoid nucleation and evolution through the surface area δA of intersection of the plane of normal $\mathbf{n}$ with the RVE surrounding a material point M, L. Kachanov (1958) introduced the continuing parameter ψ, the magnitude of which is determined as the ratio of the effective (remaining) area $\delta\widetilde{A} = \delta A - \delta A_D$ to the total (undamaged) area δA

$$\psi = \frac{\delta\widetilde{A}}{\delta A}, \qquad \psi \in [0, 1], \tag{1.6}$$

such that $\psi = 1$ corresponds to the undamaged (virgin) state, whereas the continuity decreases with damage growth to eventually reach zero for a completely damaged surface element $\delta A_D = \delta A$ (Fig. 1.2).

Considering planes of various normals $\mathbf{n}_x$ we define surface damage in an arbitrary direction x, or in the most damaged direction x_1, as:

$$D(M, \mathbf{n}, x) = \frac{\delta A_{Dx}}{\delta A}, \qquad D(M, \mathbf{n}) = \frac{\delta A_D}{\delta A}, \tag{1.7}$$

where $D = 1 - \psi = 0$ corresponds to the undamaged state of the surface element considered, and $D = 1$ to the completely damaged element (fully broken). The above definition is mainly applicable for crystalline materials in which, on the microscale, microscopic cracks develop both in metal

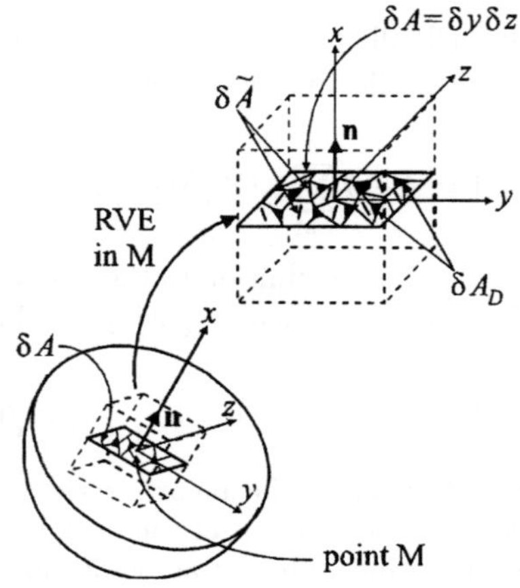

Fig. 1.2. Surface damage measure through the surface area δA of intersection of the plane of normal **n** with the RVE in a crystalline material

grains (transgranular damage) and on intergranular boundaries (intergranular damage). These microcracks have different orientations, such that the surface damage parameter also changes with the normal vector orientation when the more developed vectorial D_α, or tensorial D_{ij}, damage measures are introduced (Chap. 4).

1.3.2 Void volume fraction or porosity in ductile materials

Ductile fracture in polycrystalline metals and porous materials is, in general, the result of the following processes: growth of existing voids and cavities (if any), nucleation of new voids and their growth, and void coalescence with increase of large (visco)plastic deformation. Current void volume fraction in a RVE is defined as the ratio of the void volume to the volume of the RVE (cf. Gurson, 1977, Tvergaard, 1981, 1988, Nemes et al., 1990)

$$f = \frac{\delta V - \delta V_s}{\delta V}, \tag{1.8}$$

where δV_s denotes the volume of the solid constituents of that material element (Fig. 1.3a).

In fact, the solid with zero-void volume fraction $f = 0$ is an idealization of the voided polycrystalline material that even in the virgin state contains some voids and cavities, such that the initial void volume fraction f_0 is of the magnitude 10^{-3} to 10^{-4} (cf. Nemes et al., 1990). After nucleation the spatial distribution of voids is, roughly speaking, close to uniform, and the voids' shape may be approximated by spheres (Fig. 1.3b). The local failure

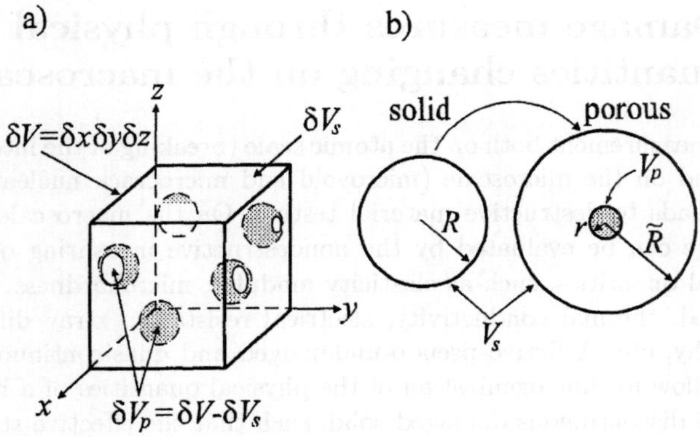

Fig. 1.3. a) Void volume fraction in the RVE in space x, y, z and b) spherical void nucleation in a spherical RVE

in the volume is a result of cavity elongation along the major tensile axis, such that two (or more) neighboring cavities coalesce when their length has grown to the magnitude of their spacing. Eventually, a slip band mechanism between the voids yields the local failure at the critical magnitude of void volume fraction of the order 0.1 to 0.2 (cf. Tvergaard, 1981).

To relate the void volume fraction f to the surface damage parameter D, consider an idealized case when a single spherical cavity of radius r is nucleated within a spherical RVE of current radius $\widetilde{R}$, the initial volume of which was R (Fig. 1.3b). Assuming no density change of the solid constituent of the RVE, the following holds:

$$f = \frac{V - V_s}{V} = \frac{r^3}{R^3 + r^3} = \frac{r^3}{\widetilde{R}^3}, \tag{1.9}$$

where $V_s = (4/3)\pi R^3$, and $V = (4/3)\pi \widetilde{R}^3$ correspond to the volume of a solid material and the volume of a RVE, respectively. Hence, when the definition of surface damage parameter in Kachanov's sense is used, we finally obtain

$$D = \frac{r^2}{\widetilde{R}^2} = \left(\frac{r^3}{R^3 + r^3}\right)^{2/3} = f^{2/3}. \tag{1.10}$$

If, on the other hand, in a heterogeneous and discontinuous spherical RVE the homogenization method is applied, the initial mass density changes from ϱ to $\widetilde{\varrho}$ with the void evolution, to yield the surface damage parameter D in terms of the initial and the current mass density

$$D = \left(1 - \frac{\widetilde{\varrho}}{\varrho}\right)^{2/3}. \tag{1.11}$$

1.4 Damage measures through physical quantities changing on the macroscale

Damage measurement both on the atomic scale (breaking of the interatomic bonds) and on the microscale (microvoid and microcrack nucleation and growth) leads to destructive material testing. On the macroscale, material damage can be evaluated by the nondestructive measuring of change of physical quantities, such as elasticity modulus, microhardness, acoustic wave speed, thermal conductivity, electrical resistance, x-ray diffraction, tomography, etc. A fictive pseudo-undamaged and quasicontinuous body concept allow for homogenization of the physical quantities of a heterogeneous and discontinuous damaged solid, such that the effective state variables and the effective physical properties may be defined in terms of the current damage state.

1.4.1 Effective stress and strain concepts and equivalence principles

Consider a one-dimensional volume element (a bar) of cross-sectional area A with a distribution of microdefects measured by the damaged surface portion A_D, loaded by the applied uniaxial stress σ. This current physical state $(\widetilde{E}, D)$ can be mapped to the fictive, pseudo-undamaged state $(E, D = 0)$ submitted to the effective stress $\widetilde{\sigma}$ such that the response remains the same, $\widetilde{\varepsilon} = \varepsilon$, (Fig. 1.4).

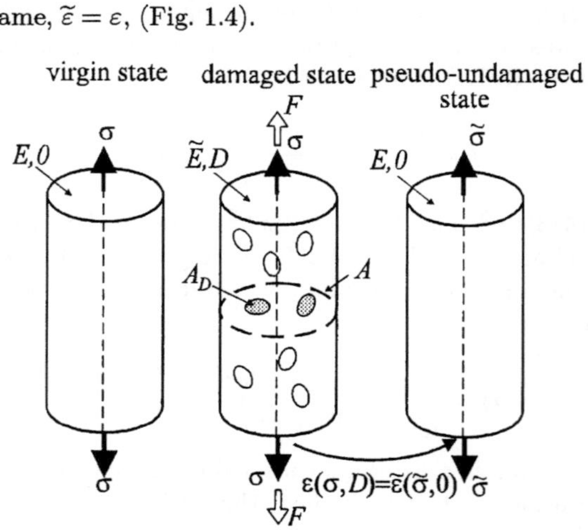

Fig. 1.4. One-dimensional effective stress concept based on strain equivalence

Hence, when the elasticity equation is furnished for both the damaged and the pseudo-undamaged material

$$\sigma = \widetilde{E}\varepsilon, \qquad \widetilde{\sigma} = E\widetilde{\varepsilon} \tag{1.12}$$

and the effective stress $\widetilde{\sigma}$ is defined by the cross-sectional area reduction in the damaged state

$$\widetilde{\sigma} = \frac{F}{A - A_D} = \frac{\sigma}{1 - D}, \qquad \widetilde{\varepsilon} = \varepsilon, \tag{1.13}$$

where $\sigma = F/A$, the following surface damage measure through the effective elasticity modulus drop with the material deterioration holds:

$$D = 1 - \frac{\widetilde{E}}{E}, \qquad \widetilde{E} = E(1 - D). \tag{1.14}$$

Note that the above definition of the uniaxial effective stress $\widetilde{\sigma}$, based on the strain equivalence principle (Rabotnov, 1968, Lemaitre, 1971), yields the linear elasticity modulus drop with damage.

Experimental validation of the formula (1.14) might be done through a series of loading/unloading tests, with the permanent strain measurement on unloading. Results for 99.9% copper at room temperature, taken after Dufailly (1980) are discussed by Lemaitre (1992) (Fig. 1.5).

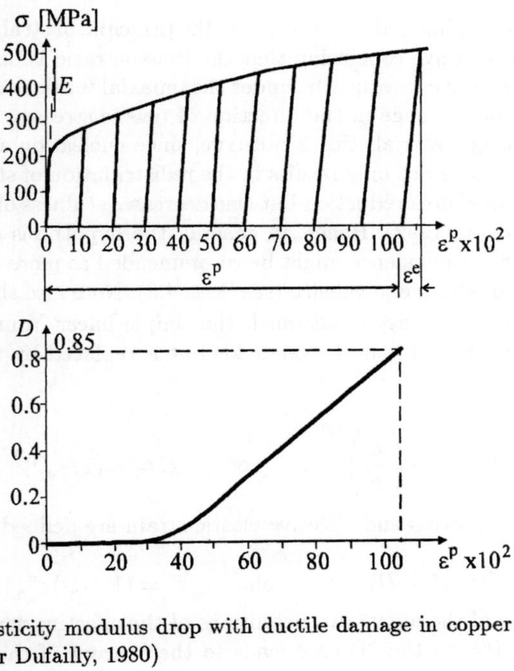

Fig. 1.5. Elasticity modulus drop with ductile damage in copper at room temperature (after Dufailly, 1980)

The linear Young's modulus drop with creep damage was also tested by Rides et al. (1989) when samples of copper were subject to constant-load creep at an elevated temperature of 300°C, and at intervals during the test they were partly unloaded and reloaded at the same rate. The results show good correlation with the formula (1.14), as shown in Fig. 1.6.

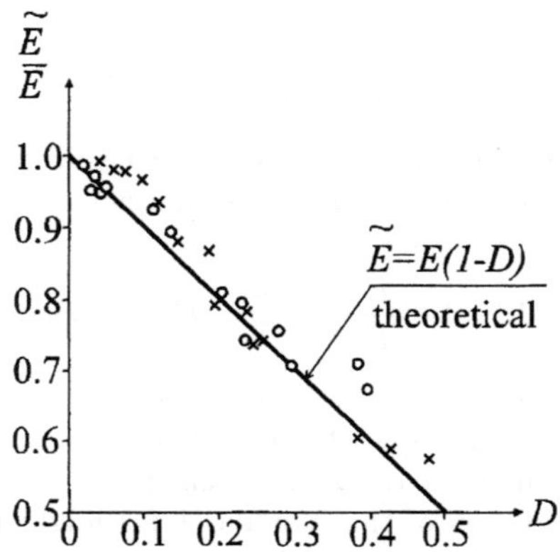

Fig. 1.6. Elasticity modulus drop with creep damage of copper at elevated temperature (after Rides et al., 1989)

It should be emphasized, however, that the principle of strain equivalence leads to the restrictive conclusion that the Poisson ratio is not affected by damage, $\widetilde{\nu} = \nu$, and consequently, under the uniaxial tension test a material suffers only from damage in the direction of tensile stresses. However, for most engineering materials this is not true, since nucleation and growth of microscopic damage not only results in the redistribution of stresses due to the cross-sectional area reduction but also decreases stiffness of the material (cf. Chow and Lu, 1992). Hence, in general, the hypothesis of the elastic (or total) energy equivalence might be recommended as more realistic than that of strain or stress equivalence (see Sect. 4.3). Note also that, when the elastic energy equivalence is assumed, the simple linear Young's modulus drop with damage (1.14) no longer holds but is replaced by the nonlinear formula

$$D = 1 - \left(\frac{\widetilde{E}}{E}\right)^{1/2}, \qquad \text{or} \qquad \widetilde{E} = E\,(1-D)^2 \tag{1.15}$$

and the effective stress and effective elastic strain are defined as follows:

$$\widetilde{\sigma} = (1-D)^{-1}\,\sigma \qquad \text{and} \qquad \widetilde{\varepsilon}^{\mathrm{e}} = (1-D)\,\varepsilon^{\mathrm{e}}. \tag{1.16}$$

Generalization of the above two definitions of the effective variables $(\widetilde{\sigma}, \widetilde{\varepsilon})$, (1.13) and (1.16), to the 3D case leads to the concept of the fourth-rank damage effect operator $\mathbf{M}(d_\alpha)$, where the damage induced anisotropy of the initially isotropic material (in a virgin state) is defined on the principal axes of damage d_α (Sect. 4.5).

In the method of strain equivalence, the effective stress tensor $\widetilde{\boldsymbol{\sigma}}$ is the stress that would have to be applied to the pseudo-undamaged material to cause the same strain tensor $\widetilde{\boldsymbol{\varepsilon}} = \boldsymbol{\varepsilon}$ as the one observed in the damaged material submitted to the current stress $\boldsymbol{\sigma}$. Hence, using Chaboche's notation (cf. Chaboche et al., 1995) the following 3D definition of the effective stress tensor holds:

$$\widetilde{\boldsymbol{\sigma}} = \mathbf{M}^{-1}(d_\alpha) : \boldsymbol{\sigma}, \tag{1.17}$$

whereas the elasticity equations furnished for both the damaged and the pseudo-undamaged material take the representation

$$\boldsymbol{\sigma} = \widetilde{\boldsymbol{\Lambda}}(d_\alpha) : \boldsymbol{\varepsilon}, \qquad \widetilde{\boldsymbol{\sigma}} = \boldsymbol{\Lambda} : \widetilde{\boldsymbol{\varepsilon}}, \tag{1.18}$$

where the fourth-rank elasticity tensor modified by damage is written as

$$\widetilde{\boldsymbol{\Lambda}}(d_\alpha) = \frac{1}{2}\left(\mathbf{M} : \boldsymbol{\Lambda} + \boldsymbol{\Lambda} : \mathbf{M}\right), \qquad \mathbf{M} = \mathbf{I} - \widehat{\mathbf{D}} \tag{1.19}$$

and $\mathbf{I}$, $\widehat{\mathbf{D}}$ are fourth-rank identity and damage tensors, respectively.

In the method of elastic energy equivalence, the effective stress $\widetilde{\boldsymbol{\sigma}}$ and effective strain $\widetilde{\boldsymbol{\varepsilon}}^{\mathrm{e}}$ are the stress and strain that would have to be applied to the pseudo-undamaged material to cause the same elastic energy as for the damaged material subjected to $\boldsymbol{\sigma}$ and $\boldsymbol{\varepsilon}^{\mathrm{e}}$. Hence, the effective stress and strain tensors are now defined as:

$$\widetilde{\boldsymbol{\sigma}} = \mathbf{M}^{-1}(d_\alpha) : \boldsymbol{\sigma} \quad \text{and} \quad \widetilde{\boldsymbol{\varepsilon}}^{\mathrm{e}} = \mathbf{M}(d_\alpha) : \boldsymbol{\varepsilon}, \tag{1.20}$$

whereas the elasticity tensor $\widetilde{\boldsymbol{\Lambda}}$ is expressed as

$$\begin{gathered} \widetilde{\boldsymbol{\Lambda}}(d_\alpha) = \mathbf{M}(d_\alpha) : \boldsymbol{\Lambda} : \mathbf{M}(d_\alpha), \\ \boldsymbol{\sigma} = \widetilde{\boldsymbol{\Lambda}} : \boldsymbol{\varepsilon}^{\mathrm{e}}, \qquad \widetilde{\boldsymbol{\sigma}} = \boldsymbol{\Lambda} : \widetilde{\boldsymbol{\varepsilon}}^{\mathrm{e}}. \end{gathered} \tag{1.21}$$

More detailed discussion of the various definitions of the damage effect operator $\mathbf{M}$ may be found in Sect. 4.4. Note that the effective stress concept should not be confused with the net-stress concept which accounts only for an area reduction (surface density of microdefects). The energy based definition leads to the Poisson' ratio varying with damage as observed for most engineering materials (cf., e.g., Murakami and Kamiya, 1997).

1.4.2 *Effect of material degradation on physical properties of damaged materials*

It is experimentally observed that damaged materials change their physical properties with damage evolution. Some of them have already been discussed: the mass density, and mechanical properties such as strength, stiffness, or compliance. The other are reported, e.g., by Lemaitre and Chaboche (1985). In what follows some of them are listed.

I. Ultrasonic wave speed drop

Longitudinal acoustic wave speed through a linear elastic medium in undamaged (virgin) and partly damaged state may be written as:

$$v^2 = \frac{E}{\varrho}\frac{1-\nu}{(1+\nu)(1-2\nu)} \qquad \text{and} \qquad \widetilde{v}^2 = \frac{\widetilde{E}}{\widetilde{\varrho}}\frac{1-\widetilde{\nu}}{(1+\widetilde{\nu})(1-2\widetilde{\nu})}. \tag{1.22}$$

Hence, when the damage definition through the drop in Young's modulus (1.14) is used, we obtain

$$D = 1 - \frac{\widetilde{\varrho}(1+\widetilde{\nu})(1-2\widetilde{\nu})(1-\nu)}{\varrho(1+\nu)(1-2\nu)(1-\widetilde{\nu})}\left(\frac{\widetilde{v}}{v}\right)^2 \tag{1.23}$$

or, when Poisson's ratio ν and the mass density ϱ change with damage may be neglected, the simplified formula holds:

$$D \approx 1 - \left(\frac{\widetilde{v}}{v}\right)^2. \tag{1.24}$$

II. Microhardness change

Assume an experimentally proved linear relationship between hardness H and the actual yield stress. When this is written for both the undamaged and partly damaged state we obtain (1D case)

$$H = k'(\sigma_y + R + X) \qquad \text{and} \qquad \widetilde{H} = k'(\sigma_y + R + X)(1-D), \tag{1.25}$$

where R and X are responsible for the isotropic and the kinematic hardening, respectively, whereas the damage affected yield stress (plasticity threshold) drops linearly with the damage increase (cf., e.g., Lemaitre, 1992). Eventually, by measuring both H and $\widetilde{H}$, the actual damage state is obtained:

$$D = 1 - \frac{\widetilde{H}}{H}. \tag{1.26}$$

III. Electric potential drop

Consider Ohm's law for the electric current through the surface A in the undamaged and the damaged state of a material

$$V = r\frac{l}{A}i \qquad \text{and} \qquad \widetilde{V} = \widetilde{r}\frac{l}{A}\widetilde{i}. \tag{1.27}$$

Introduce also the effective intensity of the electric current $\widetilde{i}$ and the effective electric resistivity $\widetilde{r}$ affected by damage:

$$\widetilde{i} = \frac{i}{1-D} \qquad \text{and} \qquad \widetilde{r} = r\left(1 + K\frac{\varrho - \widetilde{\varrho}}{\varrho}\right) = r\left(1 + KD^{3/2}\right). \tag{1.28}$$

Hence, when the effective potential difference is measured on the volume considered, the following formulas hold:

$$D = 1 - \left(1 - KD^{3/2}\right)\left(\frac{V}{\widetilde{V}}\right) \qquad \text{or} \qquad D \approx 1 - \frac{V}{\widetilde{V}}. \tag{1.29}$$

IV. Heat conductivity drop

A one-dimensional concept of the effective heat conductivity $\widetilde{\lambda}$ through the linear damaged RVE, $A_0 dx$, is based on the cross-sectional area reduction during the material degradation process to the current value $A_0(1-D)$. When Fourier's conductivity law is written for the undamaged portion of the partly damaged RVE cross-sectional area, and through the fictive pseudoundamaged equivalent homogeneous body with the defects smeared through the volume, we obtain

$$\frac{\partial}{\partial x}\left(\widetilde{\lambda}\frac{\partial T}{\partial x}\right) = \widetilde{c}_v\widetilde{\varrho}\frac{\partial T}{\partial t} \qquad \text{and} \qquad \frac{\partial}{\partial x}\left(\lambda\frac{\partial T}{\partial x}\right) = c_v\varrho\frac{\partial T}{\partial t}, \tag{1.30}$$

where the effective conductivity $\widetilde{\lambda}$ is related to the initial conductivity coefficient in a virgin body λ_0 by the simple relationship

$$\widetilde{\lambda} = \lambda_0(1-D). \tag{1.31}$$

For simplicity, the inner heat sources have been omitted, and the radiation through the damaged volume has been disregarded. In other words, this means that there is no heat flux through the fully damaged RVE. More advanced modeling, where the additional radiation term is taken into account, may be found in Sect. 5.1.

1.5 Continuum damage mechanics versus fracture mechanics

On a macroscale a structural failure mechanism may be determined by the growth of one or more macrocracks, the geometry and location of which (size, shape) is explicitly represented on the fracturing process. A crack propagation through the solid with a homogeneous microstructure under tensile stress field consists in an unstable growth of its length. In the simplest case, when linear fracture mechanics is used, the crack is assumed to be surrounded by a linear homogeneous and isotropic elastic solid, and a corresponding failure mode is perfectly brittle. On the other hand, crack propagation through the solid with a heterogeneous microstructure may be arrested, and continuum damage accumulation prior to the macrofracture may occur. Consequently, in the fracturing process in strain-controlled conditions the strong interaction between cracks is essential, and the non-local approach must be used when advanced elastic-visco-plastic material models are applicable for the solid surrounding the crack-tip (Krajcinovic, 1993). The relation between the fracture mechanics (FM) and the continuum damage mechanics (CDM) methods is a question of different characteristic sizes of microcracks and macrocracks. However, the classical characteristic parameters used by FM, such as the stress intensity factor K or J-integral, are based on the classical continuum model and, hence, both FM and CDM approaches are usually based on the local theory. It also means that a question of scale refers not to the size of crack considered but to the medium surrounding the crack (cf. Woo and Li, 1993). Nevertheless, the so-called local approach to fracture based on CDM and FEM is also used as a practical tool for coupled creep damage-fracture analysis (cf. Murakami et al., 1988, Liu et al., 1995, Murakami and Liu, 1995) or elastic-brittle damage-fracture analysis (Skrzypek et al., 1998). Let us mention, however, that additional regularization methods are often required to avoid mesh dependence of the solutions obtained in this way when the problem of stress and damage concentration at the crack tip is met (cf. Sect. 5.2).

For practical application, the scheme of the CDM and FM treatment shown in Fig. 1.7 (proposed by Chaboche, 1988), may by useful.

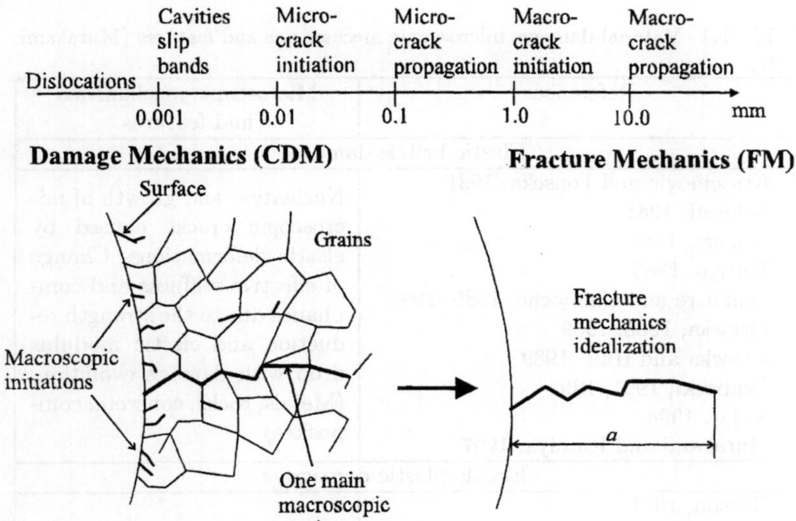

Fig. 1.7. Microcrack growth and single macrocrack initiation and propagation in a crystalline material (after Chaboche, 1988)

1.6 Classification and bibliography of material damage on the microscale

Following classification of material damage with respect to the microscopic damage characteristics (microscale) and constitutive properties of the damaged material, mainly based on the Murakami's scheme (cf. Murakami, 1987) may also be helpful in proper application of CDM modelling to damage evolution and failure analysis in structures

Table 1.1. Material damage, microscopic mechanisms and features (Murakami, 1987)

References	Microscopic mechanisms and features
Elastic-brittle damage	
Krajcinovic and Fonseka, 1981 Sidoroff, 1981 Mazars, 1985 Marigo, 1985 Lemaitre and Chaboche, 1985, 1993 Litewka, 1985, 1989 Litewka and Hult, 1989 Grabacki, 1991, 1994 Najar, 1994 Murakami and Kamiya, 1997	Nucleation and growth of microscopic cracks caused by elastic deformations. Change of effective stiffness and compliance due to the strength reduction and elastic modulus drop with damage evolution. (Metals, rocks, concrete, composites)
Elastic-plastic damage	
Gurson, 1977 Suquet, 1982 Cordebois and Sidoroff, 1982 Tvergaard, 1981, 1988 Rousselier, 1981, 1985, 1986 Lemaitre, 1984, 1985 Dragon and Chihab, 1985 Chow and Li, 1992 Voyiadjis and Kattan, 1992 Murzewski, 1992 Mou and Han, 1992 Saanouni et al., 1994 Taher, 1994	Nucleation, growth, and coalescence of microscopic voids caused by the (large) elastic-plastic deformation. Intersection of slipbands, decohesion of particles from the matrix material, cracking of particles. Void coalescence in porous media in presence of shear bands formation. (Metals, composite, polymers)
Spall damage	
Tetelman and McEvily , 1970 Gurland, 1972 Davison et al., 1977, 1978 Johnson, 1981 Grady, 1982 Perzyna, 1986 Nemes et al., 1990	Elastic and elastic-plastic damage due to impulsive loads. Propagation of shock plastic waves. Coupling between nucleation and growth of voids and stress waves. Coalescence of microcrack prior to the fragmentation process. Full separation resulting from macrocrack propagation through heavily damaged material.

Fatigue damage	
Manson, 1954 Coffin, 1954 Lemaitre, 1971 Chaboche, 1974 Manson, 1979 Lemaitre, 1992 Dufailly and Lemaitre, 1995 Skoczeń, 1996	Nucleation and growth of microscopic transgranular cracks in the vicinity of surface. High cycle fatigue (number of cycles to failure larger than 10^5): effect of macroscopic plastic strain is negligible. Very low cycle fatigue (number of cycles below 10): crack initiation in the vicinity of surface in the slip bands in grains prior to the rapid transgranular mode in the slip planes.
Creep damage	
Kachanov, 1958 Rabotnov, 1969 Leckie and Hayhurst, 1973, 1974 Hayhurst et al., 1975 Trąpczyński et al., 1981 Krajcinovic et al., 1981, 1982 Chaboche, 1979, 1981, 1988 Murakami, 1983 Hayhurst et al., 1984, 1986 Stigh, 1985 Ping Zhang and Hao Lee, 1993 Kowalewski et al., 1991a,b, 1994a,b, 1996a,b,c Needleman et al., 1995 Naumenko, 1996 H. Altenbach et al., 1990, 1997 J. Altenbach et al., 1997	Nucleation and growth of microscopic voids and cracks in metal grains (ductile transgranular creep damage at low temperatures), or on intergranular boundaries (brittle intergranular damage at high temperatures) mainly due to grain boundaries sliding and diffusion.
Creep-fatigue damage	
Chrzanowski, 1976 Lemaitre and Chaboche, 1975, 1985 Plumtree and Lemaitre, 1979 Wang, 1992 Dunne et al., 1992ab, 1994 Lin et al., 1996, 1998	Damage induced by repeated mechanical and thermal loadings at high temperature. Coupled creep-cyclic plasticity damage. Nonlinear interaction between intergranular voids and transgranular cracks. Slip bands formation due to plasticity (low temper-

	ature) combined with microcrack development due to creep (high temperature). (Metals, alloy steels, aluminum alloy, copper).
Anisotropic damage	
Sidoroff, 1981 Ladeveze, 1990 Lis, 1992 Chaboche, 1993 Chaboche et al., 1995 Chen and Chow, 1995 Voyiadjis and Venson, 1995 Litewka and Lis, 1996 Murakami and Kamiya, 1997	Damage induced anisotropy of solids or damage anisotropic materials (composites). Unilateral damage (opening/closure effect). Anisotropic elastic-brittle damage. Nonproportional and cyclic loadings. Effective state variables and damage effect tensor. (Concrete, anisotropic ceramic composites)
Corrosion damage	
Tetelman and McEvily, 1970 Knott, 1973 Schmitt and Jalinier, 1982	Pitting corrosion, intergranular corrosion, environmental degradation. Development of microcracks under stress in corrosive environments
Irradiation damage	
Tetelman and McEvily, 1970 Gittus, 1978 Tomkins, 1981 Murakami and Mizuno, 1992	Damage caused by irradiation of neutron particles and α rays. Knock-on of atoms, nucleation of voids and bubbles, swelling. Ductile behavior of creep under irradiation and brittle behavior on post-irradiation creep.
Thermo-creep damage	
Ben Hatira et al., 1994 Saanouni et al., 1994 Ganczarski and Skrzypek, 1995, 1997 Kaviany, 1997 Skrzypek and Ganczarski, 1998b	Thermo-elastic-viscoplastic damage (fully coupled approach). Damage effect on heat flux in solids. Change of temperature gradient due to damage evolution.

2

Effect of isotropic damage evolution on (visco)plasticity

2.1 Inelastic deformation processes with damage

2.1.1 Basic concepts of coupled damage – mechanical fields

Two main approaches are used to model the effect of damage evolution on the behavior of structural materials in the frame of CDM theory. In case of a weak coupling between damage and deformation processes, the effect of material damage on the elastic properties is disregarded. In this sense a coupling is established by introducing the damage variables (scalar or tensor) into the constitutive equation of the continuum solid when the effective state variables concept is used (cf. Kachanov, 1958, 1986; Rabotnov, 1968, 1969; Leckie and Hayhurst, 1973, 1974; Hayhurst, 1972, 1983, etc.). In case of a fully (strong) coupled approach, damage evolution affects both elastic properties of the material (stiffness and compliance) and inelastic response (cf. Chaboche, 1977, 1978, 1993; Cordebois and Sidoroff, 1979, 1982; Lemaitre, 1984, 1992; Litewka, 1985, 1986; Murakami and Kamiya, 1987 to mention only some of them). In this chapter the first approach is discussed when the classical strain equivalence principle is used to define the effective stress in Lemaitre's sense (Lemaitre, 1971) and the isotropic (scalar) damage variables are selected to legislate an experimentally fitted damage evolution law.

2.1.2 Creep-plasticity damage mechanisms in metals – experimental observations and general features

Two basic material damage mechanisms, ductile damage or brittle damage, can be recognized in a crystalline materials under combined creep-plasticity conditions at various temperatures. Trąpczyński et al. (1981) examined copper and aluminum alloy thin-tube specimens at 250°C and 150°C, respectively, in nonproportional loading experiments. A steady load, a single reversal of torsion, and multiple reversal of torsion were selected to follow the mechanism of microcrack and microvoid nucleation, growth, and coalescence, to eventually yield macrocrack propagation at the grain boundaries. Recently, Dunne et al. (1992) and Lin et al. (1997) examined pure copper testpieces tested to failure under condition of creep-cyclic plasticity at

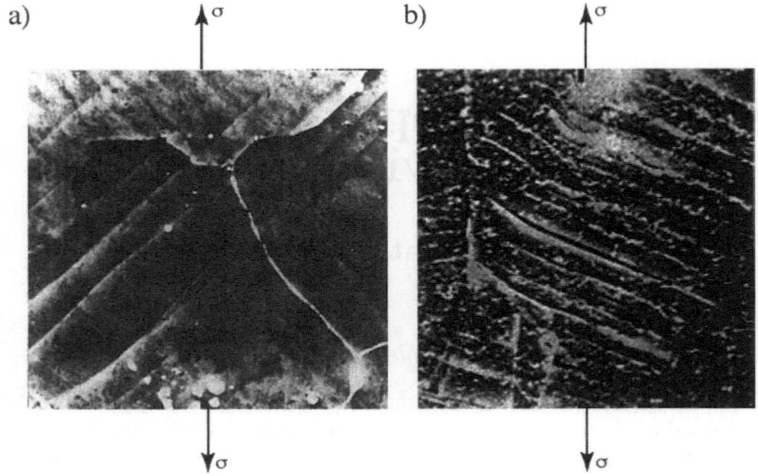

Fig. 2.1. Transgranular microcracks initiation and ductile damage growth in the sliplines regions: a) micro shear bands in Armco Iron (after Korbel et al., 1998), b) crack formed in slipbands in grains of Inconel 718 specimen tested for very low cycle fatigue (after Dufailly and Lemaitre, 1995)

room temperature, 20°C, and at 500°C.

At room temperature, crack initiation occurs in the vicinity of the surface in the slipbands of plasticity formed in the favorably oriented copper grains (of the order 0.15–0.18 mm) or subgrains (of the order 25–75 μm). They are usually oriented at 45° to the main stress direction and grow in a transgranular damage mode in the slip planes, cf. Korbel et al. (1998) (Fig. 2.1a).

A similar ductile damage mechanism, localized mainly in slipbands in the grains in Inconel 718 alloy at elevated temperature, was used by Dufailly and Lemaitre (1995) to model damage evolution in a very low cycle fatigue test (number of cycles to failure of the order of ten, or less), where the ductile damage mechanism was observed as predominant (cf. Fournier and Pineau, 1977).

At elevated temperature, the brittle intergranular microcracking process is due to microcavities which are initiated on the grain boundaries, subsequently linked to form macrocracks (Fig. 2.2). Sometimes also the saw-toothed cracks associated with a subgrain microstructure might be formed. However, strong directionality of both microcracking and macrocracking processes, both roughly perpendicular to the principal tensile stress direction, is evident.

In conclusion, the ductile or transgranular damage (or fracture) mechanism occurs at a high stress level and in the low temperature regime and

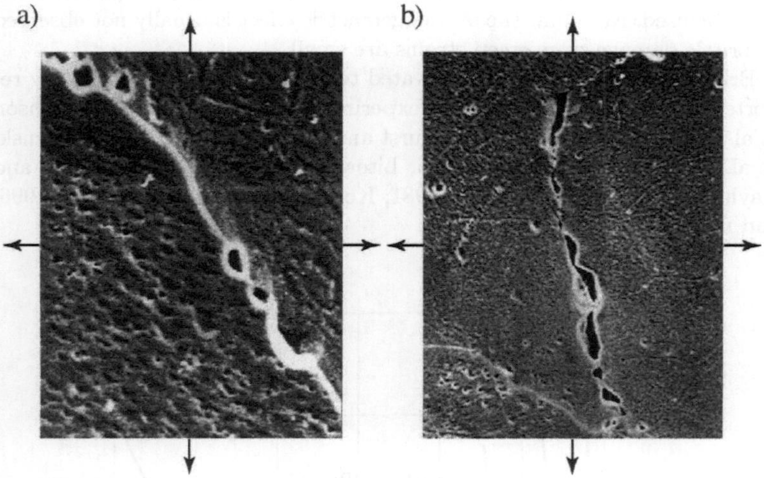

Fig. 2.2. Intergranular microcracks growth and coalescence to form cracks at grain boundary in a copper triaxial test: a) spherical grain boundary cavities, b) crack-like grain boundary cavities (after Hayhurst and Felce, 1986)

is dominated by the equivalent stress controlled damage plasticity mechanism due to the material instability from microslips at slipband regions in the grains, where material failure is initiated. This process leads to a discontinuous bifurcation of the strain-rate field that initiates the decohesion process prior to the complete failure process where the material separates with the formation of free surfaces (cf. Rudnicki and Rice, 1975; Larsson et al., 1991; Runesson et al., 1991, and others). Loss of ellipticity of the differential constitutive equations might also be considered as the initiation of material failure in this sense (cf. Shrayer and Zhou, 1995). Small voids existing in a ductile material, and their growth and coalescence, may act as an additional inhomogeneity which promotes plastic strain localization at slipbands, yielding a the failure mechanism for the material (cf., e.g., Tvergaard, 1981, 1988; Needleman et al., 1995).

In contrary, the brittle or intergranular damage (or fracture) mechanism occurs usually at a low stress level and in the high temperature regime. It is dominated by the creep micro-cavitation process on the grain boundaries which leads to the principal stress controlled micro and macrocracking process, localized mainly on the grain or subgrain boundaries. The orientation of the micro and macrocracks is selected in the damage and failure process in such a way that the normals to the average crack directions roughly coincide with the main tension direction. However, in the case of the rotating principal stress directions the microcracking process follows the main stress rotation, hence, the damage growth and accumulation process is no longer isotropic, so that the vector or tensor damage representation must be used. Note that the overall geometric effect is usually not observed in brittle damage since creep strains are small.

Brittle damage of metals at elevated temperatures has been broadly reported in the literature from the experimental point of view (cf. Johnson et al., 1956; Hayhurst, 1972; Hayhurst and Leckie, 1973, 1974; Trąpczyński et al., 1981; Murakami et al.,1985; Litewka and Hult, 1989; Othman and Hayhurst, 1990; Townley et al., 1981; Kowalewski et al., 1993, 1994, 1996, and many others).

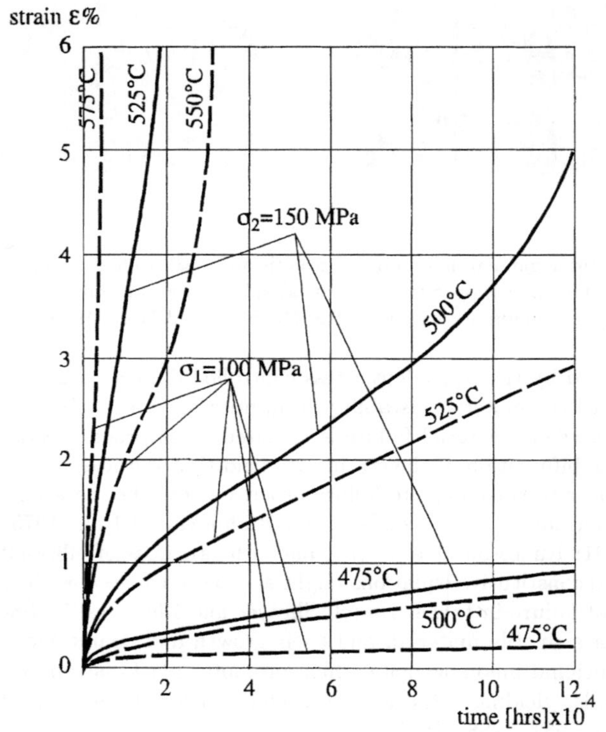

Fig. 2.3. Creep curves of 9Cr1Mo steel at constant stress levels $\sigma_1 = 100$ MPa and $\sigma_2 = 150$ MPa versus temperature (after Townley et al., 1991)

The effect of temperature on the creep curve of 9Cr1Mo steel at two constant stress levels, $\sigma_1 = 100$ MPa and $\sigma_2 = 150$ MPa, is shown in Fig.2.3 (cf. Townley et al., 1991). It is evident from the diagram that as stress and temperature increase, the time to failure decreases. So-called isostrain creep curves represent a collection of stress versus time pairs at constant stress level; the last of these curves represents rupture contour at

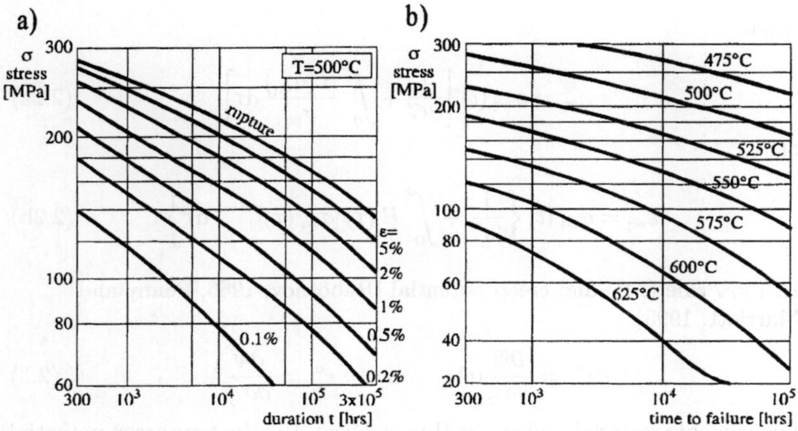

Fig. 2.4. a) Isostrain creep curves of 9Cr1Mo steel at temperature 500°C, b) rupture contours versus temperature (after Townley et al., 1991)

a given temperature that corresponds to infinite strain at rupture (Fig. 2.4, cf. Townley et al., 1991). A discussion of creep and creep-failure properties for various pressure vessel steels, according to CWST data, as well as a comparison to Odqvist's data, is given by Skrzypek (1993).

2.2 Phenomenological isotropic creep-damage models

2.2.1 Brief survey of creep constitutive equations for nondamaged materials

On the basis of the principle of strain equivalence and the effective stress concept, a simplified method to establish constitutive equations for both time-independent (plasticity) and time-dependent (creep-plasticity) materials might be proposed (cf. Lemaitre, 1971):

Any strain constitutive equation for a damaged materials may be derived in the same way as for a virgin material except that the usual stress is replaced by the effective stress.

In what follows, a brief review of creep and creep-plasticity models for nondamaged materials is presented (cf. Skrzypek, 1993).

I. Deformation or total strain (TS) theory (Rabotnov, 1948, 1966; Malinin, 1951)

$$e_{ij} = \frac{3}{2}\frac{\varepsilon_{\rm eq}\left(\sigma_{\rm eq}, t\right)}{\sigma_{\rm eq}} s_{ij}, \qquad \varepsilon_{kk} = \frac{1}{3K}\sigma_{kk}, \tag{2.1}$$

where

$$\varepsilon_{\mathrm{eq}} = \sigma_{\mathrm{eq}}(t)\left[\frac{1}{3G} + \int_0^t \frac{S(\sigma_{\mathrm{eq}})}{\sigma_{\mathrm{eq}}}\mathrm{d}\tau\right] \tag{2.2a}$$

or

$$\varepsilon_{\mathrm{eq}} = \sigma_{\mathrm{eq}}(t)\left\{\frac{1}{3G} + \int_0^t B(\tau)\left[\sigma_{\mathrm{eq}}(\tau)\right]^{m-1}\mathrm{d}\tau\right\}. \tag{2.2b}$$

II. Flow rule (FR) and creep potential (Rabotnov, 1966; Penny and Marriott, 1995)

$$\mathrm{d}\varepsilon_{ij}^{\mathrm{c}} = \frac{\partial\Psi}{\partial\sigma_{ij}}\mathrm{d}\lambda \qquad \text{or} \qquad \dot{\varepsilon}_{ij}^{\mathrm{c}} = \frac{\partial\Psi}{\partial\sigma_{ij}}\lambda. \tag{2.3}$$

For isotropic materials, when the Huber–Mises–Hencky type creep potential is applicable $\Psi(\sigma_{ij}) = (1/2)\,s_{ij}s_{ij} - (1/3)\,\sigma_0^2$, the Mises-type flow rule might be obtained:

$$\mathrm{d}\varepsilon_{ij}^{\mathrm{c}} = \frac{3}{2}\frac{\mathrm{d}\varepsilon_{\mathrm{eq}}^{\mathrm{c}}}{\sigma_{\mathrm{eq}}}s_{ij} \qquad \text{or} \qquad \dot{\varepsilon}_{ij}^{\mathrm{c}} = \frac{3}{2}\frac{\dot{\varepsilon}_{\mathrm{eq}}^{\mathrm{c}}}{\sigma_{\mathrm{eq}}}s_{ij} \tag{2.4}$$

and when the elastic strains are considered for incompressible materials the following holds:

$$\dot{e}_{ij} = \frac{1}{2G}\dot{s}_{ij} + \frac{3}{2}\frac{\dot{\varepsilon}_{\mathrm{eq}}^{\mathrm{c}}(\sigma_{\mathrm{eq}})}{\sigma_{\mathrm{eq}}}s_{ij}. \tag{2.5}$$

Specifying the equivalent stress $\sigma_{\mathrm{eq}} = \left[(3/2)\,s_{ij}s_{ij}\right]^{1/2}$ versus equivalent creep strain rates $\dot{\varepsilon}_{\mathrm{eq}}^{\mathrm{c}} = \left[(2/3)\,\dot{e}_{ij}^{\mathrm{c}}\dot{e}_{ij}^{\mathrm{c}}\right]^{1/2}$ dependence in (2.5) as $\dot{\varepsilon}_{\mathrm{eq}}^{\mathrm{c}}/\dot{\varepsilon}_c = (\sigma_{\mathrm{eq}}/\sigma_c)^n$ (cf. Odqvist and Hult, 1962) the Hooke–Norton–Odqvist flow rule is established:

$$\dot{\varepsilon}_{ij} = \frac{1+\nu}{E}\left(\dot{\sigma}_{ij} - \frac{\nu}{1+\nu}\dot{\sigma}_{kk}\delta_{ij}\right) + \frac{3}{2}\left(\frac{\sigma_{\mathrm{eq}}}{\sigma_c}\right)^{n-1}\frac{s_{ij}}{\sigma_c} \tag{2.6}$$

or when incompressibility and the power law are assumed for both elastic and creep parts $\varepsilon_{\mathrm{eq}}^{\mathrm{e}}/\varepsilon_{c0} = (\sigma_{\mathrm{eq}}/\sigma_{c0})^{n_0}$, $\dot{\varepsilon}_{\mathrm{eq}}^{\mathrm{c}}/\dot{\varepsilon}_c = (\sigma_{\mathrm{eq}}/\sigma_c)^n$ the Odqvist flow rule is furnished:

$$\frac{\mathrm{d}\varepsilon_{ij}}{\mathrm{d}t} = \frac{3}{2}\left\{\frac{\mathrm{d}}{\mathrm{d}t}\left[\left(\frac{\sigma_{\mathrm{eq}}}{\sigma_{c0}}\right)^{n_0-1}\frac{s_{ij}}{\sigma_{c0}}\right] + \left(\frac{\sigma_{\mathrm{eq}}}{\sigma_c}\right)^{n-1}\frac{s_{ij}}{\sigma_c}\right\}. \tag{2.7}$$

In the above formulas, ε_{c0}, σ_{c0}, n_0, $\dot{\varepsilon}_c$, σ_c, n are the temperature dependent material constants, whereas in (2.7) $\varepsilon_{c0} = 1$ and $\dot{\varepsilon}_c = 1$ are set.

When the time-hardening (TH) model is applied to $\dot{\varepsilon}_{\mathrm{eq}}$ and σ_{eq}, instead of $\dot{\varepsilon}$ and σ in an uniaxial case, the multiaxial time hardening creep law that accounts for both ageing and temperature dependence is established:

$$d\varepsilon_{ij}^{c} = \frac{3}{2}\frac{f_1(\sigma_{eq})}{\sigma_{eq}} s_{ij} \frac{df_2(t)}{dt} f_3(T)\, dt. \tag{2.8}$$

When the temperature effect is disregarded, two particular forms of (2.8) might be recommended:

$$\dot{\varepsilon}_{ij}^{c} = \frac{3}{2}\frac{f_1(\sigma_{eq})}{\sigma_{eq}} s_{ij} \dot{f}_2(t), \tag{2.9a}$$

$$\dot{\varepsilon}_{ij}^{c} = \frac{3}{2} C n s_{ij} \sigma_{eq}^{m-1} t^{n-1}, \tag{2.9b}$$

where in (2.9b) the functions $f_1(\sigma_{eq})$ and $\dot{f}_2(t)$ are specified in a power form that generalizes the uniaxial Nutting equation to the 3D case, $\varepsilon_{eq}^{c} = C\sigma_{eq}^{m} t^{n}$ (cf. Kraus, 1980).

III. Isotropic strain hardening (SH) theory

Multiaxial strain hardening equation that generalizes Rabotnov's 1966 concept can be presented in the form

$$\dot{p}_{ij} = \frac{3}{2}\frac{f(\sigma_{eq})}{h(q_{eq})}\frac{s_{ij}}{\sigma_{eq}}, \tag{2.10a}$$

where q_{eq} represents the length of the trajectory in the creep strain space

$$q_{eq} = \int_0^t \sqrt{\frac{2}{3}\dot{p}'_{ij}\dot{p}'_{ij}}\, dt, \tag{2.10b}$$

$\dot{\mathbf{p}} = \dot{\boldsymbol{\varepsilon}} - \dot{\boldsymbol{\varepsilon}}^{e}$, and primes stand for the deviatoric components. If the Nutting equation holds under both the uniaxial constant stress and the 3D generalization, the following equation may be furnished (cf. Kraus, 1980; Ohashi et al., 1982):

$$\dot{\varepsilon}_{ij}^{c} = \frac{3}{2} n C^{1/n} \left(\varepsilon_{eq}^{c}\right)^{(n-1)/n} \sigma_{eq}^{(m-n)/n} s_{ij}. \tag{2.11}$$

IV. Malinin–Khadjinsky creep-kinematic-hardening (CKH) theory

Malinin and Khadjinsky (1972) applied the concept of nonlinear kinematic–hardening in plasticity to the anisotropic hardening in uniaxial creep of metals at elevated temperatures (carbon steel at 455°C and aluminum alloy at 150°C) to obtain:

$$\dot{\varepsilon}^{c} = B\exp\left(\frac{|\sigma-\alpha|}{N}\right)\operatorname{sign}(\sigma-\alpha), \tag{2.12a}$$

$$\dot{\alpha} = A(|\sigma|)\,\dot{\varepsilon}^{c} - D\exp\left(\frac{|\sigma|}{N}\right)\operatorname{sign}\alpha. \tag{2.12b}$$

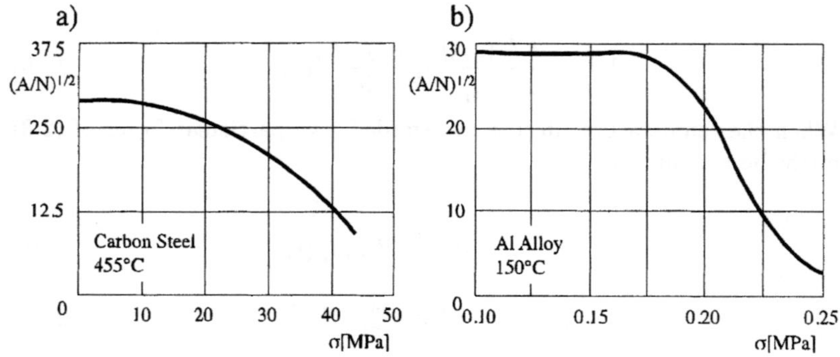

Fig. 2.5. Graphs of functions $(A/N)^{1/2}$ versus σ in the Malinin and Khadjinsky Eq. (2.12): a) 0.17% carbon steel at $T = 455°C$ (after Johnson, 1941), b) aluminum alloy at $T = 150°C$ (after Namestnikov, 1965)

The first term in (2.12b) represents the work-hardening effect, whereas the second is responsible for the thermally activated softening; $A(\sigma)$ is a function of stress, constant below the yield point and decreasing with stress above this limit; B, D, N, are the material constants (cf. Fig. 2.5, Table 2.1).

Table 2.1. Material constants for the Malinin and Khadjinisky Eq. (2.12) and the Ohashi et al. Eq. (2.13) (after Skrzypek, 1993)

Material	A[MPa]	B[h^{-1}]	D[MPah^{-1}]	N[MPa]	m[-]	n[-]
Eq. (2.12) Carbon Steel $T = 455°C$ (Johnson)	Fig. 2.5	3.53×10^{-8}	4.05×10^{-9}	4.14	–	–
Eq. (2.12) Aluminium Alloy $T = 150°C$ (Namestnikov)	Fig. 2.5	8.35×10^{-7}	3.83×10^{-5}	29.4	–	–
Eq. (2.13) Stainless Steel $T = 704°C$ (Ohashi et al.)	6.9×10^{3}	6.3×10^{-10}	0.531	29.6	3.64	3.64

The 3D generalization of Eqs. (2.12a, 2.12b) is due to Ohashi et al. (1982):

$$\dot{\boldsymbol{\varepsilon}}^{c} = B\left[\sinh\frac{J_2(\boldsymbol{\sigma}-\mathbf{X})}{N}\right]^{n}\frac{\boldsymbol{\sigma}'-\mathbf{X}'}{J_2(\boldsymbol{\sigma}-\mathbf{X})}, \tag{2.13a}$$

$$\dot{\mathbf{X}}' = A\left[J_2(\boldsymbol{\sigma})\right]\dot{\boldsymbol{\varepsilon}}^{c} - D\left[\sinh\frac{J_2(\mathbf{X})}{N}\right]^{m}\frac{\mathbf{X}'}{J_2(\mathbf{X})}. \tag{2.13b}$$

In the above absolute notation, the following definitions of the Mises-type equivalent stress $J_2(\boldsymbol{\sigma})$, the back stress (or translation tensor) $J_2(\mathbf{X})$, and the additional stress $J_2(\boldsymbol{\sigma}-\mathbf{X})$ hold:

$$\begin{gathered} J_2(\boldsymbol{\sigma}) = \left[(3/2)\,\boldsymbol{\sigma}' : \boldsymbol{\sigma}'\right]^{1/2}, J_2(\mathbf{X}) = \left[(3/2)\,\mathbf{X}' : \mathbf{X}'\right]^{1/2}, \\ J_2(\boldsymbol{\sigma}-\mathbf{X}) = \left[(3/2)\left(\boldsymbol{\sigma}' - \mathbf{X}'\right) : \left(\boldsymbol{\sigma}' - \mathbf{X}'\right)\right]^{1/2}. \end{gathered} \tag{2.13c}$$

Symbols B, D, N, n, m, are the material constants, and $A\,[.]$ is a function of the equivalent stress (cf. Table 2.1), where primes stand here for the deviatoric components.

V. Chaboche-Rousselier creep-isotropic/kinematic hardening (CIKH) theory

Chaboche (1977) applied the concept of a mixed isotropic-kinematic hardening to the creep plasticity flow rule, to obtain (cf. Chaboche and Rousselier, 1983):

$$\mathrm{d}\boldsymbol{\varepsilon}^{\mathrm{p}} = \frac{3}{2}\left\langle \frac{J_2(\boldsymbol{\sigma}-\mathbf{X}) - R - \sigma_y}{K} \right\rangle^n \frac{\boldsymbol{\sigma}' - \mathbf{X}'}{J_2(\boldsymbol{\sigma}-\mathbf{X})}\mathrm{d}t, \tag{2.14a}$$

where the following nonlinear kinematic and isotropic hardening rules hold:

$$\mathrm{d}\mathbf{X} = C\left(\frac{2}{3}a\mathrm{d}\boldsymbol{\varepsilon}^{\mathrm{p}} - \mathbf{X}\mathrm{d}p\right) - \gamma\left[J_2(\mathbf{X})\right]^{m-1}\mathbf{X}\mathrm{d}t, \tag{2.14b}$$

$$\mathrm{d}R = b\,(Q-R)\,\mathrm{d}p - \gamma R^q \mathrm{d}t. \tag{2.14c}$$

Symbol p stands for the cumulative viscoplastic strain, $\mathrm{d}p = [(2/3)\,\mathrm{d}\boldsymbol{\varepsilon}^{\mathrm{p}} : \mathrm{d}\boldsymbol{\varepsilon}^{\mathrm{p}}]^{1/2}$, and symbols n, K, σ_y, C, a, b, m, Q, γ, q are ten material coefficients, the number of which reduces to seven when two time recovery terms in (2.14b) and (2.14c) are omitted ($\gamma = 0$), or to five when, additionally, the isotropic hardening effect is disregarded ($\mathrm{d}R = 0$).

The seven-parameter CIKH theory applied to the uniaxial case reduces Eqs. (2.14) to the 1D model (cf. Chaboche and Rousselier, 1983)

$$\mathrm{d}\varepsilon^{\mathrm{p}} = \left\langle \frac{|\sigma - \alpha| - R - \sigma_y}{K} \right\rangle^n \mathrm{sign}\,(\sigma - \alpha)\,\mathrm{d}t, \tag{2.15a}$$

where

$$\begin{aligned} \mathrm{d}\alpha &= \mathrm{d}\alpha_1 + \mathrm{d}\alpha_2, \\ \mathrm{d}\alpha_1 &= C_1\left(a_1\mathrm{d}\varepsilon^{\mathrm{p}} - \alpha_1\left|\mathrm{d}\varepsilon^{\mathrm{p}}\right|\right), \\ \mathrm{d}\alpha_2 &= C_2\left(a_2\mathrm{d}\varepsilon^{\mathrm{p}} - \alpha_2\left|\mathrm{d}\varepsilon^{\mathrm{p}}\right|\right), \\ \mathrm{d}R &= b\,(Q-R)\left|\mathrm{d}\varepsilon^{\mathrm{p}}\right|. \end{aligned} \tag{2.15b}$$

Nine coefficients: n, K, σ_y, C_1, C_2, a_1, a_2, b, Q are evaluated from the tensile and relaxation tests for 316L stainless steel at room temperature:

$$\begin{array}{lllll} n = 24, & K = 151\ \text{MPa}, & \sigma_y = 82\ \text{MPa}, & C_1 = 2800, & C_2 = 25, \\ a_1 = 58\ \text{MPa}, & a_2 = 270\ \text{MPa}, & b = 8, & Q = 60\ \text{MPa}. & \end{array}$$

If, on the other hand, the kinematic-hardening effect is not taken into account, the following six-parameter (1D) NIH model might also be used for tensile tests (cf. Lemaitre and Chaboche, 1985):

$$\mathrm{d}\varepsilon^{\mathrm{p}} = \left\langle \frac{\sigma - R - \sigma_y}{K} \right\rangle^n , \tag{2.16a}$$

where

$$R\left(\varepsilon^{\mathrm{p}}\right) = Q_1\varepsilon^{\mathrm{p}} + Q_2\left[1 - \exp\left(-b\varepsilon^{\mathrm{p}}\right)\right]. \tag{2.16b}$$

Six material parameters: n, K, σ_y, Q_1, Q_2, b for 316 stainless steel at room temperature are

$$\begin{array}{lll} n = 24, & K = 151\ \text{MPa}, & \sigma_y = 84\ \text{MPa}, \\ Q_1 = 6400\ \text{MPa}, & Q_2 = 270\ \text{MPa}, & b = 25. \end{array}$$

VI. Time-independent Chaboche and Rousselier nonlinear plasticity-isotropic/kinematic hardening (PIKH) theory

For time-independent plasticity the current HMH yield surface transforms according to the following rule:

$$F^{\mathrm{Ch-R}} = J_2\left(\boldsymbol{\sigma} - \mathbf{X}\right) - R - k = 0, \tag{2.17a}$$

$$J_2\left(\boldsymbol{\sigma} - \mathbf{X}\right) = \left[(3/2)\left(\boldsymbol{\sigma}' - \mathbf{X}'\right) : \left(\boldsymbol{\sigma}' - \mathbf{X}'\right)\right]^{1/2}, \tag{2.17b}$$

where tensor $\mathbf{X}$ is a translation tensor, or a back stress tensor, that represents the current position of the yield surface (kinematic hardening effect), and scalar R, also called the drug stress, represents the size of the yield surface (isotropic hardening effect). The translation tensor $\mathbf{X}$ and the drug stress R satisfy the evolution laws:

$$\mathrm{d}\mathbf{X} = \frac{2}{3}C\left(p\right)\mathrm{d}\boldsymbol{\varepsilon}^{\mathrm{p}} - \gamma\left(p\right)\mathbf{X}\mathrm{d}p, \tag{2.18a}$$

$$\mathrm{d}R = b\left(Q - R\right)\mathrm{d}p \quad \text{or} \quad R = Q\left[1 - \exp\left(-bp\right)\right]. \tag{2.18b}$$

The nonlinear kinematic hardening rule (2.18a), in which the functions $C\left(p\right)$ and $\gamma\left(p\right)$ depend on the scalar isotropic variable called the cumulative plastic strain, $\mathrm{d}p = \left[(2/3)\,\mathrm{d}\boldsymbol{\varepsilon}^{\mathrm{p}}\!:\!\mathrm{d}\boldsymbol{\varepsilon}^{\mathrm{p}}\right]^{1/2}$, is due to Armstrong and Frederick (1966). The isotropic hardening rule (2.18b) allows for an asymptotic stabilization of the yield surface size when p tends to infinity (necessary for cyclic plasticity behavior). The normality rule associated with the current yield surface (2.17) is used to determine the plastic strain rate $\mathrm{d}\boldsymbol{\varepsilon}^{\mathrm{P}}$ with the consistency condition applied to eliminate the scalar multiplier $\mathrm{d}\lambda$:

$$\mathrm{d}\boldsymbol{\varepsilon}^{\mathrm{P}} = \mathrm{d}\lambda \frac{\partial F^{\mathrm{Ch-R}}}{\partial \boldsymbol{\sigma}} = \frac{3}{2}\mathrm{d}\lambda \frac{\boldsymbol{\sigma}' - \mathbf{X}'}{J_2(\boldsymbol{\sigma} - \mathbf{X})}, \tag{2.19a}$$

$$\mathrm{d}F^{\mathrm{Ch-R}} = \frac{\partial F}{\partial \boldsymbol{\sigma}}\mathrm{d}\boldsymbol{\sigma} + \frac{\partial F}{\partial \mathbf{X}}\mathrm{d}\mathbf{X} + \frac{\partial F}{\partial R}\mathrm{d}R = 0. \tag{2.19b}$$

Finally, the following equation is furnished

$$\mathrm{d}\boldsymbol{\varepsilon}^{\mathrm{P}} = \frac{\left\langle \frac{3}{2}\frac{(\boldsymbol{\sigma}' - \mathbf{X}'):\mathrm{d}\boldsymbol{\sigma}}{J_2(\boldsymbol{\sigma} - \mathbf{X})} \right\rangle \frac{\boldsymbol{\sigma}' - \mathbf{X}'}{J_2(\boldsymbol{\sigma} - \mathbf{X})}}{C - \frac{3}{2}\gamma \frac{\mathbf{X}:(\boldsymbol{\sigma}' - \mathbf{X}')}{J_2(\boldsymbol{\sigma} - \mathbf{X})} + b(Q - R)}, \tag{2.20}$$

where the McAuley bracket $\langle . \rangle$ is defined as: $\langle x \rangle = 0$ if $x < 0$, $\langle x \rangle = x$ if $x \geq 0$. When the indices notation is used instead of the absolute one we rewrite Eqs. (2.17)–(2.20) as follows (cf. Skrzypek, 1993):

$$\begin{aligned}
&F^{\mathrm{Ch-R}} = \sigma_{\mathrm{eq}}(\sigma_{ij} - \alpha_{ij}) - R - k = 0, \\
&\mathrm{d}\alpha_{ij} = \frac{2}{3}C(\lambda)\,\mathrm{d}\varepsilon_{ij} - \gamma(\lambda)\,\alpha_{ij}\mathrm{d}\lambda, \\
&\mathrm{d}R = b(Q - R)\,\mathrm{d}\lambda, \\
&\mathrm{d}\varepsilon^{\mathrm{P}}_{ij} = \mathrm{d}\lambda \frac{\partial F^{\mathrm{Ch-R}}}{\partial \sigma_{ij}} = \frac{3}{2}\mathrm{d}\lambda \frac{s_{ij} - \alpha'_{ij}}{\sigma_{eq}(\sigma_{ij} - \alpha_{ij})}, \\
&\mathrm{d}\lambda = \frac{1}{h}\left\langle \frac{3}{2}\frac{s_{ij} - \alpha'_{ij}}{\sigma_{eq}(s_{ij} - \alpha_{ij})}\mathrm{d}\sigma_{ij} \right\rangle, \\
&h = C - \frac{3}{2}\gamma\alpha_{ij}\frac{s_{ij} - \alpha'_{ij}}{\sigma_{eq}(\sigma_{ij} - \alpha_{ij})} + b(Q - R).
\end{aligned} \tag{2.21}$$

In the uniaxial tension/compression case Eqs. (2.21) reduce to the form:

$$\begin{aligned}
&F = |\sigma - \alpha| - R - k = 0, \\
&\mathrm{d}\alpha = C\mathrm{d}\varepsilon^{\mathrm{P}} - \gamma\alpha\,|\mathrm{d}\varepsilon^{\mathrm{P}}|, \\
&\mathrm{d}R = b(Q - R)\,|\mathrm{d}\varepsilon^{\mathrm{P}}|, \\
&\mathrm{d}\varepsilon^{\mathrm{P}} = \frac{1}{h}\left\langle \frac{3}{2}\frac{\sigma - \alpha}{|\sigma - \alpha|}\mathrm{d}\sigma \right\rangle \frac{\sigma - \alpha}{|\sigma - \alpha|} = \frac{\mathrm{d}\sigma}{h}, \\
&h = C - \gamma\alpha\,\mathrm{sign}(\sigma - \alpha) + b(Q - R).
\end{aligned} \tag{2.22}$$

2.2.2 *Single state variable creep-damage models*

In the simplest case, when the isotropic damage evolution affects only the tertiary creep phase and the primary creep phase is ignored, the creep strain-damage coupling may be established when a single scalar damage variable D (or ω) is introduced to the creep constitutive equation of a nondamaged material, and the evolution law for damage is legislated (cf. Kachanov, 1958; Rabotnov, 1969; Hayhurst, 1972, and others).

I. One-parameter uniaxial creep-damage coupling models

This approach was first proposed by Kachanov (1958) and generalized by Rabotnov (1968, 1969) (cf. Rides et al., 1989):

$$\frac{\dot{\varepsilon}^{\mathrm{c}}}{\dot{\varepsilon}_0} = \left(\frac{\sigma/\sigma_0}{1-\omega}\right)^n, \qquad \frac{\dot{\omega}}{\dot{\omega}_0} = \frac{(\sigma/\sigma_0)^\nu}{(1-\omega)^\varphi}, \tag{2.23}$$

where $\dot{\varepsilon}_0$, n and $\dot{\omega}_0$, ν, φ stand for the temperature dependent material constants in the creep law and the damage growth rule, respectively, whereas σ_0 is the reference stress. Integration of the equations (2.23) (coupled by the state variable ω) at the constant stress, $\sigma = \mathrm{const}$, and the initial condition for the damage ω and the creep strain ε^{c}, $t = 0 : \omega = \varepsilon^{\mathrm{c}} = 0$, yields:

$$\omega = 1 - \left(1 - \frac{t}{t_{\mathrm{f}}}\right)^{\frac{1}{1+\varphi}}, \qquad \frac{\varepsilon^{\mathrm{c}}}{\varepsilon_{\mathrm{f}}} = 1 - \left(1 - \frac{t}{t_{\mathrm{f}}}\right)^{\Delta}, \tag{2.24}$$

where $\Delta = (1+\varphi-n)\,/\,(1+\varphi)$, whereas symbols t_{f} and ε_{f} denote the time to failure ($\omega = 1$) and the creep strain at failure, respectively:

$$t_{\mathrm{f}} = \frac{(\sigma_0/\sigma)^\nu}{(1+\varphi)\,\dot{\omega}_0}, \qquad \varepsilon_{\mathrm{f}} = \frac{\dot{\varepsilon}_0\,(\sigma/\sigma_0)^{n-\nu}}{\dot{\omega}_0\,(1+\varphi-n)} = \frac{\dot{\varepsilon}_{\mathrm{ss}} t_{\mathrm{f}}}{\Delta}. \tag{2.25}$$

$\dot{\varepsilon}_{\mathrm{ss}} = \dot{\varepsilon}_0\,(\sigma/\sigma_0)^n$ stands here for a steady-state or a minimum creep rate (no damage effect included).

For the pure copper specimens subject to constant stress tests at temperature 300°C, when the stress $\sigma = 32.4$ MPa was used to give a failure time of the order of 15 days, Rides et al. (1989) obtained: $n = 6.56$, $\nu = 6.31$, $\varphi = 7.1$, $\sigma_0 = 300$ MPa; and $\dot{\varepsilon}_0 = 11 \times 10^{-5}\mathrm{h}^{-1}$, $\dot{\omega}_0 = 6.68 \times 10^{-4}\mathrm{h}^{-1}$(A) or $\dot{\varepsilon}_0 = 2.54 \times 10^{-5}\mathrm{h}^{-1}$, $\dot{\omega}_0 = 2.74 \times 10^{-4}\mathrm{h}^{-1}$(B); however, the model is often simplified by setting $\varphi = \nu$. Note that the second of Eqs. (2.24) describes

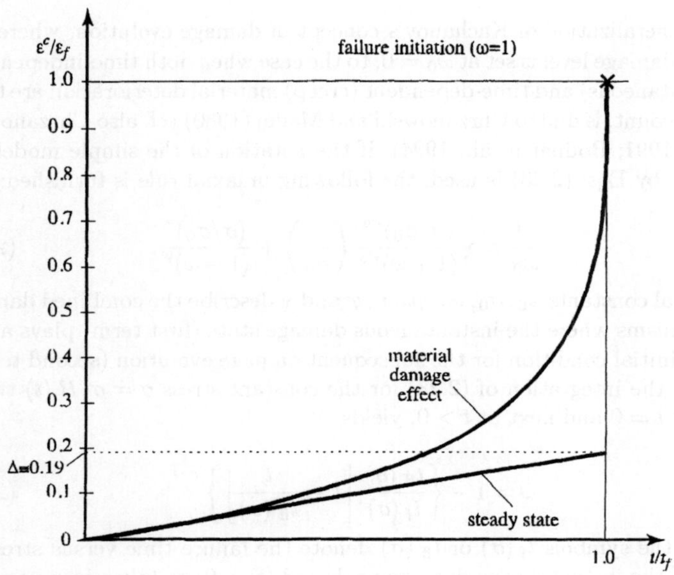

Fig. 2.6. Normalized creep strain versus normalized time for copper tested at 300°C (after Rides et al., 1989)

the growth of creep strain rate with time due to the damage evolution in the tertiary creep, the magnitude of which tends to infinity when the time to failure is reached (Fig. 2.6).

The uniaxial damage growth rule (2.23) may also be presented in an equivalent form when Kachanov's 1958, or Chaboche's 1988, notation is used

$$\frac{\mathrm{d}\psi}{\mathrm{d}t} = -C\left(\frac{\sigma_1}{\psi}\right)^r, \tag{2.26a}$$

$$\frac{\mathrm{d}D}{\mathrm{d}t} = \left(\frac{\sigma}{A}\right)^r (1-D)^{-k}, \tag{2.26b}$$

where ψ and $D = \omega$ denote the continuity and the damage, respectively, if $\psi + D = 1$ holds. The so-called life fraction rule, however, was established earlier by Robinson (1952) for steel:

$$\int_0^{t_\mathrm{R}} \frac{\mathrm{d}t}{t_\mathrm{R}(t)} = 1, \qquad t_\mathrm{R}(t) = \frac{1}{C(r+1)\left[\sigma_1(t)\right]^r}, \tag{2.27}$$

which is applicable for an arbitrarily prescribed tensile stress function $\sigma_1(t)$.

A generalization of Kachanov's concept of damage evolution, where the initial damage level is set at $\omega_0 = 0$, to the case when both time-independent (instantaneous) and time-dependent (creep) material deterioration are taken into account, is due to Chrzanowski and Madej (1980) (cf. also Chrzanowski et al., 1991; Bodnar et al., 1994). If the notation of the simple model described by Eqs. (2.23) is used, the following uniaxial rule is furnished:

$$\frac{\dot{\omega}}{\dot{\omega}_0} = \chi \frac{(\sigma/\sigma_0)^{\nu_0}}{(1+\omega)^{\varphi_0}} \left(\frac{\dot{\sigma}}{\sigma_0}\right) + \frac{(\sigma/\sigma_0)^{\nu}}{(1-\omega)^{\varphi}}. \tag{2.28}$$

Material constants $\dot{\omega}_0, \sigma_0, \nu_0, \varphi_0, \nu, \varphi$ and χ describe the combined damage mechanisms where the instantaneous damage state (first term) plays a role of the initial condition for the subsequent damage evolution (second term). Hence, the integration of (2.28) for the constant stress $\sigma = \sigma_1 H(t)$ twice, first at $t = 0$ and next at $t > 0$, yields

$$\omega = 1 - \left\{ \frac{t_{\mathrm{ff}}(\sigma)}{t_{\mathrm{f}}(\sigma)} \left[1 - \frac{t}{t_{\mathrm{ff}}(\sigma)} \right] \right\}^{\frac{1}{\varphi+1}}, \tag{2.29}$$

where the symbols $t_{\mathrm{f}}(\sigma)$ or $t_{\mathrm{ff}}(\sigma)$ denote the failure time versus stress in case of the instantaneous damage neglected ($\chi = 0$) or taken into account, respectively ($t_{\mathrm{ff}} \leq t_{\mathrm{f}}$):

$$\begin{aligned} t_{\mathrm{f}}(\sigma) &= \frac{1}{(1+\varphi)\,\dot{\omega}_0 (\sigma/\sigma_0)^{\nu}}, \\ t_{\mathrm{ff}}(\sigma) &= \left[1 - \left(\frac{\sigma}{\sigma_f} \right)^{1+\nu_0} \right]^{\frac{1+\varphi}{1+\varphi_0}} t_{\mathrm{f}}(\sigma) \end{aligned} \tag{2.30}$$

and σ_{f} stands for the instantaneous failure stress such that at $t = 0$, $\omega_0 = 1$

$$\sigma_{\mathrm{f}} = \left[\frac{1+\nu_0}{(1+\varphi_0)\,\chi\dot{\omega}_0} \right]^{\frac{1}{1+\nu_0}} \sigma_0. \tag{2.31}$$

Note that by setting $\chi = 0$ the failure stress σ_{f} tends to infinity, since (2.29) is reduced to (2.24) when $t_{\mathrm{ff}} = t_{\mathrm{f}}$. A family of damage parameters ω versus dimensionless time t/t_{f} plots is sketched in Fig.2.7.

The corresponding failure times at which $\omega = 1$ are $t_{\mathrm{ff}}/t_{\mathrm{f}} = 1$, $1/2$, $1/3$, $1/5$, $1/10$; whereas the initial damage increase with the failure time drop is $\omega_0 = 0$, 0.159, 0.240, 0.331, 0.438.

The experimental observations on metallic materials by Hayhurst et al. (1975, 1989), Othman and Hayhurst (1990), and others have shown different shapes of the normalized creep curves $\varepsilon^{\mathrm{c}}/\varepsilon_{\mathrm{f}} = f(t/t_{\mathrm{f}})$ when aluminum alloy, copper, and stainless steel specimens were tested to failure at temperatures 210°C, 250°C, and 550°C, respectively (Fig. 2.8).

In contrast to copper and aluminum alloy, where the primary creep is negligible and the tertiary creep predominates, in the case of stainless steel

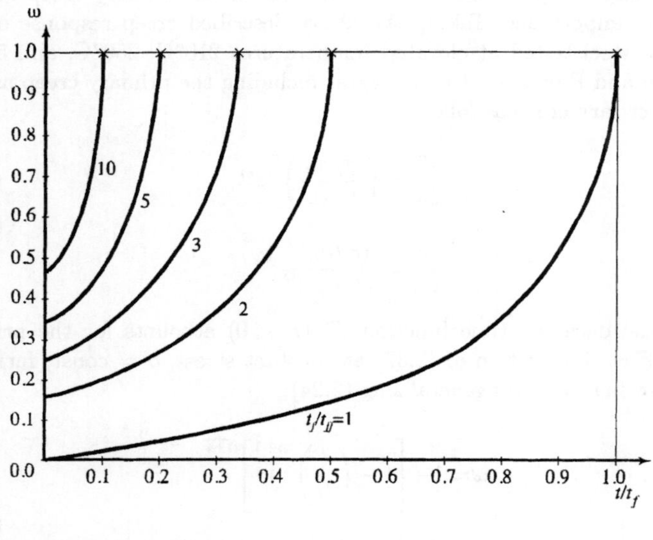

Fig. 2.7. A family of damage versus time plots (2.29) for various ratios $t_f/t_{ff} = 1, 2, 3, 5, 10$ obtained by setting $\varphi = 3$

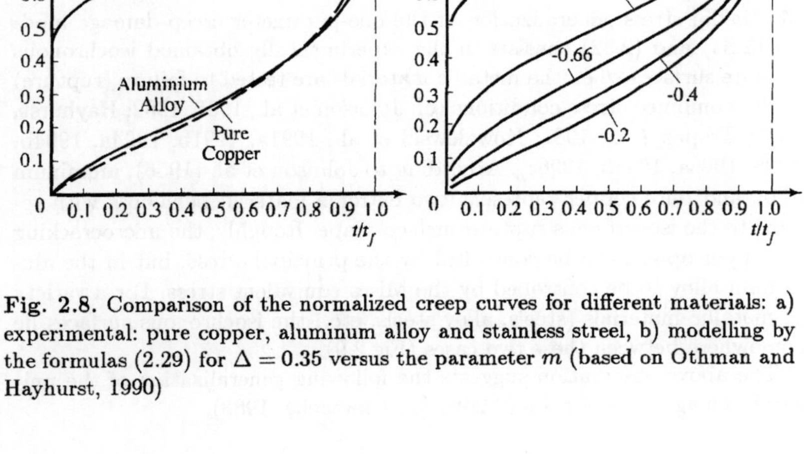

Fig. 2.8. Comparison of the normalized creep curves for different materials: a) experimental: pure copper, aluminum alloy and stainless streel, b) modelling by the formulas (2.29) for $\Delta = 0.35$ versus the parameter m (based on Othman and Hayhurst, 1990)

the primary creep manifests strongly, whereas the tertiary creep section is of less importance. Taking the above described creep response of 316 stainless steel tested at elevated temperatures 210°C, 250°C, and 550°C Othman and Hayhurst (1990) suggest including the primary creep as well as the tertiary creep as follows:

$$\frac{\dot{\varepsilon}^{\mathrm{c}}}{\dot{\varepsilon}_0} = \left(\frac{\sigma/\sigma_0}{1-\omega}\right)^n t^m, \qquad \frac{\dot{\omega}}{\dot{\omega}_0} = \frac{(\sigma/\sigma_0)^\nu}{(1-\omega)^\varphi} t^m, \tag{2.32}$$

where the decaying time function t^m $(m < 0)$ accounts for the primary creep effect. Integration of (2.32) at constant stress, $\sigma = \mathrm{const}$, furnishes the following formulas generalizing (2.24):

$$\omega = 1 - \left[1 - \left(\frac{t}{t_{\mathrm{f}}}\right)^{m+1}\right]^{\frac{1}{1+\varphi}}, \qquad \frac{\varepsilon^{\mathrm{c}}}{\varepsilon_{\mathrm{f}}} = 1 - \left[1 - \left(\frac{t}{t_{\mathrm{f}}}\right)^{m+1}\right]^{\Delta}, \tag{2.33}$$

where Δ is defined in a similar fashion as in (2.24), whereas the time to failure t_{f} and the creep strain at failure ε_{f} are:

$$t_{\mathrm{f}} = \left[\frac{(1+m)\,(\sigma_0/\sigma)^\nu}{(1+\varphi)\,\dot{\omega}_0}\right]^{\frac{1}{1+m}}, \qquad \varepsilon_{\mathrm{f}} = \frac{\dot{\varepsilon}_0\,(\sigma/\sigma_0)^{n-\nu}}{\dot{\omega}_0\,(1+\varphi-n)}. \tag{2.34}$$

II. One-parameter creep-damage models under multiaxial stress conditions

Multiaxial stress generalization of the one-parameter creep-damage models (2.23) and (2.32) consists in the experimentally obtained isochronous rupture surfaces when the metallic materials are tested to failure (rupture) under combined stress conditions (cf. Johnson et al., 1956, 1962; Hayhurst, 1972; Trąpczyński, 1981; Kowalewski et al., 1991a, 1991b, 1994a, 1994b, 1995, 1996a, 1996b, 1996c). According to Johnson et al. (1956), aluminum alloy and pure copper represent two extreme material behaviors with regard to the isochronous rupture surface shape. Roughly, the microcracking in copper appears to be controlled by the principal stress, but in the aluminum alloy to be controlled by the Mises equivalent stress. For a variety of metallic materials (steels, alloy steels, etc.) the isochronous surfaces lie somewhere between these two cases (Fig.2.9).

The above observation suggests the following generalization of the uniaxial damage growth rules (2.26b) (cf. Chaboche, 1988):

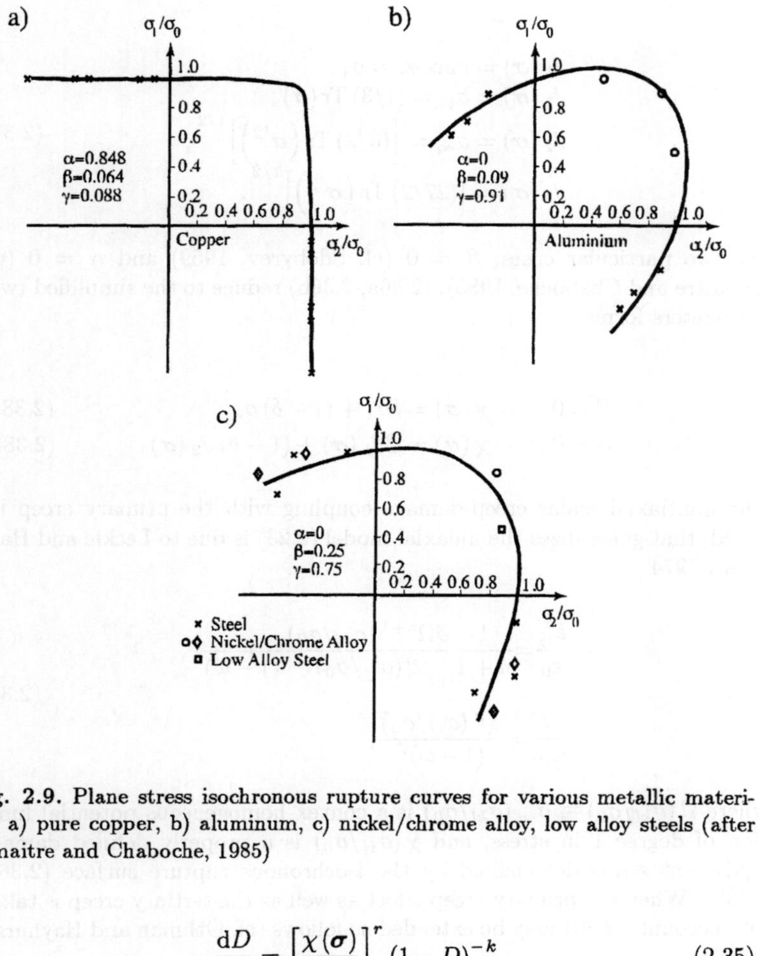

Fig. 2.9. Plane stress isochronous rupture curves for various metallic materials: a) pure copper, b) aluminum, c) nickel/chrome alloy, low alloy steels (after Lemaitre and Chaboche, 1985)

$$\frac{\mathrm{d}D}{\mathrm{d}t} = \left[\frac{\chi(\boldsymbol{\sigma})}{A}\right]^{r} (1-D)^{-k}, \qquad (2.35)$$

where the scalar function $\chi(\boldsymbol{\sigma})$, also called the damage equivalent stress, is represented as the three-parameter function of the stress invariants (Hayhurst, 1972)

$$\chi(\boldsymbol{\sigma}) = \alpha J_0(\boldsymbol{\sigma}) + 3\beta J_1(\boldsymbol{\sigma}) + (1-\alpha-\beta) J_2(\boldsymbol{\sigma}) \qquad (2.36\text{a})$$

or

$$\chi(\boldsymbol{\sigma}) = a\sigma_1 + 3b\sigma_{\mathrm{H}} + c\sigma_{\mathrm{eq}}, \qquad (2.36\text{b})$$

when Lemaitre and Chaboche's (a) or Boyle and Spence's (b) representation is used $(a+b+c=1)$ and the following definitions hold:

$$
\begin{aligned}
J_0(\boldsymbol{\sigma}) &= \max \sigma_i = \sigma_1, \\
J_1(\boldsymbol{\sigma}) &= \sigma_{\mathrm{H}} = (1/3)\,\mathrm{Tr}(\boldsymbol{\sigma}), \\
J_2(\boldsymbol{\sigma}) &= \sigma_{\mathrm{eq}} = \left[(3/2)\,\mathrm{Tr}\left(\boldsymbol{\sigma}'^2\right)\right]^{1/2}, \\
J_3(\boldsymbol{\sigma}) &= \left[(27/2)\,\mathrm{Tr}\left(\boldsymbol{\sigma}'^3\right)\right]^{1/2}.
\end{aligned}
\tag{2.37}
$$

For two particular cases, $\beta = 0$ (cf. Sdobyrev, 1959) and $\alpha = 0$ (cf. Lemaitre and Chaboche, 1985), (2.36a, 2.36b) reduce to the simplified two-parameters forms

$$
\beta = 0: \qquad \chi(\boldsymbol{\sigma}) = \delta\sigma_1 + (1-\delta)\,\sigma_{\mathrm{eq}}, \tag{2.38a}
$$

$$
\alpha = 0: \qquad \chi(\boldsymbol{\sigma}) = \beta J_1(\boldsymbol{\sigma}) + (1-\delta)\,J_2(\boldsymbol{\sigma}). \tag{2.38b}
$$

The multiaxial scalar creep-damage coupling with the primary creep ignored, that generalizes the uniaxial model (2.23) is due to Leckie and Hayhurst, 1974

$$
\begin{aligned}
\frac{\dot{\varepsilon}^{\mathrm{c}}_{ij}}{\dot{\varepsilon}_0} &= \frac{1}{n+1}\,\frac{\partial \Omega^{n+1}(\sigma_{kl}/\sigma_0)}{\partial(\sigma_{ij}/\sigma_0)}\,\frac{1}{(1-\omega)^n}, \\
\frac{\dot{\omega}}{\dot{\omega}_0} &= \frac{\chi^{\nu}(\sigma_{ij}/\sigma_0)}{(1-\omega)^{\varphi}},
\end{aligned}
\tag{2.39}
$$

where $\Omega(\sigma_{kl}/\sigma_0) \equiv \sigma_{\mathrm{eq}}(\sigma_{kl}/\sigma_0)$ is a convex homogeneous potential function of degree 1 in stress, and $\chi(\sigma_{ij}/\sigma_0)$ is a properly defined damage equivalent stress determined by the isochronous rupture surface (2.36a, 2.36b). When the primary creep effect as well as the tertiary creep is taken into account, (2.39) may be extended as follows (cf. Othman and Hayhurst, 1990):

$$
\begin{aligned}
\frac{\dot{\varepsilon}^{\mathrm{c}}_{ij}}{\dot{\varepsilon}_0} &= \frac{1}{n+1}\,\frac{\partial \Omega^{n+1}(\sigma_{kl}/\sigma_0)}{\partial(\sigma_{ij}/\sigma_0)}\,\frac{f(t)}{(1-\omega)^n}, \\
\frac{\dot{\omega}}{\dot{\omega}_0} &= \frac{\chi^{\nu}(\sigma_{ij}/\sigma_0)\,f(t)}{(1-\omega)^{\varphi}}.
\end{aligned}
\tag{2.40}
$$

A representation of a decaying time function $f(t)$, responsible for primary creep, is established to best fit the test data. In cases when the damage evolution is controlled by Mises-type equivalent stress, and the Mises-type creep potential function is used, (2.39) and (2.40) reduce in the following fashion (cf. Kowalewski et al., 1994a, b):

$$\dot{\varepsilon}^{c}_{ij} = \frac{3}{2} A \frac{\sigma_{\mathrm{eq}}^{n-1}}{(1-\omega)^n} s_{ij}, \qquad \dot{\omega} = B \frac{\sigma_{\mathrm{eq}}^{\nu}}{(1-\omega)^{\varphi}} \tag{2.41}$$

and

$$\dot{\varepsilon}^{c}_{ij} = \frac{3}{2} A \frac{\sigma_{\mathrm{eq}}^{n-1}}{(1-\omega)^n} s_{ij} t^m, \qquad \dot{\omega} = B \frac{\sigma_{\mathrm{eq}}^{\nu}}{(1-\omega)^{\varphi}} t^m, \tag{2.42}$$

where the following values of the material constants obtained for aluminum alloy tested at 150°C are: $A = 3.511 \times 10^{-29} (\mathrm{MPa})^{-n}/\mathrm{h}^{m+1}$; $B = 1.960 \times 10^{-23} (\mathrm{MPa})^{-\nu}/\mathrm{h}^{m+1}$; $n = 11.034$; $\nu = 8.220$; $\varphi = 12.107$; $m = -0.3099$; $E = 71.1 \times 10^3$ MPa.

A generalization of (2.28) to the multiaxial stress conditions can also be made as follows (cf. Bodnar et al., 1994):

$$\frac{\mathrm{d}\omega}{\mathrm{d}t} = B_0 \left(\frac{\sigma_{\mathrm{eq}_1}}{1-\omega} \right)^{\nu_0} \frac{\mathrm{d}\sigma_{\mathrm{eq}_1}}{\mathrm{d}t} + B \left(\frac{\sigma_{\mathrm{eq}_2}}{1-\omega} \right)^{\nu}, \tag{2.43}$$

where the different damage equivalent stresses σ_{eq_1} and σ_{eq_2} can be regarded as responsible for various time-independent and time-dependent damage mechanisms. When the two-parameter formula (2.38a) is used for copper, one may insert, e.g., $\delta = 0$ and $\delta = 1$, respectively, since the instantaneous damage mechanism is usually controlled by effective stress (slipbands), whereas the time-dependent microcracking may roughly be considered ascontrolled by maximum principal stress.

2.2.3 Two state variables mechanisms-based damage models

I. Two-parameter multiaxial hyperbolic sinus models for nickel and aluminum-based alloys

Othman et al. (1993) developed the mechanisms-based two state variables model in order to describe nickel-based superalloys where two physical mechanisms that operate together are included: dislocation softening (ageing) ω_1 ($0 \leq \omega_1 \leq 1$) and creep constrained cavity nucleation and growth on the grain boundaries ω_2 ($0 \leq \omega_2 \leq 0.3$). A sinh function of stress, rather than the traditionally used power law (cf. Sect. 2.2.1), is best able to represent the strain rate:

$$
\begin{aligned}
\frac{\mathrm{d}\varepsilon_{ij}^{\mathrm{c}}}{\mathrm{d}t} &= \frac{3}{2}\frac{A}{(1-\omega_2)^n}\frac{\sinh\{B\sigma_{\mathrm{eq}}[1-H(t)]\}}{(1-\omega_1)}\left\{\frac{s_{ij}}{\sigma_{\mathrm{eq}}}\right\},\\
\frac{\mathrm{d}H}{\mathrm{d}t} &= \frac{h}{\sigma_{\mathrm{eq}}}\frac{A}{(1-\omega_2)^n}\frac{\sinh\{B\sigma_{\mathrm{eq}}[1-H(t)]\}}{(1-\omega_1)}\left\{1-\frac{H(t)}{H^*}\right\},\\
\frac{\mathrm{d}\omega_1}{\mathrm{d}t} &= CA\frac{(1-\omega_1)}{(1-\omega_2)^n}\sinh\{B\sigma_{\mathrm{eq}}[1-H(t)]\},\\
\frac{\mathrm{d}\omega_2}{\mathrm{d}t} &= DA\left(\frac{\sigma_1}{\sigma_{\mathrm{eq}}}\right)^{\nu}N\frac{\sinh\{B\sigma_{\mathrm{eq}}[1-H(t)]\}}{(1-\omega_1)(1-\omega_2)^n},
\end{aligned}
\tag{2.44}
$$

where A, B, C, D, H^*, h and ν are material parameters and $n = B\sigma_{\mathrm{eq}}[1-H(t)]\coth\{B\sigma_{\mathrm{eq}}[1-H(t)]\}$. The primary creep effect is included in (2.44) through the additional variable $H(t)$ that changes from 0 to H^* (saturation) at the beginning and the end of primary phase, respectively; secondary creep is characterized by constants A and B, whereas the damage evolution in tertiary creep depends on constants C and D. Parameter N characterizes the state of loading, $N = 1$ for $\sigma_1 > 0$ and $N = 0$ for $\sigma_1 < 0$. Kowalewski et al. (1994a, b) checked the suitability of this model for predicting the tertiary creep response of aluminum alloy at 150°C to obtain: $A = 2.96 \times 10^{-11}\mathrm{h}^{-1}$; $B = 7.17 \times 10^{-2}(\mathrm{MPa})^{-1}$; $C = 35$; $D = 6.63$; $h = 1.37 \times 10^5$ MPa; $\nu = 0$.

II. Two state variables model versus stress state index

Dyson (1993) proposed a similar two state variables model based on a new mechanism of creep in particle-hardened alloy. The multiaxial generalization follows from the associated flow rule and the energy dissipation rate potential (cf. Kowalewski et al., 1994a) of nondamaged material as follows

$$
\frac{\mathrm{d}\varepsilon_{ij}^{\mathrm{c}}}{\mathrm{d}t} = \frac{\partial\Psi}{\partial\sigma_{ij}} = \frac{3}{2}A\frac{s_{ij}}{\sigma_{\mathrm{eq}}}\sinh(B\sigma_{\mathrm{eq}}); \qquad \Psi = \frac{A}{B}\cosh(B\sigma_{\mathrm{eq}}). \tag{2.45}
$$

Hence, when the two state variables ω_1 and ω_2 are introduced to model the tertiary creep softening due to dislocation mobility ageing ω_1 and grain boundary cavitation ω_2, whereas the additional state variable H stands for the primary creep effect, we obtain:

$$\begin{aligned} \frac{\mathrm{d}\varepsilon^{\mathrm{c}}_{ij}}{\mathrm{d}t} &= \frac{3}{2}\frac{A}{(1-\omega_2)^n}\left(\frac{s_{ij}}{\sigma_{\mathrm{eq}}}\right)\sinh\left\{\frac{B\sigma_{\mathrm{eq}}(1-H)}{(1-\omega_1)}\right\}, \\ \frac{\mathrm{d}H}{\mathrm{d}t} &= \frac{h}{\sigma_{\mathrm{eq}}}\frac{A}{(1-\omega_2)^n}\sinh\left\{\frac{B\sigma_{\mathrm{eq}}(1-H)}{(1-\omega_1)}\right\}\left\{1-\frac{H}{H^*}\right\}, \\ \frac{\mathrm{d}\omega_1}{\mathrm{d}t} &= \frac{K_{\mathrm{c}}}{3}(1-\omega_1)^4, \\ \frac{\mathrm{d}\omega_2}{\mathrm{d}t} &= \frac{DA}{(1-\omega_2)^n}\left(\frac{\sigma_1}{\sigma_{\mathrm{eq}}}\right)^{\nu} N\sinh\left\{\frac{B\sigma_{\mathrm{eq}}(1-H)}{(1-\omega_1)}\right\} \end{aligned} \tag{2.46a}$$

and

$$n = \frac{B\sigma_{\mathrm{eq}}(1-H)}{(1-\omega_1)}\coth\left\{\frac{B\sigma_{\mathrm{eq}}(1-H)}{(1-\omega_1)}\right\}. \tag{2.46b}$$

For aluminum alloy, six material constants that identify (2.46a, 2.46b) are obtained (cf. Kowalewski et al., 1994a): $A = 2.960\times10^{-11}\mathrm{h}^{-1}$; $B = 7.167\times 10^{-2}(\mathrm{MPa})^{-1}$; $h = 1.370\times10^5$ MPa; $H^* = 0.2032$; $K_{\mathrm{c}} = 19.310\times10^{-5}\mathrm{h}^{-1}$; $D = 6.630$.

For multiaxial rupture the constant ν, also called the stress state index, characterizes different types of stress state sensitive rupture behavior of the material considered. For example, in the case of aluminum alloy, where the damage evolution is nearly equivalent stress controlled (see Fig. 2.9), we may set $\nu = 0$. However, for other metals the magnitude of ν should be found experimentally, Fig. 2.10.

A comparison of the two-parameter model (2.46a, 2.46b) with the one-parameter model (2.41) made by Altenbach et al. (1997) by setting $\nu = 0$ shows a satisfactory coincidence on the primary creep only and, as a consequence, the one-parameter model yields an unacceptable underestimation of the failure time in plates. On the other hand, examination of the effect of the stress state index ν on the lifetime prediction and the failure mechanisms in clamped square plates under uniform pressure, done by the authors, proved the significance of the proper estimation of ν, as well as its influence on the lifetime and the failure mode of the plate.

2.2.4 Creep-cyclic plasticity damage interaction model for copper

Dunne and Hayhurst)1992a, b, 1994) developed a model based on two physical mechanisms of damage in copper validated for creep, cyclic plasticity, and creep-plasticity interaction under cyclic mechanical and cyclic thermal loading at high (500°C) and room (20°C) temperatures. The internal variable **X** models the kinematic hardening in cyclic creep-plasticity,

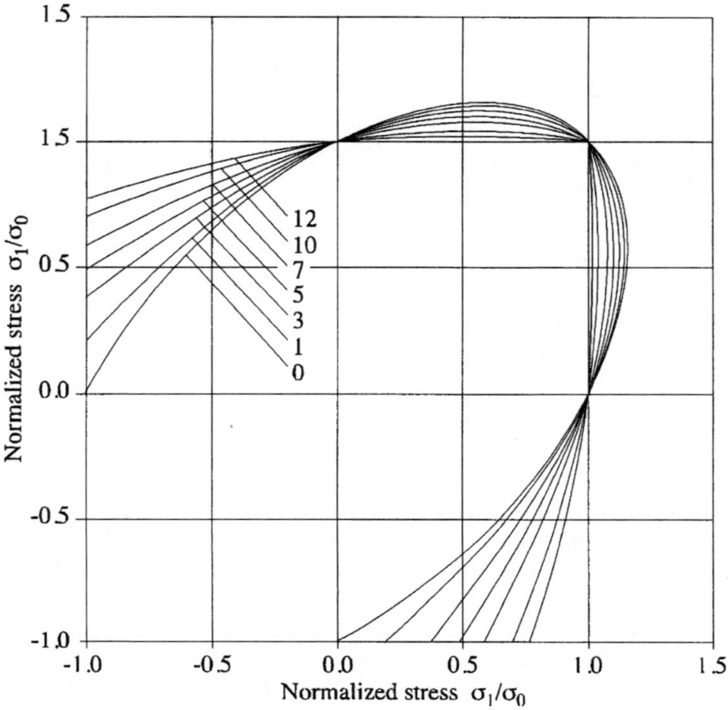

Fig. 2.10. Isochronous rupture loci for biaxial stress state versus the stress sensitivity index ν (after Kowalewski et al., 1994a)

and D is a scalar variable that accounts for a combined grain boundary time-dependent cavitation damage predominantly due in copper to principal stress controlled high temperature creep damage and the transgranular slip bands formation predominantly due to the cyclic plasticity mechanism controlled by low temperature effective stress (cf. Sect. 2.1.2 and Figs.2.1 and 2.2). The creep-cyclic plasticity damage interaction is given by

$$\begin{aligned} D^{\mathrm{c}} &= \omega_1 + \alpha_1 z(\omega_1)\,\omega_2, \\ D^{\mathrm{p}} &= \omega_2 + \alpha_2 z(\omega_1)\,\omega_1, \\ D &= D^{\mathrm{c}} + D^{\mathrm{p}}, \\ z(\omega_1) &= \frac{1}{2} + \frac{1}{\pi}\arctan \mu\,(\omega_1 - \omega_0), \end{aligned} \tag{2.46}$$

where α_1, α_2, μ, ω_0 are experimentally determined for copper (cf. Dunne and Hayhurst, 1992a), whereas ω_1 and ω_2 are the creep damage and the cyclic plasticity damage per cycle, respectively, which are controlled by the independent damage evolutions

$$\begin{aligned}\frac{\mathrm{d}\omega_1}{\mathrm{d}t} &= A\left[\delta\sigma_1 + (1-\delta)\,\sigma_{\mathrm{eq}}\right]^{\nu} / \left(1 - D^{\mathrm{c}}\right)^{\varphi}, \\ \frac{\mathrm{d}\omega_2}{\mathrm{d}N} &= \left[1 - \left(1 - D^{\mathrm{p}}\right)^{\beta+1}\right]^{\varrho}\left[\frac{A_{II}}{M\left(1 - D^{\mathrm{p}}\right)}\right]^{\beta},\end{aligned} \tag{2.47}$$

where A_{II} is the maximum effective stress range in a cycle, ϱ is defined in terms of A_{II}, M is a function of the mean stress, and β is a constant. Eventually, the creep cyclic plasticity kinematic hardening model with damage evolution included (CPKHD) is furnished (extension of the Chaboche-Rousselier theory, Eqs. 2.14a, 2.14b, 2.14c)

$$\begin{aligned}\dot{\boldsymbol{\varepsilon}}^{\mathrm{p}} &= \frac{3}{2}\left\langle\frac{J_2\left(\boldsymbol{\sigma} - \mathbf{X}\right)/\left(1 - D\right) - \sigma_y}{K}\right\rangle^{n}\frac{\boldsymbol{\sigma}' - \mathbf{X}'}{J_2\left(\boldsymbol{\sigma} - \mathbf{X}\right)}, \\ \dot{\mathbf{X}}_1 &= \tfrac{2}{3}C_1\dot{\boldsymbol{\varepsilon}}^{\mathrm{p}}\left(1 - D\right) - \gamma_1\mathbf{X}_1\dot{p} + \left(C_1'/C_1\right)\mathbf{X}_1\dot{T}, \\ \dot{\mathbf{X}}_2 &= \tfrac{2}{3}C_2\dot{\boldsymbol{\varepsilon}}^{\mathrm{p}}\left(1 - D\right) - \gamma_2\mathbf{X}_2\dot{p} + \left(C_2'/C_2\right)\mathbf{X}_2\dot{T}, \\ \mathbf{X} &= \mathbf{X}_1 + \mathbf{X}_2, \\ \boldsymbol{\sigma} &= E\left(1 - D\right)\left(\boldsymbol{\varepsilon} - \boldsymbol{\varepsilon}^{\mathrm{p}} - \boldsymbol{\varepsilon}^{\mathrm{T}}\right),\end{aligned} \tag{2.48}$$

where p is the cumulative plastic strain and $J_2\left(\boldsymbol{\sigma} - \mathbf{X}\right)$ is the effective stress given by

$$\dot{p} = \left(\frac{2}{3}\dot{\boldsymbol{\varepsilon}}^{\mathrm{p}}:\dot{\boldsymbol{\varepsilon}}^{\mathrm{p}}\right)^{1/2}, \qquad J_2 = \left[\frac{3}{2}\left(\boldsymbol{\sigma}' - \mathbf{X}'\right):\left(\boldsymbol{\sigma}' - \mathbf{X}'\right)\right]^{1/2}. \tag{2.49}$$

Symbols K, σ_y, n, C_1, C_2, γ_1, γ_2, E are temperature dependent material constants for copper (cf. Dunne and Hayhurst, 1992a, 1992b) as shown in Table 2.2, C_1' and C_2' are derivatives of $C_1$and C_2 with respect to temperature T, and $\boldsymbol{\varepsilon}^{\mathrm{T}}$ is the thermal strain given by $\boldsymbol{\varepsilon}^{\mathrm{T}} = \alpha T\mathbf{1}$.

Table 2.2. Viscoplasticity material parameters for copper (after Dunne and Hayhurst, 1992a)

Temp. T [°C]	C_1	C_2	γ_1	γ_2	K [MPa]
20	54041	721	962	1.1	4.5
50	52880	700	1000	1.1	4.5
150	45760	600	1100	1.1	4.5
250	38040	400	1300	10.0	35.0
500	28952	300	1700	35.0	20.1

Temp. T [°C]	n	E [MPa]	σ_y [MPa]		
			0.3%	0.6%	1.0%
20	2.814	96890	45	58	68
50	3.227	92106	38	60	70
150	5.34	89583	33	52	78
250	9.735	79762	13	33	45
500	7.378	63991	4	13	21

2.3 Unified thermodynamic formulation of the coupled isotropic damage-thermo-elastic (visco)plasticity

2.3.1 Kinetic law of damage evolution

I. Concept of the elastic strain energy density release rate Y^{e}

Chaboche (1976) developed a concept of the elastic strain energy release following the isotropic damage accumulation in a material, based on the effective stress using the hypothesis of strain equivalence (Sect. 1.4.1). It is based on the observation that for ductile materials continuous isotropic damage may be represented by a single scalar variable D the evolution of which is governed by the variation of the elastic strain energy $\mathrm{d}\Phi^{\mathrm{e}}/\mathrm{d}D$. In other words, in this simplified approach the variable Y^{e} associated with the isotropic damage internal variable D contains the contribution of the elastic (reversible) energy only, whereas the inelastic (irreversible) stored energy associated with the strain hardening (isotropic and kinematic) is not released by the initiation and growth of damage.

Assuming small strain theory, the total strain may be written as a sum of the elastic and the inelastic part $\boldsymbol{\varepsilon} = \boldsymbol{\varepsilon}^{\mathrm{e}} + \boldsymbol{\varepsilon}^{\mathrm{an}}$. For elastic strain the anisotropic elasticity law coupled with isotropic damage is assumed

$$\sigma_{ij} = E_{ijkl}\varepsilon^{\mathrm{e}}_{kl}\,(1 - D)\,. \tag{2.50}$$

Hence, applying (2.50) the elastic strain energy density is furnished:

$$\Phi^{\mathrm{e}} = \int \sigma_{ij} \mathrm{d}\varepsilon^{\mathrm{e}}_{ij} = \int E_{ijkl} \varepsilon^{\mathrm{e}}_{kl} (1-D) \, \mathrm{d}\varepsilon^{\mathrm{e}}_{ij} = \frac{1}{2} E_{ijkl} \varepsilon^{\mathrm{e}}_{ij} \varepsilon^{\mathrm{e}}_{kl} (1-D) \, . \quad (2.51)$$

When the constant stress condition is used,

$$\mathrm{d}\sigma_{ij} = E_{ijkl} \left[(1-D) \, \mathrm{d}\varepsilon^{\mathrm{e}}_{kl} - \varepsilon^{\mathrm{e}}_{kl} \mathrm{d}D \right] = 0, \quad (2.52)$$

a variation of the elastic strain energy due to the continuous damage growth is represented by

$$\mathrm{d}\Phi^{\mathrm{e}} = \sigma_{ij} \mathrm{d}\varepsilon^{\mathrm{e}}_{ij} = \sigma_{ij} \varepsilon^{\mathrm{e}}_{ij} \frac{\mathrm{d}D}{1-D}. \quad (2.53)$$

Eventually, applying (2.50) and (2.51), the variable Y^{e} associated with the internal (scalar) variable D is defined as follows

$$Y^{\mathrm{e}} \stackrel{\mathrm{def}}{=} \frac{1}{2} \frac{\mathrm{d}\Phi^{\mathrm{e}}}{\mathrm{d}D} \bigg|_{\sigma_{ij}=\mathrm{const}} = \frac{1}{2} \frac{\sigma_{ij} \varepsilon^{\mathrm{e}}_{ij}}{1-D} = \frac{1}{2} E_{ijkl} \varepsilon^{\mathrm{e}}_{ij} \varepsilon^{\mathrm{e}}_{kl} = \frac{\Phi^{\mathrm{e}}}{1-D}. \quad (2.54)$$

In general, a model based on the anisotropic elasticity coupled with the isotropic damage is not consistent. So, confining ourselves to the isotropic elasticity law

$$\varepsilon^{\mathrm{e}}_{ij} = \frac{1+\nu}{E} \frac{\sigma_{ij}}{1-D} - \frac{\nu}{E} \frac{\sigma_{kk}}{1-D} \delta_{ij} \quad (2.55)$$

and introducing the decomposition to deviatoric and hydrostatic terms

$$\sigma_{ij} = s_{ij} + \sigma_{\mathrm{H}} \delta_{ij}, \qquad \varepsilon^{\mathrm{e}}_{ij} = e^{\mathrm{e}}_{ij} + e^{\mathrm{e}}_{\mathrm{H}} \delta_{ij}, \quad (2.56)$$

the shape and volume change law coupled with damage is established:

$$e^{\mathrm{e}}_{ij} = \frac{1+\nu}{E} \frac{s_{ij}}{1-D}, \qquad e^{\mathrm{e}}_{\mathrm{H}} = \frac{1-2\nu}{E} \frac{\sigma_{\mathrm{H}}}{1-D}, \quad (2.57)$$

as well as the corresponding shear and hydrostatic energy portions:

$$\Phi^{\mathrm{e}} = \Phi^{\mathrm{e(S)}} + \Phi^{\mathrm{e(H)}} = \frac{1}{2} \left[\frac{1+\nu}{E} \frac{s_{ij} s_{ij}}{1-D} + 3 \frac{1-2\nu}{E} \frac{\sigma^2_{\mathrm{H}}}{1-D} \right] \quad (2.58)$$

or, equivalently,

$$\Phi^{\mathrm{e}} = \frac{\sigma^2_{\mathrm{eq}}}{2E(1-D)} \left[\frac{2}{3} (1+\nu) + 3(1-2\nu) \left(\frac{\sigma_{\mathrm{H}}}{\sigma_{\mathrm{eq}}} \right)^2 \right], \quad (2.59)$$

where σ_{eq} is used for the classical Mises-type equivalent stress $\sigma_{\mathrm{eq}} = [(3/2) \, s_{ij} s_{ij}]^{1/2}$. Substitution of (2.59) for Φ^{e} in (2.54) yields the following formula for the elastic strain energy release rate (thermodynamic force

Y^{e} associated with the isotropic damage represented by a scalar variable D):

$$Y^{\mathrm{e}} = \frac{\widetilde{\sigma}_{\mathrm{eq}}^2}{2E} R_\nu, \quad R_\nu = \frac{2}{3}(1+\nu) + 3(1-2\nu)\left(\frac{\sigma_{\mathrm{H}}}{\sigma_{\mathrm{eq}}}\right)^2, \quad \widetilde{\sigma}_{\mathrm{eq}} = \frac{\sigma_{\mathrm{eq}}}{1-D}. \tag{2.60}$$

Note that in case of three-dimensional stress state the force Y^{e} associated with the isotropic damage variable D is mainly influenced by the stress triaxiality ratio $\sigma_{\mathrm{H}}/\sigma_{\mathrm{eq}}$. In the 1D case, the stress triaxiality ratio is equal to $(\sigma_{\mathrm{H}}/\sigma_{\mathrm{eq}})^{1\mathrm{D}} = 1/3$, whereas $(R_\nu)^{1\mathrm{D}} = 1$, and it increases with the hydrostatic stress growth, as does the damage rate $\dot{D}$. At variance with the classical Mises-type equivalent stress σ_{eq} and the corresponding effective equivalent stress $\widetilde{\sigma}_{\mathrm{eq}} = \sigma_{\mathrm{eq}}/(1-D)$ the damage equivalent stress $\sigma_{\mathrm{eq}}^{\mathrm{D}}$ is furnished by equating the elastic strain energy in the 3D state $\Phi^{\mathrm{e}}(\boldsymbol{\sigma})$ with the equivalent 1D state $\Phi^{\mathrm{e}}(\sigma_{\mathrm{eq}}^{\mathrm{D}})$

$$\Phi^{\mathrm{e}}\left(\sigma_{\mathrm{eq}}^{\mathrm{D}}\right) = \frac{\left(\sigma_{\mathrm{eq}}^{\mathrm{D}}\right)^2}{2E(1-D)}, \qquad \Phi^{\mathrm{e}}(\boldsymbol{\sigma}) = Y^{\mathrm{e}}(1-D) \tag{2.61}$$

to obtain, in view of (2.60), the formula that differs from the Mises-type σ_{eq} in the function $R_\nu^{1/2}$

$$\sigma_{\mathrm{eq}}^{\mathrm{D}} = \sigma_{\mathrm{eq}} R_\nu^{1/2}. \tag{2.62}$$

For example, if plane stress is assumed, the following holds:

$$\begin{aligned} \sigma_{\mathrm{eq}} &= \left[\sigma_1^2 + \sigma_2^2 - \sigma_1\sigma_2\right]^{1/2}, \\ \sigma_{\mathrm{eq}}^{\mathrm{D}} &= \left[\frac{2}{3}(1+\nu) + 3(1-2\nu)\frac{(\sigma_1+\sigma_2)^2}{\sigma_1^2+\sigma_2^2-\sigma_1\sigma_2}\right]^{1/2} \sigma_{\mathrm{eq}}. \end{aligned} \tag{2.63}$$

II. Time-independent plasticity coupled with isotropic damage (Chaboche, 1988)

Chaboche (1988) introduced the coupled dissipative potential by an extension of the Chaboche–Rousselier nonlinear isotropic/kinematic hardening theory (2.17) to yield:

$$\begin{aligned} F^{\mathrm{Ch}} &= (\widetilde{\boldsymbol{\sigma}}, \mathbf{X}, R, D) = f(\widetilde{\boldsymbol{\sigma}}, \mathbf{X}, R) + F^{\mathrm{D}}(Y^{\mathrm{e}}) \\ &= J_2(\widetilde{\boldsymbol{\sigma}} - \mathbf{X}) - R - \sigma_y + F^{\mathrm{D}}(Y^{\mathrm{e}}), \\ J_2(\widetilde{\boldsymbol{\sigma}} - \mathbf{X}) &= \left[\frac{3}{2}\left(\frac{\sigma'_{ij}}{1-D} - X'_{ij}\right)\left(\frac{\sigma'_{ij}}{1-D} - X'_{ij}\right)\right]^{1/2} \end{aligned} \tag{2.64}$$

and the generalized normality rule (associative theory)

$$\dot{\boldsymbol{\varepsilon}}^{\mathrm{p}} = \dot{\lambda}\frac{\partial F^{\mathrm{Ch}}}{\partial \boldsymbol{\sigma}}, \quad \dot{\boldsymbol{\alpha}} = -\dot{\lambda}\frac{\partial F^{\mathrm{Ch}}}{\partial \mathbf{X}}, \quad \dot{r} = -\dot{\lambda}\frac{\partial F^{\mathrm{Ch}}}{\partial R}, \quad \dot{D} = -\dot{\lambda}\frac{\partial F^{\mathrm{Ch}}}{\partial Y^{\mathrm{e}}}, \quad (2.65)$$

where $f(\widetilde{\boldsymbol{\sigma}}, \mathbf{X}, R) = J_2(\widetilde{\boldsymbol{\sigma}} - \mathbf{X}) - R - \sigma_y = 0$ is a Mises-type partly coupled yield function and σ_y is the initial yield stress under uniaxial tension test (for a more general fully coupled case see Sect. 2.3.4(II)).

The following couples of the external (observable) state variables $(\boldsymbol{\varepsilon}, \boldsymbol{\sigma})$ and the internal state variables $(\boldsymbol{\alpha}, \mathbf{X})$, (r, R), (D, Y^{e}) are introduced to represent kinematic hardening, isotropic hardening, and isotropic damage. Note that the mechanical behavior of the damaged solid is derived from the same dissipation potential as a virgin undamaged solid, where the state stress variable $\boldsymbol{\sigma}$ is replaced by the effective stress variable $\widetilde{\boldsymbol{\sigma}} = \boldsymbol{\sigma}/(1-D)$ and the additional term $F^{\mathrm{D}}(Y^{\mathrm{e}})$ describes the damage evolution. This approach, based on the strain equivalence, ignores the damage effect on the release of inelastic stored energy so that the potential function for damage evolution depends only on the elastic energy release rate $F^{\mathrm{D}} = F^{\mathrm{D}}(Y^{\mathrm{e}})$ and variables associated with strain hardening are not affected by damage (the effective state variable $\widetilde{\mathbf{X}}$ and $\widetilde{R}$ are not built into the model).

When the generalized normality rule (2.65) is applied together with the dissipative potential (2.64) the following state equations are furnished:

$$\begin{aligned}
\dot{\varepsilon}^{\mathrm{p}}_{ij} &= \frac{3}{2}\frac{\dot{\lambda}}{1-D}\frac{\widetilde{\sigma}'_{ij} - X'_{ij}}{J_2(\widetilde{\sigma}_{ij} - X_{ij})} = \frac{3}{2}\dot{\widetilde{\lambda}}\frac{\widetilde{\sigma}'_{ij} - X'_{ij}}{J_2(\widetilde{\sigma}_{ij} - X_{ij})}, \\
\dot{\alpha}_{ij} &= \frac{3}{2}\dot{\widetilde{\lambda}}\frac{\widetilde{\sigma}'_{ij} - X'_{ij}}{J_2(\widetilde{\sigma}_{ij} - X_{ij})} = \dot{\varepsilon}^{\mathrm{p}}_{ij}(1-D), \\
\dot{r} &= \dot{\lambda} = \dot{p}(1-D), \\
\dot{D} &= -\dot{\lambda}\frac{\partial F^{\mathrm{D}}(Y^{\mathrm{e}})}{\partial Y^{\mathrm{e}}} = -\frac{\partial F^{\mathrm{D}}(Y^{\mathrm{e}})}{\partial Y^{\mathrm{e}}}\dot{p}(1-D),
\end{aligned} \quad (2.66)$$

where $\dot{p}$ denotes the cumulative plastic strain $\dot{p} = \left[(2/3)\dot{\varepsilon}^{\mathrm{p}}_{ij}\dot{\varepsilon}^{\mathrm{p}}_{ij}\right]^{1/2}$, and $\dot{\widetilde{\lambda}} = \dot{\lambda}/(1-D)$, whereas third of the Eqs. (2.66) is obtained by a scalar multiplication of the first Eq. (2.66)

$$\dot{\varepsilon}^{\mathrm{p}}_{ij}\dot{\varepsilon}^{\mathrm{p}}_{ij} = \left(\frac{3}{2}\right)^2 \frac{(\widetilde{\sigma}'_{ij} - X'_{ij})(\widetilde{\sigma}'_{ij} - X'_{ij})}{J_2^2(\widetilde{\sigma}_{ij} - X_{ij})}\left(\frac{\dot{\lambda}}{1-D}\right)^2 \quad (2.67)$$

when appropriate definitions of $\dot{p}$ and J_2 hold. The effective formula for damage evolution depends on the representation of the damage potential function F^{D}. If, following Lemaitre and Chaboche (1985), it is assumed as a square function of the elastic strain energy release rate Y^{e} (2.60),

$$F^{\mathrm{D}} = -\frac{(Y^{\mathrm{e}})^2}{2S(1-D)} \qquad \text{or} \qquad F^{\mathrm{D}} = -\frac{(Y^{\mathrm{e}})^2}{2S(1-D)} H(p-p_{\mathrm{D}}), \tag{2.68}$$

then the classical kinetic law of damage evolution is furnished (cf. Lemaitre and Chaboche, 1985):

$$\dot{D} = -\dot{\lambda}\frac{\partial F^{\mathrm{D}}}{\partial Y^{\mathrm{e}}} = \frac{Y^{\mathrm{e}}}{S}\dot{p} = \frac{\sigma_{\mathrm{eq}}^2 R_\nu}{2ES(1-D)^2}\dot{p}. \tag{2.69}$$

If, on the other hand, the damage potential function is assumed as a power function of Y^{e}, then the generalized kinetic law of damage evolution is established (cf. Germain, Nguyen and Suquet, 1983; Dufailly and Lemaitre, 1995)

$$F^{\mathrm{D}} = -\left(\frac{Y^{\mathrm{e}}}{S}\right)^s \frac{Y^{\mathrm{e}}}{(s+1)(1-D)}, \tag{2.70}$$

$$\dot{D} = \left(\frac{Y^{\mathrm{e}}}{S}\right)^s \dot{p} = \left[\frac{\sigma_{\mathrm{eq}}^2 R_\nu}{2ES(1-D)^2}\right]^s \dot{p}. \tag{2.71}$$

S, s are temperature dependent material parameters and $R_\nu(\sigma_{\mathrm{H}}/\sigma_{\mathrm{eq}})$ is given by (2.60). Equations (2.69) or (2.71) constitute the damage evolution in the frame of the kinetic law of damage based on the assumption that continuous damage manifests itself as elastic energy release, whereas the inelastic energy associated with strain hardening is not released by the damage initiation and growth. In general, this is not true, and a more extended theory based on the total energy release with damage can be developed (cf., e.g., Saanouni, Forster and Ben Hatira, 1994).

2.3.2 Application of the kinetic law of damage to plasticity, creep, and damage

I. Particular cases of the kinetic law of damage model of ductile materials

Equations (2.69) and (2.71) govern isotropic damage evolution in ductile materials, as influenced by the cumulative plastic strain $\dot{p} = \left[(2/3)\,\dot{\varepsilon}_{ij}^{\mathrm{P}}\dot{\varepsilon}_{ij}^{\mathrm{P}}\right]^{1/2}$, stress state represented by the stress triaxiality function $R_\nu = R_\nu(\sigma_{\mathrm{H}}/\sigma_{\mathrm{eq}})$, and the effective equivalent stress $\tilde{\sigma}_{\mathrm{eq}} = \sigma_{\mathrm{eq}}/(1-D)$ as a function of the cumulative strain $\tilde{\sigma}_{\mathrm{eq}}(p)$. These constitutive equations of damage hold for any loading path along which the stress triaxiality ratio $\sigma_{\mathrm{H}}/\sigma_{\mathrm{eq}}$ changes, whereas in case of proportional loadings the stress triaxiality ratio can be considered as a constant with respect to time $\sigma_{\mathrm{H}}/\sigma_{\mathrm{eq}} = \mathrm{const}$. Assuming the Ramberg–Osgood isotropic power hardening function for damage material (cf. Lemaitre and Chaboche, 1985),

$$\widetilde{\sigma}_{\text{eq}} = \frac{\sigma_{\text{eq}}}{1-D} = \sigma_{\text{s}} p^n , \tag{2.72}$$

the damage evolution equations (2.69) or (2.71) are reduced as follows (cf. Lemaitre, 1985; Dufailly–Lemaitre, 1995):

$$\mathrm{d}D = \frac{\sigma_{\text{s}}^2}{2ES} R_\nu \left(\frac{\sigma_{\text{H}}}{\sigma_{\text{eq}}}\right) p^{2n} \mathrm{d}p \tag{2.73}$$

or

$$\mathrm{d}D = \left(\frac{\sigma_{\text{s}}^2}{2ES}\right)^s \left[R_\nu \left(\frac{\sigma_{\text{H}}}{\sigma_{\text{eq}}}\right)\right]^s p^{2sn} \mathrm{d}p. \tag{2.74}$$

Equations (2.73) or (2.74) represent the ductile damage as the isotropic scalar variable D dependent on the cumulative plastic strain p, the stress triaxiality ratio $\sigma_{\text{H}}/\sigma_{\text{eq}}$, and the isotropic strain hardening exponent of the material n. Note that S, s and K, n are temperature dependent material constants (substitution $s = 1$ reduces (2.74) to the classical form (2.73)), whereas the stress triaxiality ratio $\sigma_{\text{H}}/\sigma_{\text{eq}}$, which changes for a nonproportional loading, characterizes the stress state, and $\sigma_{\text{H}}/\sigma_{\text{eq}} = 1/3$ in the 1D case. In other words, according to this model, the ductile damage, as caused by the mechanisms of microvoid nucleation, growth, and coalescence, is a plastic strain controlled mechanism with p_0 and p_{cr} (or ε_0 and ε_{cr}) corresponding to the initial damage D_0 and the threshold damage at failure D_{cr}.

Material constants in (2.73) or (2.74) are determined by the one-dimensional load test at which the following holds:

$$\frac{\sigma_{\text{H}}}{\sigma_{\text{eq}}} = \frac{1}{3}, \quad R_\nu \left(\frac{\sigma_{\text{H}}}{\sigma_{\text{eq}}}\right) = 1, \quad p = \varepsilon. \tag{2.75}$$

Integration of (2.74) for the one-dimensional case, from the initial (ε_0, D_0) to failure $(\varepsilon_{\text{cr}}, D_{\text{cr}})$ conditions, with (2.75) taken into account, yields the equation

$$D_{\text{cr}} = D_0 + \frac{1}{2ns+1}\left(\frac{\sigma_{\text{s}}^2}{2ES}\right)^s \left(\varepsilon_{\text{cr}}^{2ns+1} - \varepsilon_0^{2ns+1}\right) \tag{2.76}$$

that determines the critical damage D_{cr} in terms of the temperature dependent material constants K, n, S, s and the 1D strains at initial damage and the threshold damage at failure, ε_0 and ε_{cr}. Hence, integration of (2.74), with the simplifying assumption that the triaxiality ratio does not change in a loading process (which generally is not true) furnishes damage evolution with the cumulative plastic strain p:

$$D = D_{\text{cr}} - \frac{D_{\text{cr}} - D_0}{\varepsilon_{\text{cr}}^{2ns+1} - \varepsilon_0^{2ns+1}} \left(p_{\text{cr}}^{2ns+1} - p^{2ns+1}\right) \left[R_\nu \left(\frac{\sigma_H}{\sigma_{\text{eq}}}\right)\right]^s . \tag{2.77}$$

However, in the general case the stress triaxiality ratio $\sigma_{\rm H}/\sigma_{\rm eq}$ changes with p, hence elimination of $(\sigma_{\rm s}^2/2ES)$ from (2.76) and (2.74) leads to the general damage evolution equation applicable for any loading path:

$$\mathrm{d}D = (2ns+1)\frac{D_{\rm cr} - D_0}{\left(\varepsilon_{\rm cr}^{2ns+1} - \varepsilon_0^{2ns+1}\right)} \left[R_\nu\left(\frac{\sigma_{\rm H}}{\sigma_{\rm eq}}\right)\right]^s p^{2ns}\mathrm{d}p, \tag{2.78}$$

which holds together with the formula (2.76).

In a particular case when strain hardening is saturated, $\dot{\mathbf{X}} = \mathbf{0}$, $\dot{R} = 0$,

$$f = \widetilde{\sigma}_{\rm eq} - \sigma_{\rm s} = 0 \qquad \text{or} \qquad \frac{\sigma_{\rm eq}}{1-D} = \sigma_{\rm s}, \tag{2.79}$$

the strain density release rate approximation of (2.60) is obtained,

$$Y_{\rm s}^{\rm e} = \frac{\sigma_{\rm s}^2 R_\nu}{2E}, \tag{2.80}$$

and the corresponding generalized kinetic law of damage evolution (2.71) reduces to the simplified form

$$\dot{D} = \left(\frac{Y_{\rm s}^{\rm e}}{S}\right)^s \dot{p} = \left(\frac{\sigma_{\rm s}^2 R_\nu}{2ES}\right)^s \dot{p}. \tag{2.81}$$

Integration of (2.81) for the one-dimensional case, from the initial (ε_0, D_0) to failure $(\varepsilon_{\rm cr}, D_{\rm cr})$ conditions, yields

$$D_{\rm cr} = D_0 + \left(\frac{\sigma_{\rm s}^2}{2ES}\right)^s (\varepsilon_{\rm cr} - \varepsilon_0), \tag{2.82}$$

whereas for the three-dimensional case and constant R_ν the linear damage growth with p holds:

$$D_{\rm cr} - D = \left(\frac{\sigma_{\rm s}^2}{2ES}\right)^s \left[R_\nu\left(\frac{\sigma_{\rm H}}{\sigma_{\rm eq}}\right)\right]^s (p_{\rm cr} - p) \tag{2.83}$$

or

$$D = D_{\rm cr} - \frac{D_{\rm cr} - D_0}{\varepsilon_{\rm cr} - \varepsilon_0}(p_{\rm cr} - p)\left[R_\nu\left(\frac{\sigma_{\rm H}}{\sigma_{\rm eq}}\right)\right]^s. \tag{2.84}$$

It is easily seen that (2.83) and (2.84) describing the damage evolution for hardening saturation follow from the general damage evolution for hardening material (2.76) and (2.77), as an approximate case when the hardening exponent in (2.72) equals zero $n = 0$.

II. Creep damage of metals and polymers

In a particular case of creep damage, Benallal's equation may be employed (cf. Benallal, 1985),

$$\dot{p} = \frac{\dot{\lambda}^{\mathrm{vp}}}{1-D} = \ln\left(1 - \frac{f}{K_\infty}\right)^{-n} \tag{2.85}$$

together with the classical Lemaitre–Chaboche kinetic law of damage evolution (2.69):

$$\dot{D} = \frac{Y^{\mathrm{e}}}{S}\dot{p} = \frac{\sigma_{\mathrm{eq}}^2 R_\nu}{2ES\,(1-D)^2}\dot{p}H\,(p - p_0) \tag{2.86}$$

where, according to Chaboche (1988), a Mises-type partly coupled yield function is assumed,

$$f\,(\tilde{\boldsymbol{\sigma}}, \mathbf{X}, R) = J_2\,(\tilde{\boldsymbol{\sigma}} - \mathbf{X}) - R - \sigma_y, \tag{2.87}$$

and the Heaviside function $H\,(p - p_0)$ is introduced to account for the initial damage at $p = p_0$. In a simplified case, when Norton's creep law is applied the viscoplastic multiplier is reduced to (cf. Lemaitre, 1992)

$$\dot{p} = \frac{\dot{\lambda}^{\mathrm{vp}}}{1-D} = \left[\frac{\sigma_{\mathrm{eq}}}{K_v\,(1-D)}\right]^N \tag{2.88}$$

with K_v and N denoting the temperature dependent material parameters. Hence, by combining (2.86) with (2.88) the following damage evolution equations may be obtained:

$$(3\mathrm{D}) \quad \dot{D} = \frac{\sigma_{\mathrm{eq}}^{N+2} R_\nu}{2ESK_v^N\,(1-D)^{N+2}} H\,(p - p_0) \tag{2.89}$$

Note that in the one-dimensional case, with the new parameter A and $r = k$ introduced, Kachanov's classical equation (2.26) is recovered:

$$(1\mathrm{D}) \quad \dot{D} = \left[\frac{\sigma}{A\,(1-D)}\right]^{N+2} H\,(\varepsilon - \varepsilon_0) \tag{2.90}$$

if

$$A = \left(2ESK_v^N\right)^{\frac{1}{N+2}} \quad \text{and} \quad r = N + 2. \tag{2.91}$$

Integration of (2.90) at constant uniaxial stress $\sigma =$ const, from the time of initial damage t_0 (at $\varepsilon = \varepsilon_0$, $D = D_0$) to current time t, yields

$$D\,(t) = 1 - \left[(1 - D_0)^{N+3} - (N+3)\left(\frac{\sigma}{A}\right)^{N+2}(t - t_0)\right]^{\frac{1}{N+3}}, \tag{2.92}$$

whereas, if for the initial (t_0) and the critical (t_R) damage states $D_0 = 0$ and $D_{cr} = 1$ hold, time to rupture t_R is expressed as

$$t_R = \varepsilon_0 \left(\frac{\sigma}{K_v}\right)^{-N} + \frac{1}{N+3}\left(\frac{\sigma}{A}\right)^{-(N+2)}, \tag{2.93}$$

where, in view of (2.88), with $D = 0$ time of initial damage t_0 has been eliminated, $t_0 = \varepsilon_0 (\sigma/K_v)^{-N}$. Again, integration of (2.88) for the 1D case, $\dot{p} = \dot{\varepsilon}^{vp}$, $\sigma_{eq} = \sigma$ =const, with $D(t)$ given by (2.92) and $D_0 = 0$, determines the viscoplastic strain evolution with time $t > t_0$ (cf. Lemaitre, 1992):

$$\begin{aligned} \varepsilon^{vp} = \varepsilon_0 + \frac{1}{3}\left(\frac{\sigma}{K_v}\right)^{-N} \Bigg\{ 1 - \Bigg[1 - (N+3) \\ \times \left(\frac{\sigma}{A}\right)^{N+2} (t - t_0) \Bigg]^{\frac{3}{N+3}} \Bigg\} \left(\frac{\sigma}{A}\right)^{-(N+2)}. \end{aligned} \tag{2.94}$$

In the case of high temperature isotropic tertiary creep damage, Zhang and Lee (1993) developed the new constitutive law in the form

$$\frac{\mathrm{d}D}{\mathrm{d}t} = G\dot{\varepsilon}^{s}_{\min} (t - t_*)^{r-1} (1 - D)^{-n}. \tag{2.95}$$

If the following relationships hold:

$$s = m/n, \quad r = 1, \quad \dot{\varepsilon}^{s}_{\min} = A_1 \sigma^n \exp\left(-\frac{Q'}{RT}\right), \tag{2.96}$$

an extension of the Kachanov-type uniaxial damage evolution by the explicit absolute temperature function is recovered:

$$\text{(1D)} \quad \frac{\mathrm{d}D}{\mathrm{d}t} = A\sigma^n \exp\left(-\frac{Q}{RT}\right)(1 - D)^{-n}. \tag{2.97}$$

When the three-dimensional stress state is considered, the following holds:

$$\text{(3D)} \quad \frac{\mathrm{d}D}{\mathrm{d}t} = A\left(\sigma^{D}_{eq}\right)^n \exp\left(-\frac{Q}{RT}\right)(1 - D)^{-n}, \tag{2.98}$$

where σ^{D}_{eq} stands for the damage equivalent stress (2.62)

$$\sigma^{D}_{eq} = \sigma_{eq} R_\nu^{1/2} = \sigma_{eq}\left[\frac{2}{3}(1+\nu) + 3(1-2\nu)\left(\frac{\sigma_H}{\sigma_{eq}}\right)^2\right]^{1/2}. \tag{2.99}$$

In the above equations, t_* is the initiation time of tertiary creep, n, G, s, r, $A = GA_1^{m/n}$ are material constants, and $Q = Q'\frac{m}{n}$ is the activation energy. Integration of (2.98) for variable damage equivalent stress furnishes damage evolution with time $t > t_*$:

$$D(t) = 1 - \left\{ 1 - (n+1) \int_{t_*}^{t} \left[\sigma_{\mathrm{eq}}^{\mathrm{D}}(\tau) \right]^m A \exp\left(-\frac{Q}{RT} \right) \mathrm{d}\tau \right\}^{\frac{1}{n+1}} . \quad (2.100)$$

III. Fatigue damage

The kinetic law of damage evolution (2.69), or its generalized representation (2.71), may successfully be applied to the coupled fatigue-damage behavior. A particular form of the constitutive law of damage evolution applicable to this phenomenon depends on the cycles to failure range. A useful classification is due to Dufailly and Lemaitre (1995) (Table 2.3).

Table 2.3. Fatigue classification (after Dufailly and Lemaitre, 1995)

	Number of cycles to failure	Stress range	Strain ratio $\Delta\varepsilon^{\mathrm{p}}/\Delta\varepsilon^{\mathrm{e}}$	Energy ration $\Delta W^{\mathrm{p}}/\Delta W^{\mathrm{e}}$
High cycle fatigue HCF	$> 10^5$	$< \sigma_y$	$\cong 0$	$\cong 0$
Low cycle fatigue LCF	10^2–10^4	σ_y–σ_u	1–10	1–10
Very low cycle fatigue VLCF	1–10	close to σ_u	1–100	1–100

In the above classification the following nomenclature has been used: σ_y is the yield stress, σ_u the ultimate stress, $\Delta\varepsilon^{\mathrm{p}}$ the plastic strain amplitude, $\Delta\varepsilon^{\mathrm{e}}$ the elastic strain amplitude, ΔW^{p} the (visco)plastic (dissipative) energy per cycle, and ΔW^{e} the elastic (reversible) energy per cycle.

In case of high cycle fatigue (HCF), the average stress level on the macroscale should remain below the yield stress $\sigma < \sigma_y$, such that very small plastic strain manifests only around the microscopic defects, hence, in consequence, the dissipative energy ΔW^{p} can be disregarded when compared to the elastic strain energy ΔW^{e}. Damage in HCF tests is a strongly localized phenomenon with high stress and damage concentration, so the classical CDM method, based on the effective quasicontinuum concept of microdefects and the stress field homogenization method in RVE, should rather be replaced by a two-scale nonlocal mechanical model in which the size of the weak microplastic and damage zones is much smaller than the size of a specimen made of the elastic matrix and, hence, a direct correlation between the weak inclusions should be incorporated to the model. A number of cycles to HCF failure is assumed to be as large as 10^5.

In contrast to HCF, in the case of low cycle fatigue (LCF) the stress level is larger than the yield stress $\sigma > \sigma_y$, such that the continuum damage develops together with the cyclic plastic strain after the incubation period that precedes the nucleation and growth of microdefects. The mechanism of the ductile damage on the LCF tests is manifested through the transgranular slipband fields of plasticity developed in the large size grains, hence, the dissipative energy ΔW^{p} is of the same order as the elastic energy ΔW^{e}. In other words, for LCF tests the plasticity-damage fields involve a large volume of the specimen with weak localization, such that the classical local CDM approach is applicable as the objective method for the number of cycles to failure prediction, which customarily is supposed to be between 10^2 and 10^4.

In case of very low cycle fatigue (VLCF), the number of cycles to failure is of the order of 10. The cyclic damage mechanism is governed by the slipbands of plasticity in the grains the orientation of which is approximately inclined at 45° to the main stress, and rapid macrocrack growth in a transgranular mode in the slip planes occurs (see Fig. 2.1, after Dufailly and Lemaitre, 1995). Strain hardening saturation is reached during the first cycle, so the perfectly plastic model is justified. The number of sites of microcrack initiation is large enough to allow for damage homogenization, and no damage threshold is needed since the damage evolution starts immediately just on the first cycle (cf. Dufailly and Lemaitre, 1995).

Consider the strain controlled process for repeated cycles known as the cyclic plasticity response. It may be analyzed by the use of coupled damage-isotropic/kinematic hardening theory developed by Chaboche (1988), (2.64)–(2.65), with the kinetic law (2.69) or the generalized kinetic law (2.71) taken as the damage evolution. Two competing processes, stress amplitude growth on the nucleation period (cyclic hardening due to isotropic/kinematic hardening mechanism) and stress amplitude drop on the damage evolution period (cyclic softening due to material deterioration), result in a cyclic response as illustrated in Fig. 2.11.

Cyclic relationships between $\Delta\varepsilon^{\mathrm{p}}$ and $\Delta\sigma$ are assumed as follows (cf. Lemaitre, 1992):

$$\Delta\varepsilon^{\mathrm{p}} = \left(\frac{\Delta\sigma}{K}\right)^{M} \tag{2.101}$$

or

$$\Delta\varepsilon^{\mathrm{p}} = \left[\frac{\Delta\sigma}{K(1-D)}\right]^{M}, \tag{2.102}$$

in cases of no strain-damage or saturated plasticity-damage coupling, respectively.

In case of the one-dimensional LCF test, the kinetic damage evolution (2.69) holds

$$\mathrm{d}D = \frac{\sigma_{\mathrm{eq}}^2 R_\nu}{2ES(1-D)^2}\mathrm{d}p = \frac{\sigma^2}{2ES(1-D)^2}\left|\dot{\varepsilon}^{\mathrm{p}}\right|. \tag{2.103}$$

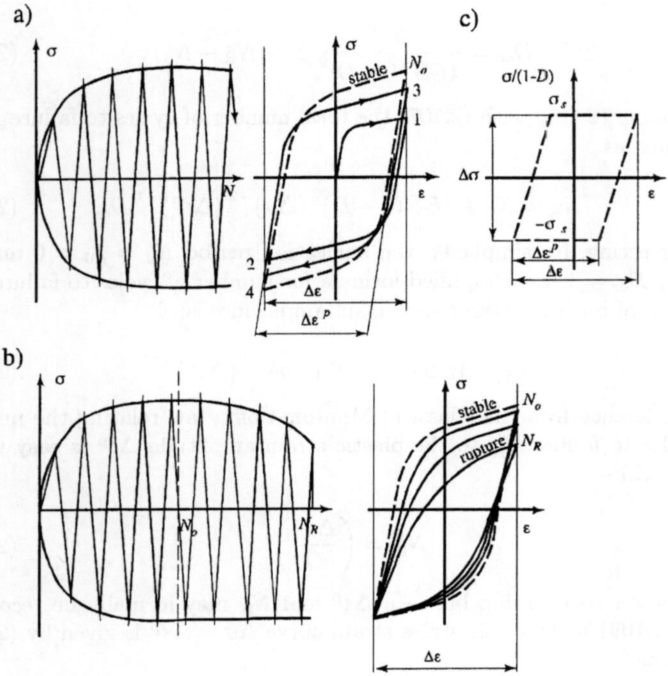

Fig. 2.11. Strain controlled cyclic hardening: a) without damage, b) with damage, c) stress–strain curve simplification for stable hysteresis loop

For the nucleation period $N < N_0$ accumulated plastic strain per cycle $(D = 0)$ is

$$\frac{\Delta p}{\Delta N} = 2\Delta\varepsilon^{\mathrm{p}} \tag{2.104}$$

and, assuming a simplified stable stress–strain loop ($\Delta\varepsilon_N^{\mathrm{p}} = \mathrm{const}$), the accumulated plastic strain to reach the nucleation limit N_0 is

$$p_0 = 2N_0\Delta\varepsilon^{\mathrm{p}} \quad \text{or} \quad N_0 = \frac{p_0}{2\Delta\varepsilon^{\mathrm{p}}}. \tag{2.105}$$

For the coupled plasticity-damage period $N_0 < N < N_{\mathrm{R}}$ assuming for plastic hardening saturation $\sigma_{\mathrm{s}} = \sigma/(1-D)$ to be constant over each cycle the damage evolution may easily be integrated over one cycle to give the damage per cycle $\Delta D/\Delta N$ (cf. Fig. 2.11c)

$$\frac{\Delta D}{\Delta N} = 2\frac{\sigma_{\mathrm{s}}^2}{2ES}\Delta\varepsilon^{\mathrm{p}} = \frac{\Delta\sigma^2}{4ES(1-D)^2}\Delta\varepsilon^{\mathrm{p}} \tag{2.106}$$

and damage per $(N_{\mathrm{R}} - N_0)$ cycles to critical damage D_{cr}

$$D_{\mathrm{cr}} = \frac{\Delta\sigma^2}{4ES\left(1-D\right)^2}\Delta\varepsilon^{\mathrm{p}}\left(N_{\mathrm{R}}-N_0\right). \tag{2.107}$$

Combining (2.105) with (2.107) the total number of cycles to failure N_{R} is furnished as

$$N_{\mathrm{R}} = N_0 + 4ES\left(1-D\right)^2\left(\Delta\sigma\right)^{-2}\left(\Delta\varepsilon^{\mathrm{p}}\right)^{-1}D_{\mathrm{cr}} \tag{2.108}$$

or, neglecting, for simplicity, the nucleation period $N_0 = p_0 = 0$ and assuming $D_{\mathrm{cr}} = 1$, the simplified formula for number of cycles to failure as a function of both the stress and strain amplitudes is:

$$N_{\mathrm{R}} = 4ES\left(1-D\right)^2\left(\Delta\sigma\right)^{-2}\left(\Delta\varepsilon^{\mathrm{p}}\right)^{-1}. \tag{2.109}$$

The difference from the classical Manson–Coffin law relating the number of cycles to failure $N_{\mathrm{R}}^{\mathrm{C}}$ to the plastic strain amplitude $\Delta\varepsilon^{\mathrm{p}}$ is easy to see (Fig. 2.12):

$$N_{\mathrm{R}}^{\mathrm{C}} = \left(\frac{\Delta\varepsilon^{\mathrm{p}}}{C}\right)^{\gamma}. \tag{2.110}$$

This power relationship between $\Delta\varepsilon^{\mathrm{p}}$ and N_{R} may formally be recovered from (2.109) if the cyclic stress strain curve $\Delta\sigma - \Delta\varepsilon^{\mathrm{p}}$ is given by (2.102) to obtain

$$N_{\mathrm{R}} = \frac{4ES}{K^2}\left(\Delta\varepsilon^{\mathrm{p}}\right)^{-\frac{M+2}{2}}. \tag{2.111}$$

For the very low cycle fatigue (VLCF) range, a big gap between the experimental results and theoretical prediction is observed.

To avoid this inconsistency, Dufailly and Lemaitre (1995) propose the generalized kinetic law of damage evolution as given by (2.71) applicable for 3D cases in the form

$$\mathrm{d}D = \left[\frac{\sigma_{\mathrm{eq}}^2 R_\nu}{2ES\left(1-D\right)^2}\right]^s \mathrm{d}p. \tag{2.112}$$

In the case of cyclic loading, assuming the plasticity criterion coupled with damage $\sigma_{\mathrm{eq}}/\left(1-D\right) = \sigma_{\mathrm{s}} = \Delta\sigma/2$ and integrating (2.112) over cycle, the damage per one cycle is obtained

$$\frac{\Delta D}{\Delta N} = \left(\frac{\sigma_{\mathrm{s}}^2 R_\nu}{2ES}\right)^s \Delta p, \tag{2.113}$$

which yields the critical damage $D_{\mathrm{cr}} = 1$ at $N = N_{\mathrm{R}}$ given by

$$N_R = \frac{\left(8ES\right)^s}{2}\left(1-D\right)^{2s}\Delta\sigma^{-2s}\left(\Delta\varepsilon^{\mathrm{p}}\right)^{-1}. \tag{2.114}$$

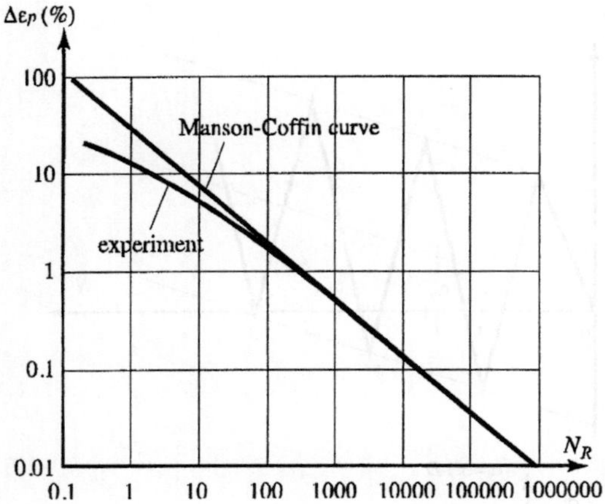

Fig. 2.12. Manson–Coffin curve for Inco 718 alloy at 550°C (after Dufailly and Lemaitre, 1995)

Note that for $s = 1$, (2.109) is again recovered.

In the fatigue-damage models discussed so far, a stable hysteresis loop reached after the incubation period was assumed. In case of the LCF or the VLCF it turns out that the hysteresis loop deforms due to cyclic hardening and simultaneously moves along the strain axis due to the progressively increasing large mean plastic strain (mixed fatigue ratchetting mode). A generalization of the Manson and Coffin law (2.110) to the case of accumulated mean plastic strain is due to Skoczeń, 1996. A push–pull loading program for a specimen made of Nickel A is sketched in Fig. 2.13.

Applying a power cyclic relationship

$$\sigma_{\mathrm{s}} = \frac{\sigma}{1-D} = \frac{1}{2}K\left(\varepsilon^{\mathrm{p}} - \varepsilon^{\mathrm{p}}_{\mathrm{min}}\right)^{1/M} \quad \text{or} \quad \Delta\varepsilon^{\mathrm{p}} = \left[\frac{\Delta\sigma}{K(1-D)}\right]^{M}, \tag{2.115}$$

the damage per cycle is traditionally obtained by integration of the kinetic damage evolution (2.103)

$$\frac{\Delta D}{\Delta N} = \frac{1}{ES}\int_{\varepsilon^{\mathrm{p}}_{\mathrm{min}}}^{\varepsilon^{\mathrm{p}}_{\mathrm{max}}} \left[\frac{1}{2}K\left(\varepsilon^{\mathrm{p}} - \varepsilon^{\mathrm{p}}_{\mathrm{min}}\right)^{1/M}\right]^{2} \mathrm{d}\varepsilon^{\mathrm{p}} = \frac{K^2}{4ES\gamma}\left(\Delta\varepsilon^{\mathrm{p}}\right)^{\gamma}, \tag{2.116}$$

where $\gamma = (M+2)/M$. When the ratchetting effect (progressively increasing mean plastic strain) was incorporated, the author arrived at the following generalization of (2.116):

$$\frac{\Delta D}{\Delta N} = \frac{K^2}{4ES\gamma}\frac{\left(\Delta\varepsilon^{\mathrm{p}}\right)^{\gamma}}{\left[1 - \frac{\varepsilon^{\mathrm{p}}_{\mathrm{m}(N)}}{\varepsilon_{\mathrm{f_o}}}\right]^{\alpha}}, \tag{2.117}$$

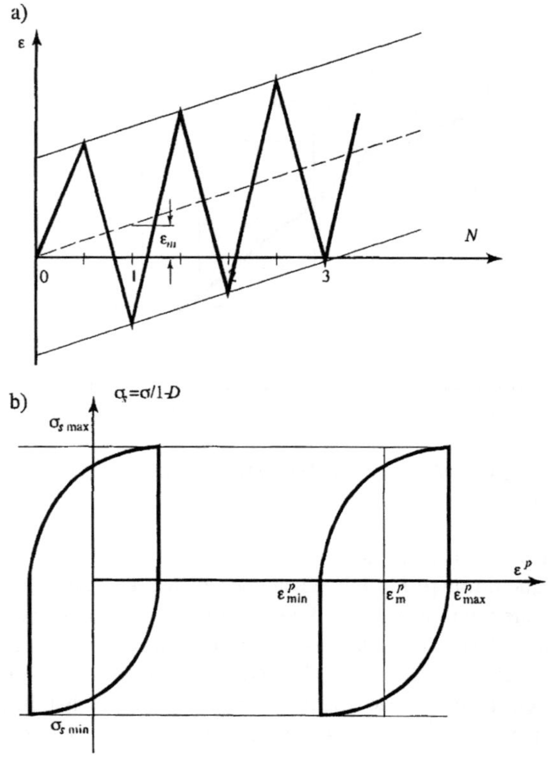

Fig. 2.13. Superimposed diameter strain (logarithmic) for linearly increasing plastic strain: a) constant ratchetting rate model, b) simplified moving hysteresis loop

where the mean plastic strain is $\varepsilon_{\mathrm{m}}^{\mathrm{p}} = \left(\varepsilon_{\mathrm{max}}^{\mathrm{p}} + \varepsilon_{\mathrm{min}}^{\mathrm{p}}\right)/2$ and $\varepsilon_{\mathrm{f_o}}$ is the tensile ductility, such that for $\varepsilon_{\mathrm{m}}^{\mathrm{p}} \to \varepsilon_{\mathrm{f_o}}$ the corrected increment of damage per cycle tends to infinity. For damage per N cycles the integration of (2.117) for a given function $\varepsilon_{\mathrm{m}}^{\mathrm{p}}(N)$ and constant $\Delta\varepsilon^{\mathrm{p}}$ over the number of cycles yields

$$D(N) = \int_0^N \frac{\Delta D}{\Delta N} \mathrm{d}N = \frac{K^2}{4ES\gamma} \left(\Delta\varepsilon^{\mathrm{p}}\right)^{\gamma} \int_0^N \left[1 - \frac{\varepsilon_{\mathrm{m}}^{\mathrm{p}}(N)}{\varepsilon_{\mathrm{fo}}}\right]^{-\alpha} \mathrm{d}N. \quad (2.118)$$

If, for simplicity, a linear function holds for mean plastic strain, $\varepsilon_{\mathrm{m}}^{\mathrm{p}}(N) = kN$, Eq. (2.118) may be integrated to arrive at

$$D(N) = \frac{K^2}{4ES\gamma} \left(\Delta\varepsilon^{\mathrm{p}}\right)^{\gamma} \frac{(1 - aN)^{1-\alpha} - 1}{a(\alpha - 1)}, \quad (2.119)$$

where $a = K/\varepsilon_{\mathrm{fo}}$, or assuming $D = D_{\mathrm{cr}}$ for $N = N_{\mathrm{R}}$, and introducing $C^{1/\gamma} = 4ESD_{\mathrm{cr}}\gamma/K^2$, a corrected number of cycles to failure for finite ratchetting is furnished

$$N_{\mathrm{R}} = \frac{1}{a}\left\{1 - \left[1 + a(\alpha - 1)\, C^{1/\gamma}\left(\Delta\varepsilon^{\mathrm{p}}\right)^{\gamma}\right]^{1/(1-\alpha)}\right\} \tag{2.120}$$

or

$$N_{\mathrm{R}} = \frac{1}{a}\left\{1 - \left[1 + a(\alpha - 1)\, N_{\mathrm{R}}^{\mathrm{C}}\right]^{1/(1-\alpha)}\right\}. \tag{2.121}$$

Note that, if $\alpha = 0$ is assumed, the classical Coffin formula $N_{\mathrm{R}} = N_{\mathrm{R}}^{\mathrm{C}}$ is recovered.

2.3.3 *Unified energy-based CDM model of ductile damage of materials*

A unified CDM model for ductile materials derived from the principles of thermodynamics is due to Mou and Han (1996). When a quasi-static loading is applied to a solid the energy transferred to the solid is either stored as elastic strain energy or dissipated by irreversible mechanisms arising from microstructural changes. The damage variable $D^{\mathrm{B}} = \ln\left(A/\widetilde{A}\right)$ is considered after Broberg (1974) as one of internal state variables which influence the Helmholtz free energy of the solid $\psi^{\mathrm{H}}\left(\boldsymbol{\varepsilon}^{\mathrm{e}}, \boldsymbol{\alpha}, r, D^{\mathrm{B}}, T\right)$

$$\psi^{\mathrm{H}}\left(\boldsymbol{\varepsilon}^{\mathrm{e}}, \boldsymbol{\alpha}, r, D^{\mathrm{B}}, T\right) = \psi^{\mathrm{e}}\left(\boldsymbol{\varepsilon}^{\mathrm{e}}, D^{\mathrm{B}}, T\right) + \psi^{\mathrm{an}}\left(\boldsymbol{\alpha}, r, T\right) \tag{2.122}$$

and the generalized thermodynamic forces $(\boldsymbol{\sigma}, \mathbf{X}, R, Y^{\mathrm{e}})$ are associated with elastic strain, kinematic hardening, isotropic hardening, and damage, respectively $\left(\boldsymbol{\varepsilon}^{\mathrm{e}}, \boldsymbol{\alpha}, r, D^{\mathrm{B}}\right)$ through

$$\boldsymbol{\sigma} = \rho\frac{\partial\psi^{\mathrm{e}}}{\partial\boldsymbol{\varepsilon}^{\mathrm{e}}}, \quad \mathbf{X} = \rho\frac{\partial\psi^{\mathrm{an}}}{\partial\boldsymbol{\alpha}}, \quad R = \rho\frac{\partial\psi^{\mathrm{an}}}{\partial r}, \quad Y^{\mathrm{e}} = -\rho\frac{\partial\psi^{\mathrm{e}}}{\partial D^{\mathrm{B}}} \tag{2.123}$$

and the specific entropy production rate can be expressed as

$$\boldsymbol{\sigma}\dot{\boldsymbol{\varepsilon}}^{\mathrm{p}} + R\dot{r} + \mathbf{X}\dot{\boldsymbol{\alpha}} + Y^{\mathrm{e}}\dot{D}^{\mathrm{B}} - q\frac{1}{T}\mathrm{grad}T \geq 0, \tag{2.124}$$

where T denotes absolute temperature, q is the heat flux vector, and A, $\widetilde{A}$ denote the initial and the fictive undamaged cross-sectional area, respectively. Also in this model the damage evolution influences the elastic energy release, whereas the inelastic energy is not affected by the continuous damage. Additionally, the hypothesis of complementary energy equivalence is employed in the derivation to obtain:

$$\psi^{\mathrm{e}} = \frac{1}{2}\widetilde{E}_{ijkl}\left(D^{\mathrm{B}}\right)\varepsilon^{\mathrm{e}}_{ij}\varepsilon^{\mathrm{e}}_{kl} = \frac{1}{2}E_{ijkl}\varepsilon^{\mathrm{e}}_{ij}\varepsilon^{\mathrm{e}}_{kl}e^{-2D^{\mathrm{B}}}, \tag{2.125}$$

$$\widetilde{\mathbf{E}} = \mathbf{E}\left(1-D\right)^2 = \mathbf{E}\exp\left(-2D^{\mathrm{B}}\right). \tag{2.126}$$

Note that (2.125) differs from previously used (2.51) in the application of energy equivalence principle and the damage measure D^{B} that differs from the Kachnov's type variable $D = 1 - \exp\left(-D^{\mathrm{B}}\right)$, $D^{\mathrm{B}} \in \langle 0, \infty\rangle$. Limiting ourselves to the isotropic elasticity law (2.88), employing (2.125), and decomposing the elastic strain energy to the shear strain energy and the volume dilation energy, we obtain

$$\begin{aligned}\varrho Y^{\mathrm{e}} = \varrho Y^{\mathrm{e}^{(\mathrm{S})}} + \varrho Y^{\mathrm{e}^{(\mathrm{H})}} &= \frac{\sigma^2_{\mathrm{eq}}}{2E\exp\left(-2D^{\mathrm{B}}\right)}\left[\frac{2}{3}\left(1+\nu\right)\right. \\ \left. + 3\left(1-3\nu\right)\left(\frac{\sigma_{\mathrm{H}}}{\sigma_{\mathrm{eq}}}\right)^2\right] &= \frac{\sigma^2_{\mathrm{eq}}R_\nu}{2E\exp\left(-2D^{\mathrm{B}}\right)},\end{aligned} \tag{2.127}$$

whereas the damage conjugate force $Y^{\mathrm{e}}\left(D^{\mathrm{B}}\right)$ is furnished now as follows:

$$Y^{\mathrm{e}}\left(D^{\mathrm{B}}\right) \stackrel{\mathrm{def}}{=} -\varrho\frac{\partial\Psi^{\mathrm{e}}}{\partial D^{\mathrm{B}}} = -\frac{\sigma^2_{\mathrm{eq}}\exp\left(\Psi^{\mathrm{e}}D^{\mathrm{B}}\right)}{E}R_\nu\left(\frac{\sigma_{\mathrm{H}}}{\sigma_{\mathrm{eq}}}\right). \tag{2.128}$$

To establish the damage evolution model, suppose there exists a dissipative potential ψ^{d} in the form (cf. Mou and Han, 1996)

$$\psi^{\mathrm{d}}\left(Y^{\mathrm{e}}, \dot{p}, p, D^{\mathrm{B}}, T\right) = CS\left(-\frac{Y^{\mathrm{e}}}{S}\right)^2\frac{\left(p_{\mathrm{cr}}-p\right)^{n-1}}{p^{2n}}e^{-2D^{\mathrm{B}}}\dot{p}, \tag{2.129}$$

where C, S, and n $(n<1)$ are temperature dependent material constants and p is the cumulative plastic strain, $\dot{p} = \left[(2/3)\,\dot{\varepsilon}^{\mathrm{p}}_{ij}\dot{\varepsilon}^{\mathrm{p}}_{ij}\right]^{1/2}$. Hence, for the constitutive equation of damage evolution the following is obtained

$$\dot{D}^{\mathrm{B}} = \frac{\partial\psi^{\mathrm{d}}}{\partial Y^{\mathrm{e}}} = -\frac{2C}{ES}\sigma^2_{\mathrm{eq}}R_\nu\left(\frac{\sigma_{\mathrm{H}}}{\sigma_{\mathrm{eq}}}\right)\frac{\left(p_{\mathrm{cr}}-p\right)^{n-1}}{p^{2n}}\dot{p} \tag{2.130}$$

or, assuming the Ramberg–Osgood hardening law $\sigma_{\mathrm{eq}} = Kp^n$, a particular representation of damage evolution holds:

$$\dot{D}^{\mathrm{B}} = -2\frac{CK^2}{ES}R_\nu\left(\frac{\sigma_{\mathrm{H}}}{\sigma_{\mathrm{eq}}}\right)\left(p_{\mathrm{cr}}-p\right)^{n-1}\dot{p} \tag{2.131}$$

or

$$\dot{D}^{\mathrm{B}} = -\frac{\sigma^2_{\mathrm{B}}}{2ES}R_\nu\left(\frac{\sigma_{\mathrm{H}}}{\sigma_{\mathrm{eq}}}\right)\left(p_{\mathrm{cr}}-p\right)^{n-1}\dot{p}, \tag{2.132}$$

where $\sigma_B^2 = 4CK^2$ is a new material constant which should be determined from the one-dimensional test, with (2.75) accounted for, to yield the critical damage

$$D_{cr} = D_0 + \frac{\sigma_B^2}{2nES}\left(\varepsilon_{cr} - \varepsilon_0\right)^n . \tag{2.133}$$

Eventually, integration of (2.132) under the assumption of constant R_ν gives the damage evolution law for proportional loading

$$D = D_{cr} - \frac{D_{cr} - D_0}{\left(\varepsilon_{cr} - \varepsilon_0\right)^n}\left(p_{cr} - p\right)^n R_\nu\left(\frac{\sigma_H}{\sigma_{eq}}\right) . \tag{2.134}$$

In the general case of nonproportional loading, (2.132) and (2.133) hold. Observe that in a particular case, $n = 1$, (2.134) reduces to (2.77) or (2.84) with $s = 1$. Damage evolution depends on the cumulative plastic strain p, the hardening exponent n, and the triaxiality ratio σ_H/σ_{eq}. The damage rate decreases as the exponent $0 < n < 1$ increases and becomes constant with p when $n \to 1$. On the other hand, the damage rate increases linearly with the triaxiality function R_ν.

2.3.4 Irreversible thermodynamics model of a coupled isotropic damage-thermoelastic-(visco)plastic material

I. General coupled state equations derived from irreversible thermodynamics

In Sect. 2.3.1 it was assumed that continuum damage evolution is manifested by elastic strain energy release only. In general, the inelastic (irreversible) energy associated with the strain hardening is also released with damage growth. A consistent unified model, based on the assumptions that variable Y associated with the isotropic damage internal variable D contains both the classical elastic (reversible) energy Y^e and the inelastic (irreversible) energy Y^{an}, was developed by Saanouni, Forster, and Ben Hatira (1994). The hypothesis of total energy equivalence is used to define the effective state variables in a fictive undamaged configuration, instead of the classical state variables in a damaged configuration (cf. Chow and Lu, 1992). Now, we introduce the mechanical flux vector $\dot{\mathbf{J}}$ and its thermodynamic conjugate force vector $\mathbf{F}$ as follows:

$$\dot{\mathbf{J}} = \left\{\dot{\boldsymbol{\varepsilon}}^p, \dot{\boldsymbol{\alpha}}, \dot{r}, \dot{D}, q\right\}^T ,$$
$$\mathbf{F} = \left\{\boldsymbol{\sigma}, \mathbf{X}, R, Y, -\frac{1}{T}\mathrm{grad}T\right\} , \tag{2.135}$$

such that the entropy production rate is written as

$$\rho\dot{\Theta} = \mathbf{F}\dot{\mathbf{J}} \geq 0. \tag{2.136}$$

The small strain is decomposed to the elastic $\boldsymbol{\varepsilon}^{\mathrm{e}}$ and inelastic $\boldsymbol{\varepsilon}^{\mathrm{an}}$ parts and the total energy equivalence is applied (independently) to the elastic Φ^{e} and the inelastic Φ^{kh} and Φ^{ih} energy portions, respectively, in the damaged and the fictive undamaged configurations:

$$\begin{aligned}
&\Phi^{\mathrm{e}}\left(\boldsymbol{\varepsilon}^{\mathrm{e}}, D\right) = \tfrac{1}{2}\boldsymbol{\sigma} : \boldsymbol{\varepsilon}^{\mathrm{e}} = \tfrac{1}{2}\widetilde{\boldsymbol{\sigma}} : \widetilde{\boldsymbol{\varepsilon}}^{\mathrm{e}},\\
&\Phi^{\mathrm{kh}}\left(\boldsymbol{\alpha}, D\right) = \tfrac{1}{2}\mathbf{X} : \boldsymbol{\alpha} = \tfrac{1}{2}\widetilde{\mathbf{X}} : \widetilde{\boldsymbol{\alpha}},\\
&\Phi^{\mathrm{ih}}\left(r, D\right) = \tfrac{1}{2}rR = \tfrac{1}{2}\widetilde{r}\widetilde{R}.
\end{aligned} \tag{2.137}$$

Hence, the couples of effective state variables are given by

$$\begin{aligned}
&\widetilde{\boldsymbol{\sigma}} = \frac{\boldsymbol{\sigma}}{g_e\left(D\right)}, \quad \widetilde{\boldsymbol{\varepsilon}}^{\mathrm{e}} = g_e\left(D\right)\boldsymbol{\varepsilon}^{\mathrm{e}};\\
&\widetilde{\mathbf{X}} = \frac{\mathbf{X}}{h_\alpha\left(D\right)}, \quad \widetilde{\boldsymbol{\alpha}}^{\mathrm{e}} = h_\alpha\left(D\right)\boldsymbol{\alpha};\\
&\widetilde{R} = \frac{R}{h_r\left(D\right)}, \quad \widetilde{r} = h_r\left(D\right)r;
\end{aligned} \tag{2.138}$$

where $g_e\left(D\right)$, $h_\alpha\left(D\right)$, $h_r\left(D\right)$ are positive decaying functions of D defined as follows:

$$g_e\left(D\right) = h_\alpha\left(D\right) = h_r\left(D\right) = \left(1 - D\right)^{1/2}. \tag{2.139}$$

Hence, the effective state variables are used in the state potential instead of the classical state variables, and the Helmholtz free energy is taken as a state potential

$$\psi^{\mathrm{H}}\left(\boldsymbol{\varepsilon}^{\mathrm{e}}, \boldsymbol{\alpha}, r, D, T\right) = \psi^{\mathrm{e}}\left(\widetilde{\boldsymbol{\varepsilon}}^{\mathrm{e}}, T\right) + \psi^{\mathrm{an}}\left(\widetilde{\boldsymbol{\alpha}}, \widetilde{r}\right), \tag{2.140}$$

where

$$\begin{aligned}
&\rho\psi^{\mathrm{e}}\left(\widetilde{\boldsymbol{\varepsilon}}^{\mathrm{e}}, T\right) = \tfrac{1}{2}\widetilde{\boldsymbol{\varepsilon}}^{\mathrm{e}} : \boldsymbol{\Lambda} : \widetilde{\boldsymbol{\varepsilon}}^{\mathrm{e}} - \left(T - T_0\right)\mathbf{k} : \widetilde{\boldsymbol{\varepsilon}}^{\mathrm{e}} - \rho c_v T\left[\log\left(\frac{T}{T_0}\right) - 1\right],\\
&\rho\psi^{\mathrm{an}}\left(\widetilde{\boldsymbol{\alpha}}, \widetilde{r}, T\right) = \tfrac{1}{3}C\widetilde{\boldsymbol{\alpha}} : \widetilde{\boldsymbol{\alpha}} + \tfrac{1}{2}Q\widetilde{r}^2.
\end{aligned} \tag{2.141}$$

In the above equations, $\boldsymbol{\Lambda}$ is the symmetric fourth-rank elastic stiffness tensor for undamaged material such that $\widetilde{\boldsymbol{\sigma}} = \boldsymbol{\Lambda} : \widetilde{\boldsymbol{\varepsilon}}^{\mathrm{e}}$ (with thermal terms omitted), C and Q denote the temperature dependent kinematic and isotropic hardening moduli, $\mathbf{k}$ is the symmetric second-order tensor of thermal conductivity, ρ is the mass density, c_v the specific heat, and T_0 the initial temperature.

At variance with the model from Sect. 2.3.1(II) (2.65) the damage affects both the elastic (reversible) and the inelastic (irreversible) energy portions and the effective state variables $\widetilde{\boldsymbol{\varepsilon}}^{\mathrm{e}}$, $\widetilde{\boldsymbol{\alpha}}$, $\widetilde{r}$ are consistently used for the state potential (free energy) of damaged material. Additionally, the inelastic flow is assumed to be initially isotropic and of the Mises-type, thermoelastic behavior is assumed to be linear and free from the inelastic effect, damage is assumed to be isotropic, and the nonlinear kinematic-isotropic hardening law is applied. Eventually, the state equations are furnished from the state potential in the following manner:

$$
\begin{aligned}
\boldsymbol{\sigma} &= \rho \frac{\partial \psi^{\mathrm{H}}}{\partial \boldsymbol{\varepsilon}^{e}} = \widetilde{\boldsymbol{\Lambda}} : \boldsymbol{\varepsilon}^{\mathrm{e}} - (T - T_0)\, \widetilde{\mathbf{k}},\\
\mathbf{X} &= \rho \frac{\partial \psi^{\mathrm{H}}}{\partial \boldsymbol{\alpha}} = \frac{2}{3} \widetilde{C} \boldsymbol{\alpha},\\
R &= \rho \frac{\partial \psi^{\mathrm{H}}}{\partial r} = \widetilde{Q} r, \qquad (2.142)\\
s &= -\rho \frac{\partial \psi^{\mathrm{H}}}{\partial T} = \frac{1}{\rho} \widetilde{\mathbf{k}} : \boldsymbol{\varepsilon}^{\mathrm{e}} + c_v \log\left(\frac{T}{T_0}\right),\\
Y &= -\rho \frac{\partial \psi^{\mathrm{H}}}{\partial D} = Y^{\mathrm{e}} + Y^{\mathrm{an}},
\end{aligned}
$$

where the elastic energy and inelastic energy release rates are given by

$$
\begin{aligned}
Y^{\mathrm{e}} &= -\rho \frac{\partial \psi^{\mathrm{e}}}{\partial D} = \frac{1}{2} \boldsymbol{\varepsilon}^{\mathrm{e}} : \boldsymbol{\Lambda} : \boldsymbol{\varepsilon}^{\mathrm{e}} - \frac{1}{2} (T - T_0) \frac{\mathbf{k}}{(1-D)^{1/2}} : \boldsymbol{\varepsilon}^{\mathrm{e}},\\
Y^{\mathrm{an}} &= -\rho \frac{\partial \psi^{\mathrm{an}}}{\partial D} = \frac{1}{3} C \boldsymbol{\alpha} : \boldsymbol{\alpha} + \frac{1}{2} Q r^2
\end{aligned}
\qquad (2.143)
$$

and effective thermo-mechanical moduli for damaged material are introduced:

$$
\widetilde{\boldsymbol{\Lambda}} = (1-D)\, \boldsymbol{\Lambda}, \quad \widetilde{C} = (1-D)\, C, \quad \widetilde{Q} = (1-D)\, Q, \quad \widetilde{\mathbf{k}} = (1-D)^{1/2}\, \mathbf{k}. \qquad (2.144)
$$

Let us mention that in this model the fully damaged RVE is free not only from the Cauchy stress $\boldsymbol{\sigma}$ but also from the internal stresses $\mathbf{X}$ and R and it is fully unable to support heat conduction. A more developed model accounting for a combined conduction/radiation heat transfer mechanism is discussed in Sect. 5.1.

II. Time-independent nonlinear isotropic/kinematic hardening coupled with isotropic damage

State equations (2.142)–(2.144), which contain scalar damage variable D representing the actual continuum damage state, require a suitable dissipation potential. In case of a plastic dissipation the plastic flow $\boldsymbol{\varepsilon}^{\mathrm{an}}=\boldsymbol{\varepsilon}^{\mathrm{p}}$ is time-independent. A consistent coupled yield function is obtained as the extension of the Chaboche–Rousselier uncoupled yield function of the virgin nonlinear isotropic/kinematic hardening material, when the classical state variables ($\boldsymbol{\sigma}$, $\mathbf{X}$, R) in (2.17) are replaced by the effective variables ($\widetilde{\boldsymbol{\sigma}}$, $\widetilde{\mathbf{X}}$, $\widetilde{R}$). In other words, the Mises-type yield function governs the plastic dissipation coupled with damage in space of effective dual variables (Saanouni, Forster, and Ben Hatira, 1994)

$$f\left(\widetilde{\boldsymbol{\sigma}},\widetilde{\mathbf{X}},\widetilde{R}\right)=J_2\left(\widetilde{\boldsymbol{\sigma}}-\widetilde{\mathbf{X}}\right)-\widetilde{R}-\sigma_y=0 \tag{2.145}$$

and the fully coupled plastic potential, which generalizes Chaboche's equation (2.64), may be written as

$$F^{\mathrm{SFB}}\left(\widetilde{\boldsymbol{\sigma}},\widetilde{\mathbf{X}},\widetilde{R},D,T\right)=f+\frac{1}{2}\frac{a}{C}J_2^2\left(\widetilde{\mathbf{X}}\right)+\frac{1}{2}\frac{b}{Q}\widetilde{R}^2+F^{\mathrm{D}}\left(Y\right), \tag{2.146}$$

where, following Germain, Nguyen, and Suquet (1983), the damage evolution potential $F^{\mathrm{D}}\left(Y\right)$ is supposed to be a power function of the total (elastic and inelastic) energy release due to damage evolution $Y=Y^{\mathrm{e}}+Y^{\mathrm{an}}$ (extension of (2.70))

$$F^{\mathrm{D}}\left(Y\right)=-\frac{S}{(s+1)}\left(\frac{Y}{S}\right)^{s+1}\frac{1}{(1-D)^{\beta}}. \tag{2.147}$$

The present unified formulation assumes the same (single) potential to describe both the plastic dissipation and the damage dissipation. In other words, in this model, which is well applicable for ductile metals, it is supposed that damage cannot initiate without plastic deformation. However, it is not true in case of brittle materials, geomaterials, or composite materials, where more advanced multisurface theory must be developed. Moreover, by contrast to the Chaboche's fully associative model (2.64)–(2.65), the present model is associative with respect to the Cauchy stress $\boldsymbol{\sigma}$, but non-associative with respect to the internal variables $\mathbf{X}$, R, Y. Hence, the state equations are obtained by the generalized normality rule (2.65), with F^{Ch} (2.64) replaced by F^{SFB} (2.146)–(2.147)

$$\dot{\varepsilon}^{\mathrm{p}}_{ij} = \dot{\lambda}\frac{\partial F^{\mathrm{SFB}}}{\partial \sigma_{ij}} = \frac{3}{2}\frac{\dot{\lambda}}{(1-D)^{1/2}}\frac{\sigma'_{ij}-X'_{ij}}{J_2\left(\sigma_{ij}-X_{ij}\right)},$$

$$\dot{\alpha}_{ij} = -\dot{\lambda}\frac{\partial F^{\mathrm{SFB}}}{\partial X_{ij}} = \frac{3}{2}\frac{\dot{\lambda}}{(1-D)^{1/2}}\frac{\sigma'_{ij}-X'_{ij}}{J_2\left(\sigma_{ij}-X_{ij}\right)} - \frac{3a}{2C}\frac{\dot{\lambda}}{(1-D)}X'_{ij},$$

$$\dot{r} = -\dot{\lambda}\frac{\partial F^{\mathrm{SFB}}}{\partial R} = \dot{\lambda}\left[\frac{1}{(1-D)^{1/2}} - \frac{b}{Q}\frac{R}{(1-D)}\right],$$

$$\dot{D} = -\dot{\lambda}\frac{\partial F^{\mathrm{SFB}}}{\partial Y} = -\dot{\lambda}\left(\frac{Y}{S}\right)^{s}\frac{1}{(1-D)^{\beta}} \tag{2.148}$$

or, by employing (2.142) the equivalent form can be furnished

$$\dot{\varepsilon}^{\mathrm{p}}_{ij} = \frac{3}{2}\dot{\tilde{\lambda}}\frac{\sigma'_{ij}-X'_{ij}}{J_2\left(\sigma_{ij}-X_{ij}\right)} = \frac{3}{2}\dot{p}\frac{\sigma'_{ij}-X'_{ij}}{J_2\left(\sigma_{ij}-X_{ij}\right)},$$

$$\dot{\alpha}_{ij} = \dot{\varepsilon}^{\mathrm{p}}_{ij} - a\dot{\lambda}\alpha_{ij} = \dot{\varepsilon}^{\mathrm{p}}_{ij} - \dot{p}\frac{3aX'_{ij}}{2C\left(1-D\right)^{1/2}}, \tag{2.149}$$

$$\dot{r} = \dot{\tilde{\lambda}}\left(1-b\tilde{r}\right) = \dot{p}\left(1-\frac{b}{Q}\tilde{R}\right),$$

$$\dot{D} = -\dot{\lambda}\left(\frac{Y}{S}\right)^{s}\frac{1}{(1-D)^{\beta}} = -\dot{p}\left(\frac{Y}{S}\right)^{s}(1-D)^{\frac{1}{2}-\beta},$$

where

$$\dot{\tilde{\lambda}} = \frac{\dot{\lambda}}{(1-D)^{1/2}} = \dot{p} = \left(\frac{2}{3}\dot{\varepsilon}^{\mathrm{p}}_{ij}\dot{\varepsilon}^{\mathrm{p}}_{ij}\right)^{1/2}. \tag{2.150}$$

It is observed that substitution of $a = b = 0$ reduces state equations (2.149) to the fully coupled linear hardening theory:

$$\dot{\varepsilon}^{\mathrm{p}}_{ij} = \frac{3}{2}\dot{p}\frac{\sigma'_{ij}-X'_{ij}}{J_2\left(\sigma_{ij}-X_{ij}\right)},$$

$$\dot{\alpha}_{ij} = \dot{\varepsilon}^{\mathrm{p}}_{ij}, \tag{2.151}$$

$$\dot{r} = \dot{p},$$

$$\dot{D} = -\dot{p}\left(1-D\right)^{1/2}\left(\frac{Y}{S}\right)^{s}\frac{1}{(1-D)^{\beta}},$$

where plastic strain $\boldsymbol{\varepsilon}^{\mathrm{p}}$ and cumulative plastic strain p may be identified with internal variables of kinematic and isotropic hardening; however, these equations differ from Chaboche's equations (2.66) where the classical effective stress concept has been used, $\widetilde{\boldsymbol{\sigma}} = \boldsymbol{\sigma}/(1-D)$, and the elastic energy release rate Y^{e} has been applied as the thermodynamic conjugate force of damage D.

III. Time-dependent viscoplastic flow coupled with isotropic damage

In case of time-dependent coupled damage-creep-isotropic/kinematic hardening material the single surface coupled viscodamage dissipation potential may be expressed as a sum of the viscoplastic and the creep-damage parts (cf. Saanouni, Forster, and Ben Hatira, 1994)

$$\Phi^{*}\left(\widetilde{\boldsymbol{\sigma}}, \widetilde{\mathbf{X}}, \widetilde{R}, D, T\right) = \Phi^{*^{\mathrm{vp}}}\left(\widetilde{\boldsymbol{\sigma}}, \widetilde{\mathbf{X}}, \widetilde{R}, T\right) + \Phi^{*^{\mathrm{D}}}\left(\boldsymbol{\sigma}, D, T\right), \tag{2.152}$$

where the viscoplastic term is represented by a following power function of f extended by the additional terms representing nonlinear hardening

$$\Phi^{*^{\mathrm{vp}}} = \frac{K}{n+1}\left\langle \frac{1}{K}\left[f + \frac{3}{4}\frac{a}{C}\widetilde{\mathbf{X}} : \widetilde{\mathbf{X}} - \frac{1}{3}aC\widetilde{\boldsymbol{\alpha}} : \widetilde{\boldsymbol{\alpha}} + \frac{1}{2}\frac{b}{Q}\widetilde{R}^{2} - \frac{1}{2}bQ\widetilde{r}^{2}\right]\right\rangle^{n+1}, \tag{2.153}$$

whereas the creep-damage term is given by

$$\Phi^{*^{\mathrm{D}}} = -Y\left\langle \frac{\chi(\boldsymbol{\sigma})}{A}\right\rangle^{r}(1-D)^{-k}. \tag{2.154}$$

In the above equations, the function $f\left(\widetilde{\boldsymbol{\sigma}}, \widetilde{\mathbf{X}}, \widetilde{R}\right)$ denotes the Mises-type isotropic/kinematic hardening yield function as defined by (2.145); a, b, C, Q are temperature dependent hardening parameters, K and n characterize creep behavior of the material, a scalar function $\chi(\boldsymbol{\sigma})$ represents the Hayhurst-type damage equivalent stress as defined by (2.36) and A, r, k characterize creep-damage under multiaxial stress according to Chaboche, (2.35).

Hence, the following definitions hold:

$$\begin{aligned}
&f\left(\widetilde{\boldsymbol{\sigma}}, \widetilde{\mathbf{X}}, \widetilde{R}\right) = J_2\left(\widetilde{\boldsymbol{\sigma}} - \widetilde{\mathbf{X}}\right) - \widetilde{R} - \sigma_y > 0,\\
&\chi(\boldsymbol{\sigma}) = \alpha J_0(\boldsymbol{\sigma}) + 3\beta J_1(\boldsymbol{\sigma}) + (1-\alpha-\beta)\, J_2(\boldsymbol{\sigma}),\\
&J_0(\boldsymbol{\sigma}) = \max \sigma_i, \quad J_1(\boldsymbol{\sigma}) = \frac{\sigma_{ii}}{3}, \quad J_2(\boldsymbol{\sigma}) = \left[\left(\frac{3}{2}\right)\sigma'_{ij}\sigma'_{ij}\right]^{1/2},\\
&J_2\left(\widetilde{\boldsymbol{\sigma}} - \widetilde{\mathbf{X}}\right) = \left[\frac{3}{2}\left(\sigma'_{ij} - X'_{ij}\right)\left(\sigma'_{ij} - X'_{ij}\right)\right]^{1/2},
\end{aligned} \tag{2.155}$$

and symbols denoted by primes $(\,)'$ represent deviatoric components, whereas the tilde $(\widetilde{\ })$ refers to the effective state variables, (2.138). Note that in view of the effective state variables definitions (2.138), and the state equations (2.142), the additional terms in the MacAuley bracket $\langle . \rangle$, Eq. (2.153), which was introduced in order to describe a nonlinear hardening effect on the damage-viscoplastic dissipation potential, may be reduced to the linear hardening case if the following holds:

$$\widetilde{\mathbf{X}} = \frac{2}{3} C \widetilde{\boldsymbol{\alpha}} \quad \text{or} \quad \mathbf{X} = \frac{2}{3} \widetilde{C} \boldsymbol{\alpha}; \quad \widetilde{R} = Q \widetilde{r} \quad \text{or} \quad R = \widetilde{Q} r \tag{2.156}$$

and

$$\widetilde{C} = (1 - D)\, C, \quad \widetilde{Q} = (1 - D)\, Q. \tag{2.157}$$

Eventually, if the generalized normality rule is applied to the single coupled viscoplastic-damage potential, (2.152)–(2.154), the state equations are established as:

$$\begin{aligned}
\dot{\varepsilon}_{ij}^{\mathrm{vp}} &= \frac{\partial \Phi^*}{\partial \sigma_{ij}} = \frac{3}{2} \frac{\langle f/K \rangle^n}{(1-D)^{1/2}} \frac{\sigma'_{ij} - X'_{ij}}{J_2(\sigma_{ij} - X_{ij})}, \\
\dot{\alpha}_{ij} &= -\frac{\partial \Phi^*}{\partial X_{ij}} = \frac{3}{2} \frac{\langle f/K \rangle^n}{(1-D)^{1/2}} \left[\frac{\sigma'_{ij} - X'_{ij}}{J_2(\sigma_{ij} - X_{ij})} - \frac{a}{C} \frac{X_{ij}}{(1-D)^{1/2}} \right], \\
\dot{r} &= -\frac{\partial \Phi^*}{\partial R} = \frac{\langle f/K \rangle^n}{(1-D)^{1/2}} \left[1 - \frac{b}{Q} \widetilde{R} \right], \\
\dot{D} &= -\frac{\partial \Phi^*}{\partial Y} = \left[\frac{\chi(\sigma_{ij})}{A} \right]^r (1-D)^{-k}
\end{aligned} \tag{2.158}$$

or, in an equivalent form,

$$\begin{aligned}
\dot{\varepsilon}_{ij}^{\mathrm{vp}} &= \frac{3}{2} \dot{\widetilde{\lambda}}^{\mathrm{vp}} \frac{\sigma'_{ij} - X'_{ij}}{J_2(\sigma_{ij} - X_{ij})}, \\
\dot{\alpha}_{ij} &= \dot{\varepsilon}_{ij}^{\mathrm{vp}} - a \dot{\lambda}^{\mathrm{vp}} \alpha_{ij}, \\
\dot{r} &= \dot{\widetilde{\lambda}}^{\mathrm{vp}} (1 - b\widetilde{r}), \\
\dot{D} &= \left[\frac{\chi(\sigma_{ij})}{A} \right]^r (1-D)^{-k},
\end{aligned} \tag{2.159}$$

where

$$\dot{\lambda}^{vp} = \left\langle \frac{f}{K} \right\rangle^n, \quad \dot{\tilde{\lambda}}^{vp} = \frac{\dot{\lambda}^{vp}}{(1-D)^{1/2}}; \tag{2.160}$$

and the MacAuley bracket $\langle F \rangle$ is defined as: $\langle F \rangle = 0$ if $F < 0$, whereas $\langle F \rangle = F$ if $F > 0$. Note that (2.159 and 2.160) represent the time-dependent fully coupled visco-plastic-damage state equations, whereas (2.149 and 2.150) describe the time-independent coupled plastic-damage dissipation although, the representation for the state variables $(\dot{\boldsymbol{\varepsilon}}^{vp}, \dot{\boldsymbol{\alpha}}, \dot{r})$ and $(\dot{\boldsymbol{\varepsilon}}^{p}, \dot{\boldsymbol{\alpha}}, \dot{r})$ is in both cases (i.e., both (2.159) and (2.149)) analogous, although different definitions for $\dot{\lambda}$ and $\dot{\lambda}^{vp}$ hold.

3

Three-dimensional anisotropic damage representation

3.1 Damage anisotropy

3.1.1 Directional nature of damage

In Sect. 2.1.2, two basic damage mechanisms in crystalline metallic materials were distinguished.

The first damage mechanism, called the ductile or transgranular damage mode, is predominant at room (or low) temperature and high stress level tests when the slipbands of plasticity are formed in favorably oriented grains. The microslips are inclined roughly at 45° to the main stress direction and the coupled damage-(visco)plasticity mechanism may approximately be described by the isotropic (scalar) damage internal variable D, the evolution of which may be governed by the elastic energy release rate (Lemaitre and Chaboche, 1985) or the total (elastic and inelastic) energy release (Saanouni, Forster, and Ben Hatira, 1994) in a more general case. The material instability from microslips initiation eventually yields a discontinuous bifurcation of the velocity field (cf. Runesson et al., 1991; Shrayer and Zhou, 1995). The plastic strain localization in zones of microvoid concentration leads to a failure mode with material separation and the formation of free surfaces (decohesion) on the macrolevel. The macrocracks are formed in a transgranular mode with a preferable inclination that coincides with the directions of slipbands of plasticity (Fig. 2.1).

The second damage mechanism, usually identified for simplicity with brittle or intergranular damage, is representative for high temperature but rather low stress level loading conditions. It is mainly based on the microcracking process initiated at the grain (or subgrain) boundaries, and it is recognized to be controlled by the maximal stress-rather than the effective stress, such that the normal to the microcrack direction coincides with the principal stress direction at the point considered (Fig. 2.2). The macrocracking process may be observed at selected grain boundaries to result from coalescence of microcracks of similar average orientation. No, or negligibly small, plastic deformations precede the damage evolution, hence; pure brittle failure mechanism occurs. The discontinuous and heterogeneous damaged solid is approximated by the pseudo-undamaged continuum by the use of the couples of effective state variables, the definitions of which depend on the equivalence principles employed. In such a case, however, the damage evolution in the elastic-brittle or creep materials is no longer isotropic; hence, unlike the ductile damage phenomenon, brittle damage behavior is anisotropic in nature, so that the description by scalar internal variable(s) is insufficient. The essentially anisotropic description of damage in the elastic-brittle or creep solids by the development of distributed and oriented microscopic cracks require damage variables ranging from a vector to second or higher-rank tensors (Vakulenko and Kachanov, 1971; Krajcinovic et al., 1981, 1983, 1993; Murakami et al., 1981, 1983, 1987, 1988; Murakami and Kamiya, 1997 and others).

The damage anisotropy is easily observed in metal specimens subjected to creep under nonproportional loading conditions. Microstructural observations by Trąpczyński, Hayhurst, and Leckie (1981) allowed the identification of two classes of metallic materials: copper-like, where cavitation takes place on grain boundaries essentially perpendicular to the maximum principal stress, and aluminum alloy-like, where grain boundary cavitation is much more isotropically distributed (cf. also Hayhurst and Felce, 1986). The complexity of the damage accumulation depends on the loading path or on the rotation of principal stress axes with respect to material fibers. Thin cylindrical copper tubes were tested to failure at a temperature of 250°C under the following programs:

i. steady load (constant principal stress direction),

ii. single reverse torsion, steady tension (single principal axes jump),

iii. multiple reverse torsion, steady tension (multiple principal stress axes rotations).

In the case of steady load, the majority of cracks are found on planes perpendicular to the maximum principal tension stress σ_1 (Fig. 3.1a). In the single reverse torsion steady tension test, two failure planes may be observed, each corresponding to the principal stress plane (Fig. 3.1b). In the case of a multiple-reverse torsion steady tension test, the crack planes of different orientation within the angle between two principal stress planes might be recognized (Fig. 3.1c). For the lifetime and deformation prediction, the single damage variable theory by Leckie and Hayhurst (1974) (2.39) was employed for both copper and aluminum specimens under nonproportional loadings. For aluminum alloy, the lifetime and deformation prediction were in sufficient accord with the experimental results. However, for copper, strain rate discrepancies with a factor of two were reached.

3.1.2 Damage variables review

The crucial problem for continuum damage mechanics is the proper and accurate modeling of material damage. In all cases of various equivalence principles it is assumed that in a quasicontinuum the true distribution of

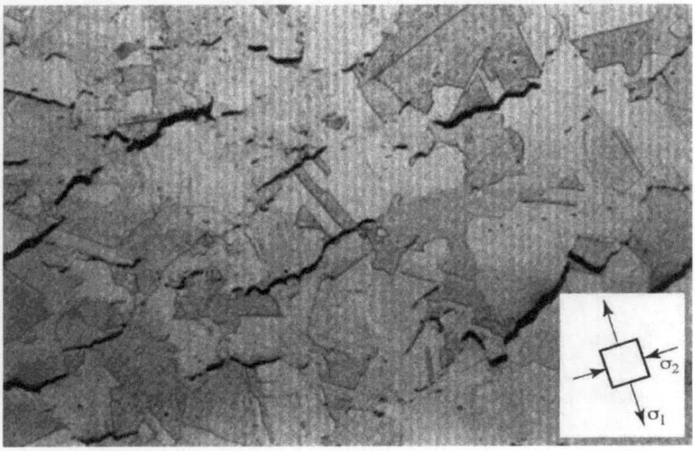

Fig. 3.1a. Mid-thickness micrograph of a copper tube tested to failure under steady load (magnification ×75) (after Trąpczyński et al., 1981)

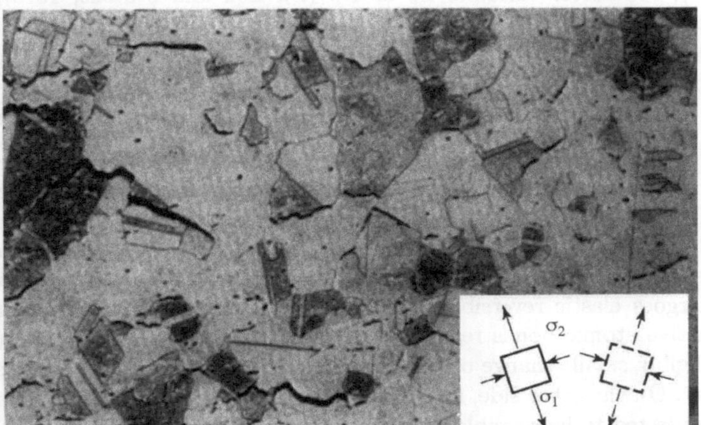

Fig. 3.1b. Mid-thickness micrograph of a copper tube tested to failure under single reverse torsion, steady tension loading (magnification ×65) (after Trąpczyński et al., 1981)

Fig. 3.1c. Mid-thickness micrograph of a copper tube tested to failure under multiple torsion steady tension (magnification ×65) (after Trąpczyński et al., 1981)

defects is smeared out and homogenized by properly defined internal variables that characterize damage: the scalar variables ω or D (Kachanov, 1958), the vector variables ω_ν or D_ν (Davison and Stevens, 1973), the second-rank tensor variables $\boldsymbol{\Omega}$, $\mathbf{D}$ (Vakulenko and Kachanov, 1971; Murakami and Ohno, 1981), or the fourth-rank tensor variables $\widehat{\mathbf{D}}$ (Chaboche, 1982; Krajcinovic, 1989; etc.). In general, damage may be characterized by the set $\mathcal{D}$ of scalars, vectors, and/or second, fourth or higher-rank tensors that function as internal variables $\mathcal{D} = \left\{D, D_\nu, \mathbf{D}, \widehat{\mathbf{D}}, \ldots\right\}$. The extended damage variables review used to describe the damage process is presented in Table 3.1.

Roughly speaking, in a ductile deformation process of crystalline materials, a flow of mass through the lattice takes place, at which the lattice undergoes elastic reversible deformation only, whereas the total number of active atomic bonds remains approximately constant. Hence, no (or a negligibly small) change of the effective material properties is assumed to occur. On the other side, in a brittle deformation process the lattice itself is subjected to irreversible changes resulting from breaking of the atomic bonds and, hence, a progressive material degradation through strength and stiffness reduction takes place. This fully coupled CDM approach to the elastic-brittle damage or creep damage, when the damage evolution influences both the stress and strain state and also the elastic properties, leads to the concept of fourth-rank elasticity tensors modified by damage $\mathcal{D}$, stiffness $\widetilde{\mathbf{\Lambda}}(\mathcal{D})$, or compliance $\widetilde{\mathbf{\Lambda}}^{-1}(\mathcal{D})$:

Table 3.1. Damage variables review (cf. Murakami, 1987; Skrzypek and Ganczarski, 1998a)

Reference	Material damage
Scalar damage variables	
Kachanov, 1958	creep (isotropic)
Kachanov, 1974	creep (anisotropic)
Rabotnov, 1968 (1969)	creep (anisotropic)
Martin–Leckie, 1972	creep (anisotropic)
Hayhurst–Leckie, 1973	creep (anisotropic)
Davison et al., 1977	spalling, elastic
Gurson, 1977	elastic-plastic
Trączyński et al., 1981	creep (nonproportional)
Lemaitre–Chaboche, 1978	creep (anisotropic)
Chaboche, 1988	creep, fatigue, ductile, brittle, anisotropic
Lemaitre, 1987, 1992	general
Rides et al., 1989	effect of creep damage on elastic properties
Randy–Cozzarelli, 1988	propagation of rupture
Murakami–Mizuno, 1992	creep under irradiation
Zheng–Lee, 1993	creep, high temperature
Scalar: $\omega(x)$ [Kachanov] (isotropic) continuity: $\psi = \dfrac{A_{\text{ef}}}{A_0}$ damage: $\omega = 1 - \psi$ effective stress: $\widetilde{\sigma} = \dfrac{\sigma}{\psi} = \dfrac{\sigma}{1-\omega}$	
$\varrho(\mathbf{n})$ – nonuniform distribution of damage (defined on a unit sphere) [Krajcinovic]: $\varrho(\mathbf{n}) = \varrho_0$ isotropic $\varrho_0 = \int_{4\pi} \varrho(\mathbf{n})\mathrm{d}A$ nearly isotropic	
Vector damage variables	
Davison–Stevens, 1973	spalling, elastic
Kachanov, 1974, 1986	creep
Krajcinovic–Fonseka, 1981	elastic-brittle
Krajcinovic, 1983	general, creep
Singh–Digby, 1989	brittle solid, anisotropic
Lubarda–Krajcinovic, 1993	general
Vector: $\boldsymbol{\omega}(x)$ [Kachanov]: $\boldsymbol{\omega} = \omega_\nu \boldsymbol{\nu}$ $\boldsymbol{\psi} = \psi_\nu \boldsymbol{\nu}$ $\widetilde{\sigma}_\nu = \dfrac{\sigma_\nu}{\psi_\nu}$	

Second-rank damage tensor	
Murzewski,1957,1958,1992	quasi-homogeneous metals (stochastic approach)
Rabotnov, 1969	creep
Vakulenko–Kachanov, 1971	plastic-brittle
Murakami–Ohno, 1981	creep
M. Kachanov, 1972	elastic
Dragon–Mróz, 1979	brittle-plastic
Cordebois–Sidoroff, 1982	elastic, elastic-plastic
Betten, 1983	general, creep
Litewka, 1985, 1987, 1989	creep (anisotropic)
Kondaurov, 1988	elastic (orthotropic)
Murakami, 1983, 1987, 1988	creep, fatigue
Karihaloo–Fu, 1989	concrete
Chow–Lu, 1992	anisotropic, elastic-plastic
Lis, 1992	(nonproportional loading)
Chaboche, 1993	damage induced elastic
Lis–Litewka, 1996	anisotropy
Zheng–Betten, 1996	effective stress review
Murakami–Kamiya, 1997	elastic-brittle anisotropy
Skrzypek et al., 1998a	elastic-brittle anisotropy

Second-rank tensor [Murakami–Ohno]: $\boldsymbol{\Omega} = \sum_{i=1}^{3} \Omega_i \mathbf{n}_i \mathbf{n}_i$

$\boldsymbol{\Omega} = \mathbf{1} - \boldsymbol{\Psi}$, $\widetilde{\boldsymbol{\sigma}} = \frac{1}{2}[(\mathbf{1} - \boldsymbol{\Omega})^{-1} : \boldsymbol{\sigma} + \boldsymbol{\sigma} : (\mathbf{1} - \boldsymbol{\Omega})^{-1}]$

$\varrho(\mathbf{n}) = \varrho_{kl} n_k n_l$ second rank-crack density tensor [Krajcinovic]:

$D_{ij} = \int_{4\pi} \varrho(\mathbf{n}) n_i n_j \mathrm{d}A$

Fourth-rank damage tensor	
Chaboche, 1982	creep (anisotropic)
Leckie–Onat, 1981	creep
Simo–Ju, 1987	general
Chow–Wang, 1987	general, anisotropic
Krajcinovic, 1989	general, anisotropic
Lubarda–Krajcinovic, 1993	general, anisotropic
Schiesse, 1994	elastic-plastic, anisotropic
Chen–Chow, 1995	damage effect tensor, anisotropic
Voyiadjis–Park, 1996	anisotropic damage, plasticity
Qi–Bertram, 1997	single crystal, anisotropic

Fourth-rank tensor [Chaboche]:

$\widehat{\mathbf{D}}$-fourth-rank damage tensor

$\widetilde{\mathbf{E}}$-fourth-rank elastic tensor, $\widetilde{\mathbf{E}}(\widehat{\mathbf{D}}) = (\mathbf{I} - \widehat{\mathbf{D}}) : \mathbf{E}$, $\widetilde{\boldsymbol{\sigma}} = (\mathbf{I} - \widehat{\mathbf{D}})^{-1} : \boldsymbol{\sigma}$

$\widetilde{\boldsymbol{\sigma}} = \mathbf{M}(\widehat{\mathbf{D}}) : \boldsymbol{\sigma}$, $\mathbf{M}(\widehat{\mathbf{D}})$ – fourth-rank damage effect tensor

$\varrho(\mathbf{n}) = \varrho_{ijkl} n_i n_j n_k n_l$ fourth-rank crack density tensor [Krajcinovic]

$\widehat{D}_{ijkl} = \int_{4\pi} \varrho(\mathbf{n}) n_i n_j n_k n_l \mathrm{d}A$

Eighth-rank damage tensor	
Chaboche, 1981	creep, fatigue

$$\boldsymbol{\sigma} = \widetilde{\boldsymbol{\Lambda}}(\mathcal{D}) : \boldsymbol{\varepsilon}^{\mathrm{e}} \qquad \text{or} \qquad \boldsymbol{\varepsilon}^{\mathrm{e}} = \widetilde{\boldsymbol{\Lambda}}^{-1}(\mathcal{D}) : \boldsymbol{\sigma}, \tag{3.1}$$

where $\mathcal{D}$ stands for properly selected damage variables (cf. Litewka, 1985, 1989; Chen and Chow, 1995; Murakami and Kamiya, 1997, etc.). A general concept of the fourth-rank damage effect tensor $\mathbf{M}(\mathcal{D})$ that transforms the Cauchy stress tensor in a damaged configuration $\boldsymbol{\sigma}$ to the effective (conjugate) Cauchy stress tensor in an equivalent fictive pseudo-undamaged solid $\widetilde{\boldsymbol{\sigma}}$, based on the appropriate damage equivalence hypothesis (strain or stress or complementary energy or total energy equivalence) takes into account the fully anisotropic nature of damage in the form (cf. Chow and Lu, 1992; Zheng and Betten, 1996)

$$\widetilde{\boldsymbol{\sigma}} = \mathbf{M}(\mathcal{D}) : \boldsymbol{\sigma} \qquad \text{or} \qquad \widetilde{\boldsymbol{\sigma}} = \widetilde{\boldsymbol{\sigma}}(\boldsymbol{\sigma}, \mathcal{D}). \tag{3.2}$$

$\mathbf{M}(\mathcal{D})$ is an isotropic fourth-rank tensor-valued function of the damage state variable $\mathcal{D}$, and the effective stress tensor $\widetilde{\boldsymbol{\sigma}}(\boldsymbol{\sigma}, \mathcal{D})$ is an isotropic second-rank tensor-valued function of $\boldsymbol{\sigma}$ and $\mathcal{D}$ (damage isotropy principle), the representation of which depends on the equivalence principle adopted.

3.2 Second-rank damage tensors

In order to develop the orthotropic damage theory we postulate that the damage state is sufficiently described by the second-rank damage tensor $\mathbf{D}$ as defined by Murakami and Ohno (1981):

$$\mathbf{D} = \sum_{i=1}^{3} D_i \mathbf{n}_i \otimes \mathbf{n}_i, \tag{3.3}$$

where D_i and $\mathbf{n}_i$ are principal values and the unit vector of principal directions of the tensor $\mathbf{D}$. D_i may be interpreted here as the ratio of area reduction in the plane perpendicular to $\mathbf{n}_i$ caused by the development of damage $D_i = \delta A_{\mathrm{D}i}/\delta A_i$ (cf. Fig. 3.2).

The assumed property of symmetry with respect to three planes determined by the normals n_1, n_2, and n_3 reflects damage orthotropy. In other words, the area reduction in the directions of damage orthotropy can be expressed in terms of the principal damage components D_1, D_2, D_3, respectively.

Transformation of the area element δA to $\delta \widetilde{A}$ is described as:

$$(I_{ki} - D_{ki})\, n_k \delta A = \widetilde{n}_i \delta \widetilde{A}, \tag{3.4}$$

or

$$(1 - D_1)\, \delta A_1 = \delta \widetilde{A}_1, \quad (1 - D_2)\, \delta A_2 = \delta \widetilde{A}_2, \quad (1 - D_3)\, \delta A_3 = \delta \widetilde{A}_3. \tag{3.5}$$

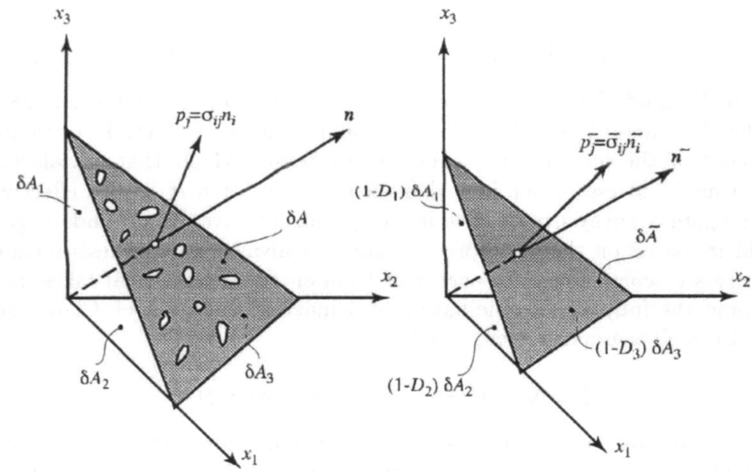

Fig. 3.2. Schematics of the RVE transformation from the current (damaged) to the equivalent (pseudo-undamaged) configuration

By equating the tractions p_j through δA and $\widetilde{p}_j$ through $\delta\widetilde{A}$ the effective stress is furnished:

$$\sigma_{ij}n_i\delta A = \widetilde{\sigma}_{ij}\widetilde{n}_i\delta\widetilde{A} \tag{3.6}$$

or

$$\sigma_{ij}n_i\delta A = \widetilde{\sigma}_{ij}\left(I_{ki} - D_{ki}\right)n_k\delta A. \tag{3.7}$$

The above is equivalent to the following definition of the asymmetric effective stress:

$$\widetilde{\boldsymbol{\sigma}}^{\mathrm{as}} = \boldsymbol{\sigma} : (\mathbf{1} - \mathbf{D})^{-1}. \tag{3.8}$$

However, only the symmetric part of (3.8) accounts for the constitutive equations, so

$$\widetilde{\boldsymbol{\sigma}} = \frac{1}{2}\left[\boldsymbol{\sigma} : (\mathbf{1} - \mathbf{D})^{-1} + (\mathbf{1} - \mathbf{D})^{-1} : \boldsymbol{\sigma}\right]. \tag{3.9}$$

The following review of various effective stress concepts is due to Zheng and Betten (1996):

1. Isotropic damage (Lemaitre and Chaboche, 1978)

$$\widetilde{\boldsymbol{\sigma}} = \boldsymbol{\sigma}\,(1 - D)^{-1}; \tag{3.10}$$

2. Asymmetric effective stress tensor (Murakami and Ohno, 1981)
$$\tilde{\boldsymbol{\sigma}}^{\text{as}} = \boldsymbol{\sigma} : (\mathbf{1} - \mathbf{D})^{-1}; \tag{3.11}$$

3. Symmetric part of the asymmetric effective stress tensor (Murakami, 1988)
$$\tilde{\boldsymbol{\sigma}} = \frac{1}{2}\left[\boldsymbol{\sigma} : (\mathbf{1} - \mathbf{D})^{-1} + (\mathbf{1} - \mathbf{D})^{-1} : \boldsymbol{\sigma}\right]; \tag{3.12}$$

4. Alternative representation of the symmetric effective stress tensors applied to elasticity, plasticity, and ductile damage (Chow and Wang, 1987; Cordebois and Sidoroff, 1992)
$$\tilde{\boldsymbol{\sigma}} = (\mathbf{1} - \mathbf{D})^{-1/2} : \boldsymbol{\sigma} : (\mathbf{1} - \mathbf{D})^{-1/2}; \tag{3.13}$$

5. Pseudo-net-stress tensor (Betten, 1986)
$$\tilde{\boldsymbol{\sigma}} = (\mathbf{1} - \mathbf{D})^{-1} : \boldsymbol{\sigma} : (\mathbf{1} - \mathbf{D})^{-1}; \tag{3.14}$$

6. General representation of the effective stress tensor by a linear transformation between the Cauchy stress and the effective Cauchy stress tensors, by the use of a fourth-rank damage effect tensor (Chow and Lu, 1992)
$$\tilde{\boldsymbol{\sigma}} = \mathbf{M}(\mathbf{D}) : \boldsymbol{\sigma}. \tag{3.15}$$

In a particular case when the Cauchy stress $\boldsymbol{\sigma}$ and the second-rank damage tensor $\mathbf{D}$ are coaxial in their principal directions, or, in other words, rotation of principal axes of the stress (and damage) tensor is excluded, they both are commutable $\boldsymbol{\sigma} : \mathbf{D} = \mathbf{D} : \boldsymbol{\sigma}$ and, as the consequence, the model (3) and (4) reduce to the simplified form
$$\tilde{\boldsymbol{\sigma}} = (\mathbf{1} - \mathbf{D})^{-1} : \boldsymbol{\sigma} = \boldsymbol{\sigma} : (\mathbf{1} - \mathbf{D})^{-1}. \tag{3.16}$$
However, in general, the above does not hold when current principal directions of the stress tensor α_i and of the damage tensor β_i do not coincide if the principal stress axes rotate (e.g., due to a shear effect) and, hence, the principal axes of damage follow them (cf. Skrzypek and Ganczarski, 1998). Additionally, when the damage is not highly developed, the difference between the models (3) and (4) is negligible (Zheng and Betten, 1996) and, since $(\mathbf{1} - \mathbf{D})$ and $(\mathbf{1} - \mathbf{D})^{1/2}$ are both positive definite second-rank symmetric tensors, there is no essential difference between models (4) and (5).

Let us also mention another definition of the second-rank damage tensor $\mathbf{D}^*$ which is due to Vakulenko and Kachanov (1971) and applied by Litewka (1985, 1987, 1989). The concept is restricted to the case of the regularly damaged material possessing three mutually perpendicular planes of orthotropy defined by the unit normal vectors $\mathbf{n}_1$, $\mathbf{n}_2$, $\mathbf{n}_3$ for which the principal damage components are defined as the ratios of corresponding damaged δA_{D} to residual (undamaged) $\delta A_{\mathrm{R}} = \delta A - \delta A_{\mathrm{D}}$ portions of the surface elements $D_i^* = \delta A_{\mathrm{D}i}/\delta A_{\mathrm{R}i}$ $(i = 1, 2, 3)$:

$$\mathbf{D}^* = \sum_{i=1}^{3} D_i^* \mathbf{n}_i \otimes \mathbf{n}_i, \tag{3.17}$$

where

$$D_i^* = \frac{D_i}{1 - D_i}, \qquad D_i^* \in \langle 0, \infty), \qquad D_i \in \langle 0, 1\rangle. \tag{3.18}$$

An alternative way to define the second-rank damage tensor is to include the microcracks morphology. Following the concepts of Vakulenko and Kachanov (1971), also extended by Kachanov (1980) the so-called damage descriptor through crack opening displacement is evaluated in the volume of RVE

$$D_{ij} = \frac{1}{V}\int_S \mathbf{u}_i \mathbf{n}_i \mathrm{d}S, \tag{3.19}$$

where $\mathbf{n}$ is a unit normal vector to the crack surface, $\mathbf{u}$ is the displacement jump across the crack surface, V is the volume of the RVE and the integration is done over all crack surfaces, S. In the particular cases of the 3D penny-shaped cracks or the 2D slit cracks of characteristic size r_k (3D crack radius) or a_k (2D crack half-length), respectively, the average crack density second-rank tensors over volume V or area A are furnished (cf. Lacy et al., 1997):

$$D_{ij} = \frac{1}{V}\sum_{k=1}^{N} r_k^3 \mathbf{n}_i^k \mathbf{n}_j^k \tag{3.20}$$

or

$$D_{ij} = \frac{1}{A}\sum_{k=1}^{N} a_k^2 \mathbf{n}_i^k \mathbf{n}_j^k. \tag{3.21}$$

3.3 Strain, stress, and energy based CDM models

Consider a damaged solid in a current configuration, the mechanical state of which is defined by the couple of external state variables $(\boldsymbol{\varepsilon}, \boldsymbol{\sigma})$, where $\boldsymbol{\varepsilon}$ is the small strain tensor and its associated variable $\boldsymbol{\sigma}$ is the Cauchy stress tensor. Introduce, next, a fictive pseudo-undamaged state characterized by the effective state variables $(\tilde{\boldsymbol{\varepsilon}}, \tilde{\boldsymbol{\sigma}})$, the definition of which depends on the damage equivalence principle. In Sect. 1.4.1, the hypotheses of strain equivalence (Chaboche, 1978) and elastic energy equivalence (Cordebois and Sidoroff, 1979) whare applied to the 1D case to yield the formulas (1.13) and (1.16), respectively. Let us discuss now the various damage equivalence principles more systematically, to generalize the above definitions to the 3D case.

3.3.1 Principle of strain equivalence – the effective stress concept

The hypothesis of strain equivalence states:

The strain associated with a damaged state under the applied stress $\boldsymbol{\sigma}$ *is equivalent to the strain associated with the undamaged state under the effective stress* $\tilde{\boldsymbol{\sigma}}$ (Fig. 3.3).

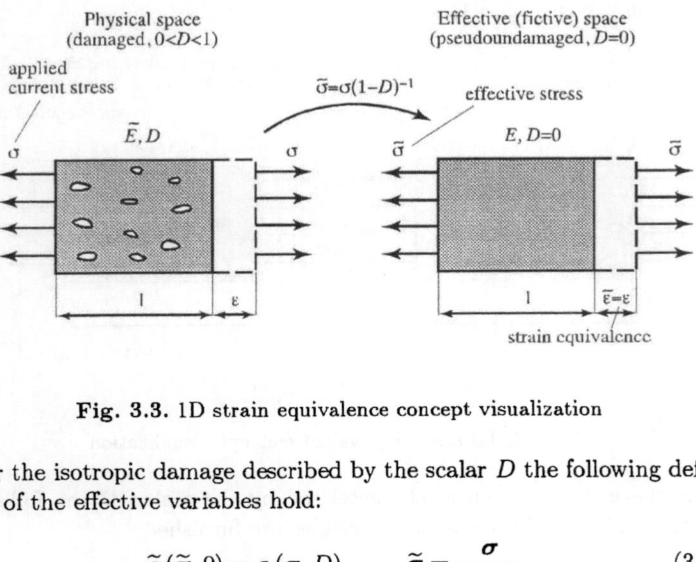

Fig. 3.3. 1D strain equivalence concept visualization

For the isotropic damage described by the scalar D the following definitions of the effective variables hold:

$$\tilde{\boldsymbol{\varepsilon}}(\tilde{\boldsymbol{\sigma}}, 0) = \boldsymbol{\varepsilon}(\boldsymbol{\sigma}, D), \qquad \tilde{\boldsymbol{\sigma}} = \frac{\boldsymbol{\sigma}}{1-D}. \tag{3.22}$$

When the fourth-rank damage effect tensor $\mathbf{M}_{\mathrm{Ch}}(\mathbf{D})$ which characterizes the anisotropic damage is used, the general transformation of the Cauchy stress tensor $\boldsymbol{\sigma}$ into the effective stress tensor $\tilde{\boldsymbol{\sigma}}$ may be introduced in case of the anisotropic damage $\mathbf{M}_{\mathrm{Ch}} = \mathbf{M}_{\mathrm{Ch}}(\mathbf{D})$

$$\tilde{\boldsymbol{\sigma}}(t) = \mathbf{M}_{\mathrm{Ch}}^{-1} : \boldsymbol{\sigma}(t) \tag{3.23}$$

and for the isotropic damage $\mathbf{M}_{\mathrm{Ch}}(D) = (1-D)\mathbf{I}$

$$\widetilde{\boldsymbol{\sigma}}(t) = \frac{\boldsymbol{\sigma}(t)}{1 - D(t)}, \tag{3.24}$$

where $\mathbf{M}_{\mathrm{Ch}}(\mathbf{D})$ is a fourth-rank damage effect tensor which characterizes damage and $\mathbf{I}$ is a fourth-rank identity tensor. In other words, the effective stress $\widetilde{\boldsymbol{\sigma}}$ expresses the stress that would have to be applied to the fictive pseudo-undamaged material to cause the same strain tensor that is observed in the damaged material sustained to current stress $\boldsymbol{\sigma}$ (Simo and Ju, 1987).

3.3.2 Principle of stress equivalence – the effective strain concept

The hypothesis of stress equivalence says:

The stress associated with a damaged state under the applied strain $\boldsymbol{\varepsilon}$ is equivalent to the stress associated with the undamaged state under the effective strain $\widetilde{\boldsymbol{\varepsilon}}$ (Fig. 3.4).

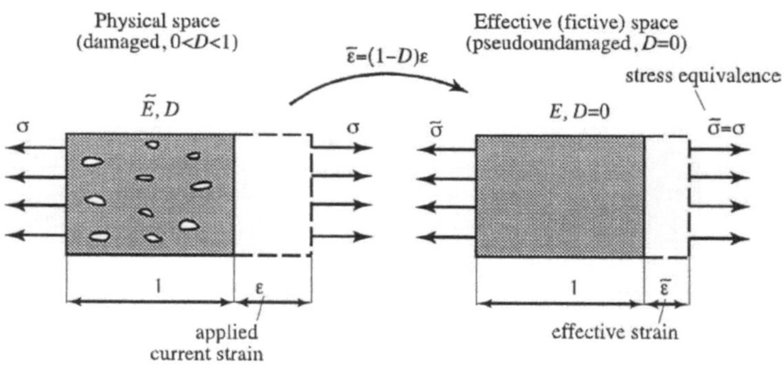

Fig. 3.4. 1D stress equivalent concept visualization

For the isotropic damage characterized by the scalar D the following (dual) definitions of the effective variables are furnished:

$$\widetilde{\boldsymbol{\sigma}}(\widetilde{\boldsymbol{\varepsilon}}, 0) = \boldsymbol{\sigma}(\boldsymbol{\varepsilon}, D), \qquad \widetilde{\boldsymbol{\varepsilon}} = (1 - D)\,\boldsymbol{\varepsilon}. \tag{3.25}$$

In a general case of the anisotropic damage characterized by the fourth-rank damage effect tensor $\mathbf{M}_{\mathrm{Ch}}(\mathbf{D})$ the transformation from the damaged space to the pseudo-undamaged space is obtained:
in case of the anisotropic damage $\mathbf{M}_{\mathrm{Ch}} = \mathbf{M}_{\mathrm{Ch}}(\mathbf{D})$,

$$\widetilde{\boldsymbol{\varepsilon}}(t) = \mathbf{M}_{\mathrm{Ch}} : \boldsymbol{\varepsilon}(t) \tag{3.26}$$

and, for the isotropic damage $\mathbf{M}_{\mathrm{Ch}}(D) = (1 - D)\,\mathbf{I}$,

$$\widetilde{\boldsymbol{\varepsilon}}(t) = [1 - D(t)]\, \boldsymbol{\varepsilon}(t). \tag{3.27}$$

3.3.3 Generalized principle of strain equivalence – the generalized effective stress concept

Taher, Baluch, and Al-Gadhib (1994) developed the generalized effective stress concept for time-independent isotropic elasto-plastic damage. In this concept, three scalar generalized, total, elastic, and plastic damage variables D^{t}, D^{e}, and D^{p} are defined by the fourth-rank secant moduli degradation tensors $\widetilde{\mathbf{\Lambda}}(t)$, $\widetilde{\mathbf{E}}(t)$, and $\widetilde{\mathbf{P}}(t)$ as the result of damage evolution

$$\begin{aligned} \boldsymbol{\sigma} &= \widetilde{\mathbf{\Lambda}}(D^{\mathrm{t}}) : \boldsymbol{\varepsilon}, & \widetilde{\mathbf{\Lambda}}(t) &= [1 - D^{\mathrm{t}}(t)]\, \mathbf{\Lambda}, \\ \boldsymbol{\sigma} &= \widetilde{\mathbf{E}}(D^{\mathrm{e}}) : \boldsymbol{\varepsilon}^{\mathrm{e}}, & \widetilde{\mathbf{E}}(t) &= [1 - D^{\mathrm{e}}(t)]\, \mathbf{E}, \\ \boldsymbol{\sigma} &= \widetilde{\mathbf{P}}(D^{\mathrm{p}}) : \boldsymbol{\varepsilon}^{\mathrm{p}}, & \widetilde{\mathbf{P}}(t) &= [1 - D^{\mathrm{p}}(t)]\, \mathbf{P}, \end{aligned} \tag{3.28}$$

where $\mathbf{\Lambda}$, $\mathbf{E}$ and $\mathbf{P}$ denote the initial values of $\widetilde{\mathbf{\Lambda}}(t)$, $\widetilde{\mathbf{E}}(t)$, and $\widetilde{\mathbf{P}}(t)$, respectively (Fig. 3.5).

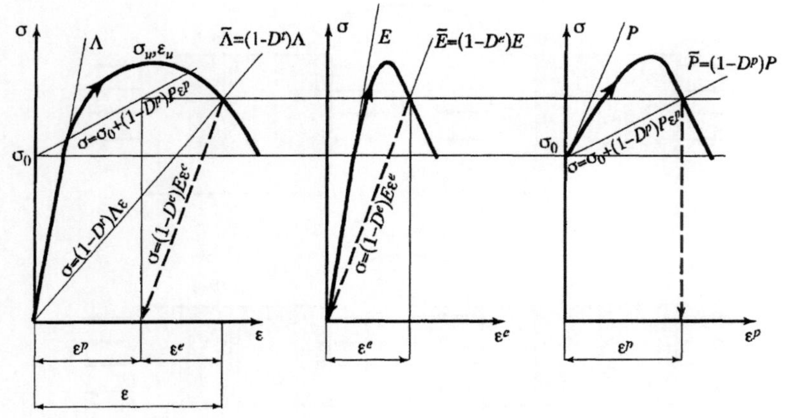

Fig. 3.5. Total uniaxial strain split into the elastic and plastic components and the secant moduli $\widetilde{\Lambda}$, $\widetilde{E}$, and $\widetilde{P}$ from damage D^{t}, D^{e}, and D^{p} (after Taher et al., 1994)

Applying the strain equivalence principle (3.22) and (3.25) to the total, elastic, and plastic strains independently, the following relationships hold:

$$\begin{aligned} \boldsymbol{\varepsilon} &= \widetilde{\mathbf{\Lambda}}^{-1} : \boldsymbol{\sigma} = \mathbf{\Lambda}^{-1} : \widetilde{\boldsymbol{\sigma}}, \\ \boldsymbol{\varepsilon}^{\mathrm{e}} &= \widetilde{\mathbf{E}}^{-1} : \boldsymbol{\sigma} = \mathbf{E}^{-1} : \widetilde{\boldsymbol{\sigma}}^{\mathrm{e}}, \\ \boldsymbol{\varepsilon}^{\mathrm{p}} &= \widetilde{\mathbf{P}}^{-1} : \left(\boldsymbol{\sigma} - \boldsymbol{\sigma}^{0}\right) = \widetilde{\mathbf{P}}^{-1} : \left(\widetilde{\boldsymbol{\sigma}}^{\mathrm{p}} - \widetilde{\boldsymbol{\sigma}}^{0}\right). \end{aligned} \tag{3.29}$$

Combining (3.28) with (3.29) the generalized effective stress tensors $\tilde{\boldsymbol{\sigma}}$ (total), $\tilde{\boldsymbol{\sigma}}^{\mathrm{e}}$ (elastic), $\tilde{\boldsymbol{\sigma}}^{\mathrm{p}}$ (plastic), and $\tilde{\boldsymbol{\sigma}}^{0}$ (initial plastic) are furnished:

$$\tilde{\boldsymbol{\sigma}} = \frac{\boldsymbol{\sigma}}{1 - D^{\mathrm{t}}}, \qquad \tilde{\boldsymbol{\sigma}}^{\mathrm{e}} = \frac{\boldsymbol{\sigma}}{1 - D^{\mathrm{e}}},$$
$$\tilde{\boldsymbol{\sigma}}^{\mathrm{p}} = \frac{\boldsymbol{\sigma}}{1 - D^{\mathrm{p}}}, \qquad \tilde{\boldsymbol{\sigma}}^{0} = \frac{\boldsymbol{\sigma}^{0}}{1 - D^{\mathrm{p}}}. \tag{3.30}$$

Only in a special case $D^{\mathrm{t}} = D^{\mathrm{e}} = D^{\mathrm{p}} = D$ the definitions (3.30) coincide with (3.22) and (3.25) (cf. Simo and Ju, 1987). One-dimensional schematics for the generalized strain equivalence principle are sketched in Fig. 3.6.

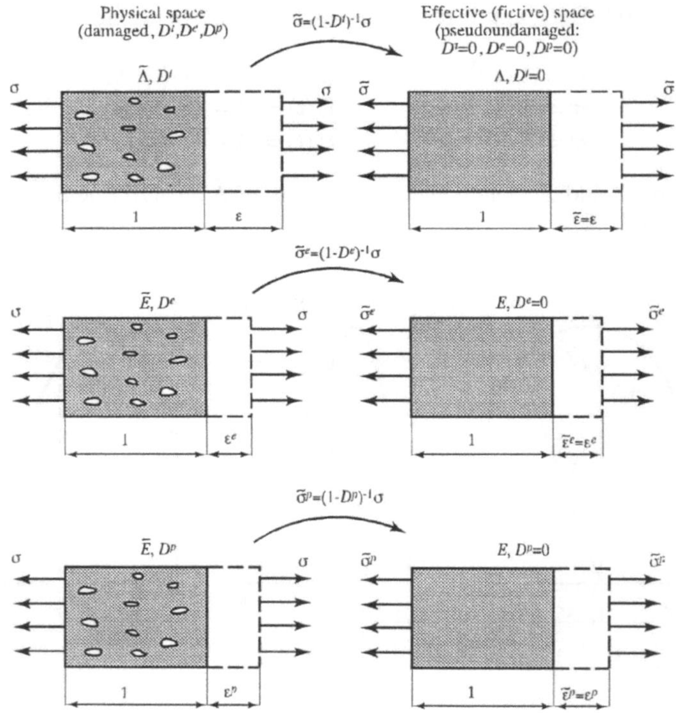

Fig. 3.6. 1D generalized stress concept visualization (after Taher et al., 1994)

Inspection of the evolution of three generalized damage variables, D^{t}, D^{e}, and D^{p}, as defined by (3.28), for two different materials, a brittle one (concrete) and a ductile one (copper 99.9%), shows the essential differences (Fig. 3.7).

Loosely speaking, in case of the brittle material under compression, Fig. 3.7a, the damage process may be approximately characterized by the single damage variable D^{t} which is intermediate to the elastic, and plastic

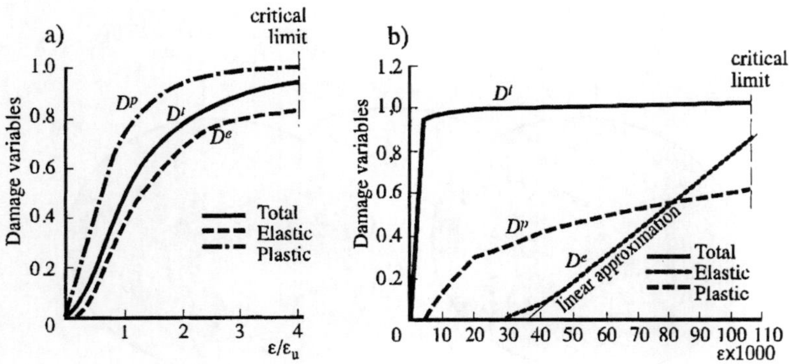

Fig. 3.7. Evolution of generalized damage variables D^{t}, D^{e}, and D^{p} in a) concrete under compression and b) copper under tension ($\varepsilon/\varepsilon_{\mathrm{u}}$ is the strain over the peak strain ratio, Fig. 3.5) (after Taher et al., 1994)

variables, D^{e} and D^{p}. For ductile materials the damage evolution cannot be described by a single damage variable following uncoupling between the total, elastic and plastic stiffness degradation as shown in Fig. 3.7b. The elastic damage variable $D^{\mathrm{e}}(\varepsilon)$ is approximately the linear function (as given by (2.76) for the Lemaitre theory, with $n = 0$), whereas the total damage variable rapidly increases to evolve asymptotically to unity, and the plastic variable significantly contributes to the damage evolution.

3.3.4 Principle of the complementary elastic energy equivalence

In the strain or the stress damage equivalent configurations (Sects. 3.3.1 or 3.3.2), stiffness reduction due to microcracks or microvoids growth affects the effective stress or the effective strain distribution, respectively, whereas the strain or the stress remain unchanged. These simplified models do not properly describe real irreversible thermodynamic material degradation processes, as reported in Sect. 2.3. Cordebois and Sidoroff (1979) postulated use of complementary elastic energy equivalence in order to define the fictive pseudo-undamaged equivalent configuration and the corresponding effective variables $\tilde{\boldsymbol{\sigma}}$ and $\tilde{\boldsymbol{\varepsilon}}$. The complementary elastic energy of the pseudo-undamaged solid $\tilde{\Phi}^{\mathrm{e}}$ was obtained directly from the virgin undamaged one Φ^{e}, except that the stress and strain variables $\boldsymbol{\sigma}$ and $\boldsymbol{\varepsilon}$ are replaced by the effective variables $\tilde{\boldsymbol{\sigma}}$ and $\tilde{\boldsymbol{\varepsilon}}$:

$$\Phi^{\mathrm{e}}(\boldsymbol{\sigma}, \mathcal{D}) = \tilde{\Phi}^{\mathrm{e}}(\tilde{\boldsymbol{\sigma}}, 0), \quad \boldsymbol{\varepsilon}^{e} = \frac{\partial \Phi^{\mathrm{e}}}{\partial \boldsymbol{\sigma}}, \tag{3.31}$$

where $\Phi^{\mathrm{e}} = (1/2)\,\boldsymbol{\sigma} : \boldsymbol{\varepsilon}^{\mathrm{e}}$ and $\tilde{\Phi}^{\mathrm{e}} = (1/2)\,\tilde{\boldsymbol{\sigma}} : \tilde{\boldsymbol{\varepsilon}}^{\mathrm{e}}$ whereas $\mathcal{D}$ represents a set of damage variables (Fig. 3.8).

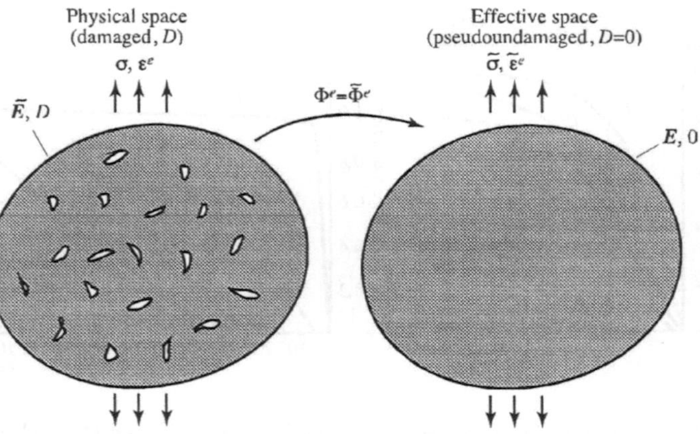

Fig. 3.8. 3D elastic energy equivalence

Applying Eqs. (3.31) to the damage coupled elasticity

$$\boldsymbol{\sigma} = \widetilde{\mathbf{E}} : \boldsymbol{\varepsilon}^{\mathrm{e}} \quad \text{and} \quad \widetilde{\boldsymbol{\sigma}} = \mathbf{E} : \widetilde{\boldsymbol{\varepsilon}}^{\mathrm{e}} \tag{3.32}$$

the following definitions of the effective variables $\widetilde{\boldsymbol{\sigma}}$, $\widetilde{\boldsymbol{\varepsilon}}^{\mathrm{e}}$ are obtained:

$$\widetilde{\boldsymbol{\sigma}} = \left(\mathbf{I} - \widehat{\mathbf{D}}\right)^{-1} : \boldsymbol{\sigma}, \quad \widetilde{\boldsymbol{\varepsilon}}^{\mathrm{e}} = \left(\mathbf{I} - \widehat{\mathbf{D}}\right) : \boldsymbol{\varepsilon}^{\mathrm{e}}, \tag{3.33}$$

where $\mathbf{I}$ and $\widehat{\mathbf{D}}$ are fourth-rank identity and damage tensors, respectively, whereas $\widehat{\mathbf{D}}$ is related to fourth-rank elasticity tensors $\mathbf{E}$ and $\widetilde{\mathbf{E}}$ of the damage equivalent (fictive) and the current (physical) state of the solid through

$$\widehat{\mathbf{D}} = \mathbf{I} - \widetilde{\mathbf{E}}^{1/2} : \mathbf{E}^{1/2}. \tag{3.34}$$

In a more general representation, when a fourth-rank damage effect tensor $\mathbf{M}(\mathcal{D})$ is used, the effective variables $\widetilde{\boldsymbol{\sigma}}$, $\widetilde{\boldsymbol{\varepsilon}}^{\mathrm{e}}$ are

$$\widetilde{\boldsymbol{\sigma}} = \mathbf{M}(\mathcal{D}) : \boldsymbol{\sigma}, \quad \widetilde{\boldsymbol{\varepsilon}}^{\mathrm{e}} = \mathbf{M}^{-1}(\mathcal{D}) : \boldsymbol{\varepsilon}^{\mathrm{e}}, \tag{3.35}$$

where $\mathcal{D}$ denotes a properly selected damage variable D, $\mathbf{D}$ or $\widehat{\mathbf{D}}$, scalar, second-rank tensor, or fourth-rank tensor, respectively. Note that in the energy based damage equivalence model the microcrack and/or microvoid growth influences both the stress and the strain distribution, which is more realistic than in the strain or stress damage equivalence postulate where the local stiffness drop results in a local stress decrease or local strain increase, exclusively. Nevertheless, it is limited as it does not allow for the physically adequate description of phenomena other than damage coupled elasticity (cf. Chow and Lu, 1992).

3.3.5 Principle of the total (elastic and anelastic) energy equivalence

For the description of anelastic material response behaviour affected by anisotropic damage, Chow and Lu (1992) extended the hypothesis of complementary elastic energy equivalence, due to Cordebois and Sidoroff, by including in the first law of thermodynamics of a material that undergoes progressive deterioration under infinitesimal deformation the inelastic energy terms, to yield

$$\mathrm{d}\Phi = \mathrm{d}\Phi^{\mathrm{e}} + \mathrm{d}\Phi^{\mathrm{p}} + \mathrm{d}\Phi^{\mathrm{d}}, \tag{3.36}$$

where $\mathrm{d}\Phi = \boldsymbol{\sigma} : \mathrm{d}\boldsymbol{\varepsilon}$ is the infinitesimal work of the applied stresses, and $\mathrm{d}\Phi^{\mathrm{e}}$, $\mathrm{d}\Phi^{\mathrm{p}}$, and $\mathrm{d}\Phi^{\mathrm{d}}$ denote the elastic (reversible) energy, the work done on (visco)plastic (irreversible) infinitesimal deformation and the work associated with damage nucleation and growth, respectively, (Fig. 3.9).

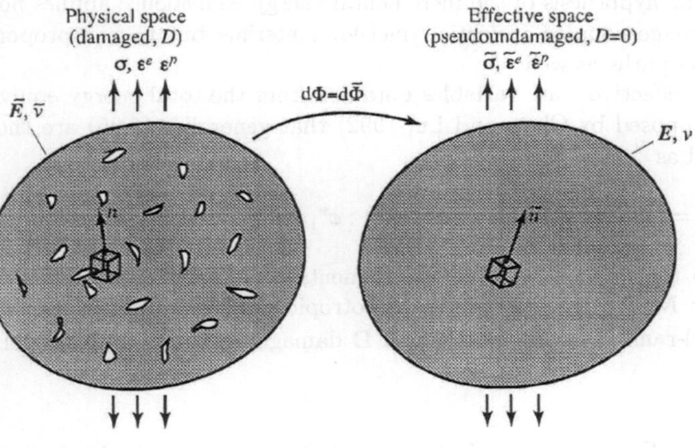

Fig. 3.9. 3D infinitesimal total (elastic and anelastic) energy equivalence (after Chow and Lu, 1992)

The total energy equivalence postulates that:
There exists a pseudo-undamaged (homogeneous) solid made of the virgin material in the sense that the total work done by the external tractions on infinitesimal deformations during the same loading history as that for the real, damaged (heterogeneous) solid is not changed

(cf. Chow and Lu, 1992). Because in a fictive configuration $\mathrm{d}\widetilde{\Phi}^{\mathrm{d}} = 0$, the following therefore holds:

$$\mathrm{d}\Phi = \mathrm{d}\widetilde{\Phi} \tag{3.37}$$

or

$$\mathrm{d}\Phi^{\mathrm{e}} + \mathrm{d}\Phi^{\mathrm{d}} = \mathrm{d}\widetilde{\Phi}^{\mathrm{e}} \quad \text{and} \quad \mathrm{d}\Phi^{\mathrm{p}} = \mathrm{d}\widetilde{\Phi}^{\mathrm{p}}, \tag{3.38}$$

where

$$\mathrm{d}\widetilde{\Phi} = \widetilde{\boldsymbol{\sigma}} : \mathrm{d}\widetilde{\boldsymbol{\varepsilon}}, \quad \mathrm{d}\widetilde{\Phi}^{e} = \frac{1}{2}\left(\widetilde{\boldsymbol{\sigma}} : \mathrm{d}\widetilde{\boldsymbol{\varepsilon}}^{e} + \mathrm{d}\widetilde{\boldsymbol{\sigma}} : \widetilde{\boldsymbol{\varepsilon}}^{e}\right), \quad \mathrm{d}\widetilde{\Phi}^{p} = \widetilde{\boldsymbol{\sigma}} : \mathrm{d}\widetilde{\boldsymbol{\varepsilon}}^{p}. \tag{3.39}$$

In an equivalent form, (3.37)–(3.39) may be written as follows:

$$\begin{aligned} &\boldsymbol{\sigma} : \mathrm{d}\boldsymbol{\varepsilon} = \widetilde{\boldsymbol{\sigma}} : \mathrm{d}\widetilde{\boldsymbol{\varepsilon}}, \\ &\tfrac{1}{2}\left(\boldsymbol{\sigma} : \mathrm{d}\boldsymbol{\varepsilon}^{\mathrm{e}} + \mathrm{d}\boldsymbol{\sigma} : \boldsymbol{\varepsilon}^{\mathrm{e}}\right) + \mathrm{d}\Phi^{\mathrm{d}} = \tfrac{1}{2}\left(\widetilde{\boldsymbol{\sigma}} : \mathrm{d}\widetilde{\boldsymbol{\varepsilon}}^{\mathrm{e}} + \mathrm{d}\widetilde{\boldsymbol{\sigma}} : \widetilde{\boldsymbol{\varepsilon}}^{\mathrm{e}}\right), \\ &\boldsymbol{\sigma} : \mathrm{d}\boldsymbol{\varepsilon}^{\mathrm{p}} = \widetilde{\boldsymbol{\sigma}} : \mathrm{d}\widetilde{\boldsymbol{\varepsilon}}^{\mathrm{p}}, \end{aligned} \tag{3.40}$$

where the state variables on the left-hand side of (3.40) refer to the physical (damaged) space, and the effective state variables on the right-hand side to the energy equivalent fictive (pseudo-undamaged) one (Fig. 3.9). Note that the hypothesis of the incremental energy equivalence applies not only to damage coupled inelastic (ductile) materials but to non-proportional loading paths as well.

The effective state variables obtained from the total energy equivalence (as proposed by Chow and Lu, 1992) that generalize (3.35) are then furnished as

$$\widetilde{\boldsymbol{\sigma}} = \mathbf{M}(\mathcal{D}) : \boldsymbol{\sigma}, \quad \widetilde{\boldsymbol{\varepsilon}}^{\mathrm{e}} = \mathbf{M}^{-1}(\mathcal{D}) : \boldsymbol{\varepsilon}^{\mathrm{e}}, \quad \mathrm{d}\widetilde{\boldsymbol{\varepsilon}}^{\mathrm{p}} = \mathbf{M}^{-1}(\mathcal{D}) : \mathrm{d}\boldsymbol{\varepsilon}^{\mathrm{p}}, \tag{3.41}$$

where the explicit form of the elements of a fourth-rank damage effect tensor $\mathbf{M}(\mathcal{D})$ depends on the anisotropic damage representation by the second-rank $\mathbf{D}$ or the fourth-rank $\widehat{\mathbf{D}}$ damage tensor components (cf. Sect. 3.4).

3.3.6 Comparison of strain versus energy equivalence in the damage evolution with strain for aluminum alloy 2024–T3 under uniaxial tension

Mapping of the stress–strain curve $\boldsymbol{\sigma}(\boldsymbol{\varepsilon})$ to the effective stress–strain curve $\widetilde{\boldsymbol{\sigma}}(\widetilde{\boldsymbol{\varepsilon}})$ depends on the damage equivalence principle used. Chow and Wang (1987) measured the effective Young's modulus $\widetilde{E}$ and the effective Poisson's ratio $\widetilde{\nu}$ for a ductile aluminum alloy 2024–T3 tensile specimen, based on which damage components were calculated from the energy and the stress or strain equivalence.

I. 1D energy equivalence concept

The matrix representation of the 1D energy based effective state variables (3.35) is

$$\left\{\begin{array}{c}\widetilde{\sigma}_1\\0\\0\end{array}\right\}=\left[\begin{array}{ccc}\dfrac{1}{1-D_1}&0&0\\&\dfrac{1}{1-D_2}&0\\&&\dfrac{1}{1-D_2}\end{array}\right]\left\{\begin{array}{c}\sigma_1\\0\\0\end{array}\right\},$$

$$\left\{\begin{array}{c}\widetilde{\varepsilon}_1^{\mathrm{e}}\\\widetilde{\varepsilon}_2^{\mathrm{e}}\\\widetilde{\varepsilon}_3^{\mathrm{e}}\end{array}\right\}=\left[\begin{array}{ccc}1-D_1&0&0\\&1-D_2&0\\&&1-D_2\end{array}\right]\left\{\begin{array}{c}\varepsilon_1^{\mathrm{e}}\\\varepsilon_2^{\mathrm{e}}\\\varepsilon_3^{\mathrm{e}}\end{array}\right\},\tag{3.42}$$

where Hooke's law written for the fictive (pseudo-undamaged) and true (damaged) solid is given by

$$\left\{\begin{array}{c}\widetilde{\sigma}_1\\0\\0\end{array}\right\}=\frac{E}{(1+\nu)(1-2\nu)}\left[\begin{array}{ccc}1-\nu&\nu&\nu\\&1-\nu&\nu\\&&1-\nu\end{array}\right]\left\{\begin{array}{c}\widetilde{\varepsilon}_1^{\mathrm{e}}\\-\nu\widetilde{\varepsilon}_1^{\mathrm{e}}\\-\nu\widetilde{\varepsilon}_1^{\mathrm{e}}\end{array}\right\},$$

$$\left\{\begin{array}{c}\sigma_1\\0\\0\end{array}\right\}=\frac{\widetilde{E}}{(1+\widetilde{\nu})(1-2\widetilde{\nu})}\left[\begin{array}{ccc}1-\widetilde{\nu}&\widetilde{\nu}&\widetilde{\nu}\\&1-\widetilde{\nu}&\widetilde{\nu}\\&&1-\widetilde{\nu}\end{array}\right]\left\{\begin{array}{c}\varepsilon_1^{\mathrm{e}}\\-\widetilde{\nu}\varepsilon_1^{\mathrm{e}}\\-\widetilde{\nu}\varepsilon_1^{\mathrm{e}}\end{array}\right\}\tag{3.43}$$

or, in the dual form,

$$\left\{\begin{array}{c}\widetilde{\varepsilon}_1^{\mathrm{e}}\\\widetilde{\varepsilon}_2^{\mathrm{e}}=-\nu\widetilde{\varepsilon}_1^{\mathrm{e}}\\\widetilde{\varepsilon}_3^{\mathrm{e}}=-\nu\widetilde{\varepsilon}_1^{\mathrm{e}}\end{array}\right\}=\frac{1}{E}\left[\begin{array}{ccc}1&-\nu&-\nu\\&1&-\nu\\&&1\end{array}\right]\left\{\begin{array}{c}\widetilde{\sigma}_1\\0\\0\end{array}\right\},$$

$$\left\{\begin{array}{c}\varepsilon_1^{\mathrm{e}}\\\varepsilon_2^{\mathrm{e}}=-\widetilde{\nu}\varepsilon_1^{\mathrm{e}}\\\varepsilon_3^{\mathrm{e}}=-\widetilde{\nu}\varepsilon_1^{\mathrm{e}}\end{array}\right\}=\frac{1}{\widetilde{E}}\left[\begin{array}{ccc}1&-\widetilde{\nu}&-\widetilde{\nu}\\&1&-\widetilde{\nu}\\&&1\end{array}\right]\left\{\begin{array}{c}\sigma_1\\0\\0\end{array}\right\}.\tag{3.44}$$

Hence, after the following rearrangement,

$$\begin{aligned}\widetilde{\sigma}_1&=\frac{\sigma_1}{1-D_1}=E\widetilde{\varepsilon}_1^{\mathrm{e}}\to\frac{\sigma_1}{1-D_1}=E(1-D_1)\varepsilon_1^{\mathrm{e}}=E(1-D_1)\frac{\sigma_1}{\widetilde{E}},\\\widetilde{\varepsilon}_2^{\mathrm{e}}&=-\nu\widetilde{\varepsilon}_1^{\mathrm{e}}=(1-D_2)\varepsilon_2^{\mathrm{e}}\to-\frac{\nu}{E}\frac{\sigma_1}{1-D_1}=(1-D_2)\left(-\frac{\widetilde{\nu}}{\widetilde{E}}\right)\sigma_1,\end{aligned}\tag{3.45}$$

the two damage components D_1 and D_2, related to $\widetilde{E}$ and $\widetilde{\nu}$ are obtained

$$\begin{aligned}D_1&=1-\left(\frac{\widetilde{E}}{E}\right)^{1/2},\\D_2&=1-\frac{\nu}{\widetilde{\nu}}\left(\frac{\widetilde{E}}{E}\right)^{1/2}=1-\frac{\nu}{\widetilde{\nu}}(1-D_1).\end{aligned}\tag{3.46}$$

II. 1D elastic strain equivalence concept

The elastic strain equivalence require the following representation of the effective variables $\widetilde{\boldsymbol{\sigma}}^*$, $\widetilde{\boldsymbol{\varepsilon}}^*$:

$$
\begin{aligned}
\left\{\begin{array}{c} \widetilde{\sigma}_1^* \\ 0 \\ 0 \end{array}\right\} &= \left[\begin{array}{ccc} \dfrac{1}{1-D_1^*} & 0 & 0 \\ 0 & 1 & 0 \\ 0 & 0 & 1 \end{array}\right] \left\{\begin{array}{c} \sigma_1 \\ 0 \\ 0 \end{array}\right\}, \\
\left\{\begin{array}{c} \widetilde{\varepsilon}_1^{e*} \\ \widetilde{\varepsilon}_2^{e*} \\ \widetilde{\varepsilon}_3^{e*} \end{array}\right\} &= \left[\begin{array}{ccc} 1 & 0 & 0 \\ 0 & 1 & 0 \\ 0 & 0 & 1 \end{array}\right] \left\{\begin{array}{c} \varepsilon_1^{e} \\ \varepsilon_2^{e} \\ \varepsilon_3^{e} \end{array}\right\}
\end{aligned}
\tag{3.47}
$$

and Hooke's law referred to the pseudo-undamaged and damaged state is given by

$$
\begin{aligned}
\left\{\begin{array}{c} \widetilde{\sigma}_1^* \\ 0 \\ 0 \end{array}\right\} &= \frac{E}{(1+\nu)(1-2\nu)} \left[\begin{array}{ccc} 1-\nu & \nu & \nu \\ & 1-\nu & \nu \\ & & 1-\nu \end{array}\right] \left\{\begin{array}{c} \widetilde{\varepsilon}_1^{e*} \\ -\nu\widetilde{\varepsilon}_1^{e*} \\ -\nu\widetilde{\varepsilon}_1^{e*} \end{array}\right\}, \\
\left\{\begin{array}{c} \sigma_1 \\ 0 \\ 0 \end{array}\right\} &= \frac{\widetilde{E}}{(1+\widetilde{\nu}^*)(1-2\widetilde{\nu}^*)} \left[\begin{array}{ccc} 1-\widetilde{\nu}^* & \widetilde{\nu}^* & \widetilde{\nu}^* \\ & 1-\widetilde{\nu}^* & \widetilde{\nu}^* \\ & & 1-\widetilde{\nu}^* \end{array}\right] \left\{\begin{array}{c} \varepsilon_1^{e} \\ -\widetilde{\nu}^*\varepsilon_1^{e} \\ -\widetilde{\nu}^*\varepsilon_1^{e} \end{array}\right\}
\end{aligned}
\tag{3.48}
$$

or

$$
\begin{aligned}
\left\{\begin{array}{c} \widetilde{\varepsilon}_1^{e*} \\ \widetilde{\varepsilon}_2^{e*} = -\nu\widetilde{\varepsilon}_1^{e*} \\ \widetilde{\varepsilon}_3^{e*} = -\nu\widetilde{\varepsilon}_1^{e*} \end{array}\right\} &= \frac{1}{E}\left[\begin{array}{ccc} 1 & -\nu & -\nu \\ & 1 & -\nu \\ & & 1 \end{array}\right] \left\{\begin{array}{c} \widetilde{\sigma}_1^* \\ 0 \\ 0 \end{array}\right\}, \\
\left\{\begin{array}{c} \varepsilon_1^{e} \\ -\widetilde{\nu}^*\varepsilon_1^{e} \\ -\widetilde{\nu}^*\varepsilon_1^{e} \end{array}\right\} &= \frac{1}{\widetilde{E}}\left[\begin{array}{ccc} 1 & -\widetilde{\nu}^* & -\widetilde{\nu}^* \\ & 1 & -\widetilde{\nu}^* \\ & & -\widetilde{\nu}^* \end{array}\right] \left\{\begin{array}{c} \sigma_1 \\ 0 \\ 0 \end{array}\right\}.
\end{aligned}
\tag{3.49}
$$

Finally, after a simple transformation the following is obtained:

$$
\begin{aligned}
\widetilde{\sigma}_1^* &= \frac{\sigma_1}{1-D_1^*} = E\widetilde{\varepsilon}_1^{e*} \rightarrow \frac{\sigma_1}{1-D_1^*} = \frac{E}{\widetilde{E}}\sigma_1, \\
\widetilde{\varepsilon}_2^{e*} &= -\frac{\nu}{E}\widetilde{\sigma}_1^* = -\frac{\widetilde{\nu}^*}{\widetilde{E}}\sigma_1 \rightarrow -\frac{\nu}{E}\frac{\sigma_1}{1-D_1^*} = -\frac{\widetilde{\nu}^*}{\widetilde{E}}\sigma_1,
\end{aligned}
\tag{3.50}
$$

and a single damage component D_1^* related to the Young's moduli ratio $\widetilde{E}/E$ is recovered, (1.14), whereas Poisson's ratio $\widetilde{\nu}^*$ does not change,

$$
D_1^* = 1 - \frac{\widetilde{E}}{E}, \qquad \widetilde{\nu}^* = \nu \quad (!),
\tag{3.51}
$$

which is far from the general experimental observation that anisotropic damage propagates not only in the direction of main stress but also in the transverse one.

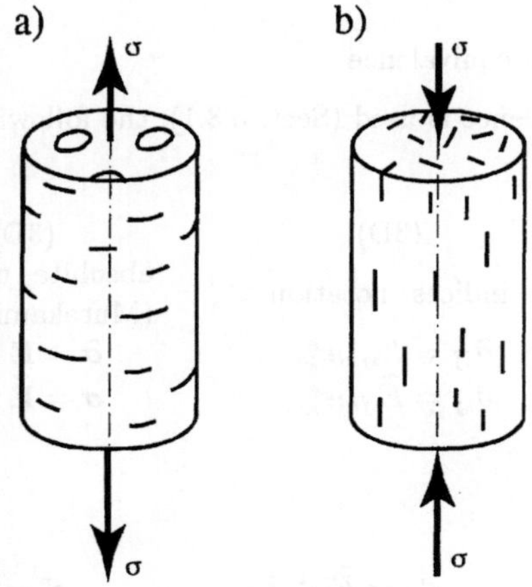

Fig. 3.10. Particular microcracking orientations in elastic-brittle rock-like solids: a) planar transverse isotropy under uniaxial tension and b) cylindrical transverse isotropy under uniaxial compression (after Chaboche, 1993)

For example, a broadly reported specific cracking phenomena of both planar transverse isotropy and cylindrical transverse isotropy (Fig. 3.10) in the elastic-brittle rock-like, ceramic, or concrete solids, under uniaxial tension and uniaxial compression, respectively, cannot be adequately described by the use of the strain equivalence hypothesis (cf. Chaboche, 1993; Chaboche, Lesne and Maire, 1995). Also, the damage–strain relations of high-strength concrete and the Young's modulus drop and Poisson's ratio increase with strain, under uniaxial tension and compression, Fig. 4.3, show the strain equivalence limitations (cf. Murakami and Kamiya, 1997).

3.4 Fourth-rank damage effect tensors

3.4.1 Strain, stress and energy based damage tensor representations

It has been shown in the previous section that a selection of the damage equivalence principle may lead to the different damage descriptions that should follow experimental observations of different materials. In this section, the fourth-rank damage tensor representations are derived from the strain, stress, and elastic energy equivalences, all applied to the 3D case, when both indices and absolute notation are used. A simple 1D case, if the damage variable reduces to a scalar one, is attached in parallel to a general damage tensors application.

I. Principle of strain equivalence

When strain equivalence is used (Sect. 3.3.1), the following holds

(1D)	(3D) indices notation	(3D) absolute notation (Murakami, 1987)	
$\widetilde{\sigma} = E\varepsilon^{\mathrm{e}},$	$\widetilde{\sigma}_{ij} = E_{ijkl}\varepsilon^{\mathrm{e}}_{kl},$	$\widetilde{\boldsymbol{\sigma}} = \mathbf{E} : \boldsymbol{\varepsilon}^{\mathrm{e}},$	(3.52)
$\sigma = \widetilde{E}\varepsilon^{\mathrm{e}},$	$\sigma_{ij} = \widetilde{E}_{ijkl}\varepsilon^{\mathrm{e}}_{kl},$	$\boldsymbol{\sigma} = \widetilde{\mathbf{E}} : \boldsymbol{\varepsilon}^{\mathrm{e}}$	

or

$$\varepsilon^{\mathrm{e}} = \widetilde{E}^{-1}\sigma, \qquad \varepsilon^{\mathrm{e}}_{kl} = \widetilde{E}^{-1}_{klij}\sigma_{ij}, \qquad \boldsymbol{\varepsilon}^{\mathrm{e}} = \widetilde{\mathbf{E}}^{-1} : \boldsymbol{\sigma}, \tag{3.53}$$

where $\widetilde{E}_{klij}$, $\widetilde{\mathbf{E}}$ denote the fourth-rank elasticity tensors modified by damage and $\widetilde{\sigma}_{ij}$, $\widetilde{\boldsymbol{\sigma}}$ are the strain equivalent effective stress tensors

$$\widetilde{\sigma} = \underbrace{E\widetilde{E}^{-1}}\sigma, \qquad \widetilde{\sigma}_{ij} = \underbrace{E_{ijrs}\widetilde{E}^{-1}_{rskl}}\sigma_{kl}, \qquad \widetilde{\boldsymbol{\sigma}} = \underbrace{\mathbf{E} : \widetilde{\mathbf{E}}^{-1}} : \boldsymbol{\sigma}. \tag{3.54}$$

If the fourth-rank identity tensors I_{ijkl} or $\mathbf{I}$ and the fourth-rank damage tensors $\widehat{D}_{ijkl}$ or $\widehat{\mathbf{D}}$ are introduced, the following formulas that define the effective stress tensors in terms of the actual damage state are furnished:

$$\widetilde{\sigma} = (1-D)^{-1}\sigma, \qquad \widetilde{\sigma}_{ij} = (I_{ijkl} - \widehat{D}_{ijkl})^{-1}\sigma_{kl}, \qquad \widetilde{\boldsymbol{\sigma}} = (\mathbf{I} - \widehat{\mathbf{D}})^{-1} : \boldsymbol{\sigma}. \tag{3.55}$$

Hence, the damage tensors' representations are obtained

$$D = 1 - \widetilde{E}E^{-1}, \qquad \widehat{D}_{ijkl} = I_{ijkl} - \widetilde{E}_{ijrs}E^{-1}_{rskl}, \qquad \widehat{\mathbf{D}} = \mathbf{I} - \widetilde{\mathbf{E}} : \mathbf{E}^{-1} \tag{3.56}$$

or

$$\widetilde{E}(D) = (1-D)E, \qquad \widetilde{E}_{ijkl} = (I_{ijmn} - \widehat{D}_{ijmn}) \times E_{mnkl}, \qquad \widetilde{\mathbf{E}}(\widehat{\mathbf{D}}) = (\mathbf{I} - \widehat{\mathbf{D}}) : \mathbf{E}. \tag{3.57}$$

II. Principle of stress equivalence

When stress equivalence is used (Sect. 3.3.2), a similar derivation yields:

$$\begin{array}{lll} \text{(1D)} & \text{(3D) indices notation} & \text{(3D) absolute notation} \\ \widetilde{\varepsilon}^{e} = E^{-1}\sigma, & \widetilde{\varepsilon}^{e}_{ij} = E^{-1}_{ijkl}\sigma_{kl}, & \widetilde{\boldsymbol{\varepsilon}}^{e} = \mathbf{E}^{-1} : \boldsymbol{\sigma}, \\ \varepsilon^{e} = \widetilde{E}^{-1}\sigma, & \varepsilon^{e}_{ij} = \widetilde{E}^{-1}_{ijkl}\sigma_{kl}, & \boldsymbol{\varepsilon}^{e} = \widetilde{\mathbf{E}}^{-1} : \boldsymbol{\sigma} \end{array} \tag{3.58}$$

or

$$\begin{array}{lll} \sigma = \widetilde{E}\varepsilon^{e}, & \sigma_{kl} = \widetilde{E}_{klij}\varepsilon^{e}_{ij}, & \boldsymbol{\sigma} = \widetilde{\mathbf{E}} : \boldsymbol{\varepsilon}^{e}, \\ \widetilde{\varepsilon}^{e} = \underbrace{E^{-1}\widetilde{E}}\varepsilon^{e}, & \widetilde{\varepsilon}^{e}_{ij} = \underbrace{E^{-1}_{ijrs}\widetilde{E}_{rskl}}\varepsilon^{e}_{kl}, & \widetilde{\boldsymbol{\varepsilon}}^{e} = \underbrace{\mathbf{E}^{-1} : \widetilde{\mathbf{E}}} : \boldsymbol{\varepsilon}^{e}, \end{array} \tag{3.59}$$

where $\widetilde{\varepsilon}^{e}_{ij}$, $\widetilde{\boldsymbol{\varepsilon}}^{e}$ denote the stress equivalent effective strain tensors

$$\widetilde{\varepsilon}^{e} = (1 - D)\varepsilon^{e}, \qquad \widetilde{\varepsilon}^{e}_{ij} = (I_{ijkl} - \widehat{D}_{ijkl})\varepsilon^{e}_{kl}, \qquad \widetilde{\boldsymbol{\varepsilon}}^{e} = (\mathbf{I} - \widehat{\mathbf{D}}) : \boldsymbol{\varepsilon}^{e}. \tag{3.60}$$

Hence, formulas analogous to (3.56) are furnished:

$$D = 1 - E^{-1}\widetilde{E}, \qquad \widehat{D}_{ijkl} = I_{ijkl} - E^{-1}_{ijrs}\widetilde{E}_{rskl}, \qquad \widehat{\mathbf{D}} = \mathbf{I} - \mathbf{E}^{-1} : \widetilde{\mathbf{E}}. \tag{3.61}$$

III. Principle of elastic energy equivalence

In a similar fashion, if elastic energy equivalence (Sect. 3.3.4) is postulated, the respective transformations may be performed:

$$\begin{array}{lll} \text{(1D)} & \text{(3D) indices notation} & \text{(3D) absolute notation} \\ \widetilde{\sigma} = E\widetilde{\varepsilon}^{e}, & \widetilde{\sigma}_{ij} = E_{ijkl}\widetilde{\varepsilon}^{e}_{kl}, & \widetilde{\boldsymbol{\sigma}} = \mathbf{E} : \widetilde{\boldsymbol{\varepsilon}}^{e}, \\ \sigma = \widetilde{E}\varepsilon^{e}, & \sigma_{ij} = \widetilde{E}_{ijkl}\varepsilon^{e}_{kl}, & \boldsymbol{\sigma} = \widetilde{\mathbf{E}} : \boldsymbol{\varepsilon}^{e}, \end{array} \tag{3.62}$$

$$\begin{array}{lll} \Phi^{e}(\sigma, \mathcal{D}) & \Phi^{e}(\sigma_{ij}, \mathcal{D}) & \Phi^{e}(\boldsymbol{\sigma}, \mathcal{D}) \\ = \widetilde{\Phi}^{e}(\widetilde{\sigma}, 0), & = \widetilde{\Phi}^{e}(\widetilde{\sigma}_{ij}, 0), & = \widetilde{\Phi}^{e}(\widetilde{\boldsymbol{\sigma}}, 0), \end{array} \tag{3.63}$$

$$\begin{array}{lll} \widetilde{\Phi}^{e}(\sigma, \mathcal{D}) = \frac{1}{2}\widetilde{\sigma}\widetilde{\varepsilon}^{e}, & \widetilde{\Phi}^{e} = \frac{1}{2}\widetilde{\sigma}_{ij}\widetilde{\varepsilon}^{e}_{ij}, & \widetilde{\Phi}^{e} = \frac{1}{2}\widetilde{\boldsymbol{\sigma}} : \widetilde{\boldsymbol{\varepsilon}}^{e}, \\ \Phi^{e} = \frac{1}{2}\sigma\varepsilon^{e}, & \Phi^{e} = \frac{1}{2}\sigma_{ij}\varepsilon^{e}_{ij}, & \Phi^{e} = \frac{1}{2}\boldsymbol{\sigma} : \boldsymbol{\varepsilon}^{e}, \end{array} \tag{3.64}$$

$$\widetilde{\sigma}\widetilde{\varepsilon}^{\mathrm{e}} = \sigma\varepsilon^{\mathrm{e}}, \qquad \widetilde{\sigma}_{kl}\widetilde{\varepsilon}^{\mathrm{e}}_{kl} = \sigma_{rs}\varepsilon^{\mathrm{e}}_{rs}, \qquad \widetilde{\boldsymbol{\sigma}} : \widetilde{\boldsymbol{\varepsilon}}^{\mathrm{e}} = \boldsymbol{\sigma} : \boldsymbol{\varepsilon}^{\mathrm{e}}, \tag{3.65}$$

$$\begin{aligned} &\widetilde{\sigma} = E\widetilde{\varepsilon}^{\mathrm{e}} && \widetilde{\sigma}_{ij} = E_{ijkl}\widetilde{\varepsilon}^{\mathrm{e}}_{kl} && \widetilde{\boldsymbol{\sigma}} = \mathbf{E} : \widetilde{\boldsymbol{\varepsilon}}^{\mathrm{e}} \\ &= E\widetilde{\sigma}^{-1}\sigma\widetilde{E}^{-1}\sigma, && = E_{ijkl}\widetilde{\sigma}^{-1}_{kl}\sigma_{rs} && = \mathbf{E} : \widetilde{\boldsymbol{\sigma}}^{-1} : \boldsymbol{\sigma} \\ & && \widetilde{E}^{-1}_{rsmn}\sigma_{mn}, && : \widetilde{\mathbf{E}}^{-1} : \boldsymbol{\sigma}, \end{aligned} \tag{3.66}$$

$$\widetilde{\sigma}^2 = E\widetilde{E}^{-1}\sigma^2, \qquad \widetilde{\sigma}_{ik}\widetilde{\sigma}_{kj} = E_{ijrs}\widetilde{E}^{-1}_{rskl}\sigma_{kp}\sigma_{pl}, \qquad \widetilde{\boldsymbol{\sigma}}^2 = \mathbf{E} : \widetilde{\mathbf{E}}^{-1} : \boldsymbol{\sigma}^2, \tag{3.67}$$

$$\widetilde{\sigma} = \underbrace{E^{1/2}\widetilde{E}^{-1/2}}\sigma, \qquad \widetilde{\sigma}_{ij} = \underbrace{E^{1/2}_{ijrs}\widetilde{E}^{-1/2}_{rskl}}\sigma_{kl}, \qquad \widetilde{\boldsymbol{\sigma}} = \underbrace{\mathbf{E}^{1/2} : \widetilde{\mathbf{E}}^{-1/2}} : \boldsymbol{\sigma}. \tag{3.68}$$

Hence, the effective state variables $(\widetilde{\sigma}, \widetilde{\varepsilon}^{\mathrm{e}})$ or $(\widetilde{\sigma}_{ij}, \widetilde{\varepsilon}^{\mathrm{e}}_{ij})$ or $(\widetilde{\boldsymbol{\sigma}}, \widetilde{\boldsymbol{\varepsilon}}^{\mathrm{e}})$ corresponding to the fictive (pseudo-undamaged) elastic energy equivalent configuration (Fig. 3.8) are defined

$$\begin{aligned} &\widetilde{\sigma} = (1-D)^{-1}\sigma, && \widetilde{\sigma}_{ij} = (I_{ijkl} - \widehat{D}_{ijkl})^{-1}\sigma_{kl}, && \widetilde{\boldsymbol{\sigma}} = (\mathbf{I} - \widehat{\mathbf{D}})^{-1} : \boldsymbol{\sigma}, \\ &\widetilde{\varepsilon}^{\mathrm{e}} = (1-D)\varepsilon^{\mathrm{e}}, && \widetilde{\varepsilon}^{\mathrm{e}}_{ij} = (I_{ijkl} - \widehat{D}_{ijkl})\varepsilon^{\mathrm{e}}_{kl}, && \widetilde{\boldsymbol{\varepsilon}}^{\mathrm{e}} = (\mathbf{I} - \widehat{\mathbf{D}}) : \boldsymbol{\varepsilon}^{\mathrm{e}}, \\ &D = 1 - \widetilde{E}^{1/2}E^{-1/2}, && \widehat{D}_{ijkl} = I_{ijkl} - \widetilde{E}^{1/2}_{ijrs}E^{-1/2}_{rskl}, && \widehat{\mathbf{D}} = \mathbf{I} - \widetilde{\mathbf{E}}^{1/2} \\ & && && : \mathbf{E}^{-1/2}. \end{aligned} \tag{3.69}$$

When a fourth-rank damage effect tensor $\mathbf{M}(\widehat{\mathbf{D}})$ is used (also see Sect. 3.4.2), the mapping of the state variables $(\sigma, \varepsilon^{\mathrm{e}})$ or $(\sigma_{ij}, \varepsilon^{\mathrm{e}}_{ij})$ or $(\boldsymbol{\sigma}, \boldsymbol{\varepsilon}^{\mathrm{e}})$ from the physical (damaged) space to the fictive (pseudo-undamaged) one $(\widetilde{\sigma}, \widetilde{\varepsilon}^{\mathrm{e}})$ or $(\widetilde{\sigma}_{ij}, \widetilde{\varepsilon}^{\mathrm{e}}_{ij})$ or $(\widetilde{\boldsymbol{\sigma}}, \widetilde{\boldsymbol{\varepsilon}}^{\mathrm{e}})$ is established:

$$\begin{aligned} &\widetilde{\sigma} = M(D)\sigma, && \widetilde{\sigma}_{ij} = M_{ijkl}(\widehat{D}_{ijkl})\sigma_{kl}, && \widetilde{\boldsymbol{\sigma}} = \mathbf{M}(\widehat{\mathbf{D}}) : \boldsymbol{\sigma}, \\ &\widetilde{\varepsilon}^{\mathrm{e}} = M^{-1}(D)\varepsilon^{\mathrm{e}}, && \widetilde{\varepsilon}^{\mathrm{e}}_{ij} = M^{-1}_{ijkl}(\widehat{D}_{ijkl})\varepsilon^{\mathrm{e}}_{kl}, && \widetilde{\boldsymbol{\varepsilon}}^{\mathrm{e}} = \mathbf{M}^{-1}(\widehat{\mathbf{D}}) : \boldsymbol{\varepsilon}^{\mathrm{e}}, \end{aligned} \tag{3.70}$$

where

$$\begin{aligned} &M(D) = (1-D)^{-1} && M_{ijkl}(\widehat{D}_{ijkl}) && \mathbf{M}(\widehat{\mathbf{D}}) = (\mathbf{I} - \widehat{\mathbf{D}})^{-1}. \\ & && = (I_{ijkl} - \widehat{D}_{ijkl})^{-1}, && \end{aligned} \tag{3.71}$$

The formulas (3.70) and (3.71) may also be interpreted as linear transformations of the Cauchy stress tensor σ_{ij} or $\boldsymbol{\sigma}$ and the elastic strain tensors ε_{ij} or $\boldsymbol{\varepsilon}$ to the effective Cauchy stress tensors $\widetilde{\sigma}_{ij}$ or $\widetilde{\boldsymbol{\sigma}}$ and the effective elastic strain tensor $\widetilde{\varepsilon}^{\mathrm{e}}_{ij}$ or $\widetilde{\boldsymbol{\varepsilon}}^{\mathrm{e}}$ through the fourth-rank damage effect tensors $M_{ijkl}(\widehat{D}_{ijkl})$ or $\mathbf{M}(\widehat{\mathbf{D}})$ and $M^{-1}_{ijkl}(\widehat{D}_{ijkl})$ or $\mathbf{M}^{-1}(\widehat{\mathbf{D}})$, respectively.

3.4.2 Matrix representations of the damage effect tensors expressed in terms of the second-rank damage tensors

I. Matrix transformation between stress and effective stress vectors

In general, a damage effect tensor $\mathbf{M}$ is a fourth-rank symmetric tensor, according to the definition of effective stress, and it may be defined in terms of a fourth-rank symmetric damage tensor $\widehat{\mathbf{D}}$ (cf. Sect. 3.4.1). In other words, a linear transformation is assumed between the Cauchy stress tensor $\boldsymbol{\sigma}$ and the effective Cauchy stress tensor $\widetilde{\boldsymbol{\sigma}}$ such that

$$\widetilde{\sigma}_{ij} = M_{ijkl}(\widehat{D}_{ijkl})\sigma_{kl} \quad \text{or} \quad \widetilde{\boldsymbol{\sigma}} = \mathbf{M}(\widehat{\mathbf{D}}) : \boldsymbol{\sigma}, \tag{3.72}$$

where a symmetrized effective stress tensor $\widetilde{\sigma}_{ij}$ is used (cf. Sect. 3.2), though the effective Cauchy stress tensor needs not to be symmetric in a more general case under this transformation. On the other hand, a second-rank symmetric damage tensor $\mathbf{D}$ is often employed instead of a fourth-rank $\widehat{\mathbf{D}}$, as its elements are easier to measure. Due to the symmetry of both stress and effective stress tensors, the fourth-rank tensor M_{ijkl} can be represented by a 6×6 matrix and, hence, the above tensor transformation can be replaced by the following matrix form transformation:

$$\begin{Bmatrix} \widetilde{\sigma}_{11} \\ \widetilde{\sigma}_{22} \\ \widetilde{\sigma}_{33} \\ \widetilde{\sigma}_{23} \\ \widetilde{\sigma}_{31} \\ \widetilde{\sigma}_{12} \end{Bmatrix} = \begin{bmatrix} M_{1111} & M_{1122} & M_{1133} \\ M_{2211} & M_{2222} & M_{2233} \\ M_{3311} & M_{3322} & M_{3333} \\ M_{2311} & M_{2322} & M_{2333} \\ M_{3111} & M_{3122} & M_{3133} \\ M_{1211} & M_{1222} & M_{1233} \end{bmatrix}$$

$$\left. \begin{matrix} M_{1123} & M_{1131} & M_{1112} \\ M_{2223} & M_{2231} & M_{2212} \\ M_{3323} & M_{3331} & M_{3312} \\ M_{2323} & M_{2331} & M_{2312} \\ M_{3123} & M_{3131} & M_{3112} \\ M_{1223} & M_{1231} & M_{1212} \end{matrix} \right] \begin{Bmatrix} \sigma_{11} \\ \sigma_{22} \\ \sigma_{33} \\ \sigma_{23} \\ \sigma_{31} \\ \sigma_{12} \end{Bmatrix}. \tag{3.73}$$

II. 6×6 matrix representations of $[\mathbf{M}(\mathbf{D})]$

When the second-rank symmetric damage tensor $\mathbf{D}$ is used the elements of a fourth-rank damage effect tensor $\mathbf{M}(\mathbf{D})$ and its 6×6 matrix representation $[\mathbf{M}(\mathbf{D})]$ may be constructed in several ways. Some of them are listed below.

A. (Lekhnitskii, 1981; Chen and Chow, 1995)

$$\begin{aligned} &\mathbf{M}_1^2(\mathbf{D}) = \mathbf{P}^{-1}(\mathbf{D}), \\ &P_{ijkl} = \frac{1}{2}\left[(I_{ik} - D_{ik})(I_{jl} - D_{jl}) + (I_{il} - D_{il})(I_{jk} - D_{jk})\right], \end{aligned} \tag{3.74}$$

where

$$\mathbf{1} = \left\{ \begin{array}{ccc} 1 & 0 & 0 \\ 0 & 1 & 0 \\ 0 & 0 & 1 \end{array} \right\}, \quad \mathbf{D} = \left\{ \begin{array}{ccc} D_{11} & D_{12} & D_{13} \\ D_{21} & D_{22} & D_{23} \\ D_{31} & D_{32} & D_{33} \end{array} \right\}, \quad \begin{array}{l} D_{12} = D_{21} \\ D_{13} = D_{31} \\ D_{23} = D_{32} \end{array} \tag{3.75}$$

and, due to the symmetry, a vector representation of $\boldsymbol{\sigma}$ and $\mathbf{D}$ is employed:

$$\begin{aligned} \{\boldsymbol{\sigma}\} &= \{\sigma_{11}, \sigma_{22}, \sigma_{33}, \sigma_{23}, \sigma_{31}, \sigma_{12}\}^{\mathrm{T}}, \\ \{\mathbf{D}\} &= \{D_{11}, D_{22}, D_{33}, D_{23}, D_{31}, D_{12}\}^{\mathrm{T}}. \end{aligned} \tag{3.76}$$

Hence, for a matrix representation of $\mathbf{P}$, we obtain:

$$[\mathbf{P}(\mathbf{D})] = \left[\begin{array}{llllll} (1-D_{11})^2 & D_{12}^2 & D_{13}^2 & 2D_{12}D_{13} & -2(1-D_{11})D_{13} & -2(1-D_{11})D_{12} \\ D_{21}^2 & (1-D_{22})^2 & D_{23}^2 & -2(1-D_{22})D_{23} & 2D_{23}D_{21} & -2(1-D_{22})D_{21} \\ D_{31}^2 & D_{32}^2 & (1-D_{33})^2 & -2(1-D_{33})D_{32} & -2(1-D_{33})D_{31} & 2D_{32}D_{31} \\ D_{21}D_{31} & -(1-D_{22})D_{32} & -(1-D_{33})D_{23} & (1-D_{22})(1-D_{33})+D_{23}D_{32} & D_{23}D_{31}-D_{21}(1-D_{33}) & D_{21}D_{32}-D_{31}(1-D_{22}) \\ -(1-D_{11})D_{31} & D_{32}D_{12} & -(1-D_{33})D_{13} & D_{32}D_{13}-D_{12}(1-D_{33}) & (1-D_{33})(1-D_{11})+D_{31}D_{13} & D_{12}D_{31}-D_{32}(1-D_{11}) \\ -(1-D_{11})D_{21} & -(1-D_{22})D_{12} & D_{13}D_{23} & D_{12}D_{23}-D_{13}(1-D_{22}) & D_{13}D_{21}-D_{23}(1-D_{11}) & (1-D_{11})(1-D_{22})+D_{12}D_{21} \end{array} \right]. \tag{3.77}$$

Note that, when constructing the above matrix, the 18 right-side elements have been multiplied by the factor 2 because, due to the symmetry of the stress tensor $\boldsymbol{\sigma}$, a six-element vectorial representation is used instead of nine-element one,

$$\{\boldsymbol{\sigma}\} = [\sigma_{11}, \sigma_{22}, \sigma_{33}, \sigma_{23}, \sigma_{32}, \sigma_{31}, \sigma_{13}, \sigma_{12}, \sigma_{21}]^T, \tag{3.78}$$

and, as a consequence, a 6×6 matrix (pseudo-symmetric) is defined instead of a 9×9 matrix in a more general case.

For instance:

$$
\begin{aligned}
P_{1111} &= \frac{1}{2}\,[(1-D_{11})(1-D_{11}) \\
&\quad + (1-D_{11})(1-D_{11})] = (1-D_{11})^2, \\
P_{2323} &= 2 \times \frac{1}{2}\,[(1-D_{22})(1-D_{33}) \\
&\quad + (-D_{23})(-D_{32})] = (1-D_{22})(1-D_{33}) + D_{23}D_{32}, \\
&\text{etc.}
\end{aligned}
\tag{3.79}
$$

A pseudo-symmetry of the matrix $[\mathbf{P}(\mathbf{D})]$ may be visualized as

$$
[P] = \begin{bmatrix} [A] & 2\,[C] \\ [C]^{\mathrm{T}} & [B] \end{bmatrix}, \tag{3.80}
$$

where

$$
[A] = \begin{bmatrix} (1-D_{11})^2 & D_{12}^2 & D_{13}^2 \\ & (1-D_{22})^2 & D_{23}^2 \\ & \text{symmetry} & (1-D_{33})^2 \end{bmatrix}, \tag{3.81}
$$

$$
[B] = \begin{bmatrix} (1-D_{22})(1-D_{33}) + D_{23}^2 & D_{23}D_{13} - D_{12}(1-D_{33}) & D_{12}D_{23} - D_{13}(1-D_{22}) \\ D_{23}D_{13} - D_{12}(1-D_{33}) & (1-D_{33})(1-D_{11}) + D_{13}^2 & D_{12}D_{13} - D_{23}(1-D_{11}) \\ D_{12}D_{23} - D_{13}(1-D_{22}) & D_{12}D_{13} - D_{23}(1-D_{11}) & (1-D_{11})(1-D_{22}) + D_{12}^2 \end{bmatrix}, \tag{3.82}
$$

$$
[C] = \begin{bmatrix} D_{12}D_{13} & -(1-D_{11})D_{13} & -(1-D_{11})D_{12} \\ -(1-D_{22})D_{23} & D_{12}D_{23} & -(1-D_{22})D_{12} \\ -(1-D_{33})D_{23} & -(1-D_{33})D_{13} & D_{23}D_{13} \end{bmatrix}. \tag{3.83}
$$

Similar results may be obtained when a general matrix transformation formulas are used as equivalent to the tensor rule (Lekhnitskii, 1981):

Tensor transformation rule (m, n, r, $p \rightarrow$ summation from 1 to 3)

$$
T'_{ijkl} = T_{mnrp} l_{im} l_{jn} l_{kr} l_{lp}, \tag{3.84}
$$

Matrix transformation rule (m, $n \rightarrow$ summation from 1 to 6)

$$
\widehat{T}'_{ij} = \widehat{T}_{mn} q_{im} q_{jn}. \tag{3.85}
$$

Symbols q_{ij} are defined as

i/j	1	2	3
1	l_{11}^2	l_{12}^2	l_{13}^2
2	l_{21}^2	l_{22}^2	l_{23}^2
3	l_{31}^2	l_{32}^2	l_{33}^2
4	$2l_{31}l_{21}$	$2l_{32}l_{22}$	$2l_{33}l_{23}$
5	$2l_{31}l_{11}$	$2l_{32}l_{12}$	$2l_{33}l_{13}$
6	$2l_{21}l_{11}$	$2l_{12}l_{22}$	$2l_{13}l_{23}$
i/j	4	5	6
1	$l_{12}l_{13}$	$l_{13}l_{11}$	$l_{12}l_{11}$
2	$l_{23}l_{22}$	$l_{23}l_{21}$	$l_{22}l_{21}$
3	$l_{33}l_{32}$	$l_{33}l_{31}$	$l_{32}l_{31}$
4	$l_{33}l_{22} + l_{32}l_{23}$	$l_{33}l_{21} + l_{31}l_{23}$	$l_{31}l_{22} + l_{32}l_{21}$
5	$l_{33}l_{12} + l_{32}l_{13}$	$l_{33}l_{11} + l_{31}l_{13}$	$l_{31}l_{12} + l_{32}l_{11}$
6	$l_{13}l_{22} + l_{12}l_{23}$	$l_{13}l_{21} + l_{11}l_{23}$	$l_{11}l_{22} + l_{12}l_{21}$

where

$$[q_{ij}] = [I_{ij} - D_{ij}] = \begin{bmatrix} 1-D_{11} & -D_{12} & -D_{13} \\ -D_{21} & 1-D_{22} & -D_{23} \\ -D_{31} & -D_{32} & 1-D_{33} \end{bmatrix}. \tag{3.86}$$

B. (Chen and Chow, 1995)

The second matrix representation of $\mathbf{M}_2(\mathbf{D})$ is as follows:

$$\begin{aligned} \mathbf{M}_2(\mathbf{D}) &= \left[\mathbf{I} - \widehat{\mathbf{D}}(\mathbf{D})\right]^{-1}, \\ \widehat{D}_{ijkl} &= \frac{1}{4}(I_{ik}D_{jl} + I_{il}D_{jk} + I_{jk}D_{il} + I_{jl}D_{ik}), \end{aligned} \tag{3.87}$$

where $\mathbf{I}$ is the fourth-rank identity tensor and $\widehat{\mathbf{D}}(\mathbf{D})$ denotes the fourth-rank damage tensor whose elements are defined by the components of the second-rank damage tensor $\mathbf{D}$, to yield the following matrix representation of $[\widehat{\mathbf{D}}(\mathbf{D})]$:

$$
\left[\widehat{\mathbf{D}}(\mathbf{D})\right] = \begin{bmatrix} D_{11} & 0 & 0 & 0 & D_{13} & D_{12} \\ 0 & D_{22} & 0 & D_{23} & 0 & D_{21} \\ 0 & 0 & D_{33} & D_{32} & D_{31} & 0 \\ 0 & \dfrac{D_{32}}{2} & \dfrac{D_{23}}{2} & \dfrac{(D_{33}+D_{22})}{2} & \dfrac{D_{21}}{2} & \dfrac{D_{31}}{2} \\ \dfrac{D_{31}}{2} & 0 & \dfrac{D_{13}}{2} & \dfrac{D_{12}}{2} & \dfrac{(D_{11}+D_{33})}{2} & \dfrac{D_{32}}{2} \\ \dfrac{D_{21}}{2} & \dfrac{D_{12}}{2} & 0 & \dfrac{D_{13}}{2} & \dfrac{D_{23}}{2} & \dfrac{(D_{11}+D_{22})}{2} \end{bmatrix} . \tag{3.88}
$$

C. (Chen and Chow, 1995)

A third matrix representation of $\mathbf{M}_3(\mathbf{D})$ by the fourth-rank tensor $\widehat{\mathbf{\Phi}}(\mathbf{D})$ whose components are defined by the second-rank tensor $\mathbf{\Phi}(\mathbf{D})$ may also be used:

$$
\begin{aligned}
&\mathbf{M}_3(\mathbf{D}) = \widehat{\mathbf{\Phi}}(\mathbf{D}), \\
&\widehat{\Phi}_{ijkl} = \frac{1}{4}(I_{ik}\Phi_{jl} + I_{il}\Phi_{jk} + I_{jk}\Phi_{il} + I_{jl}\Phi_{ik}), \\
&\mathbf{\Phi} = (\mathbf{1} - \mathbf{D})^{-1};
\end{aligned} \tag{3.89}
$$

$$
\left[\widehat{\Phi}(\Phi)\right] = \begin{bmatrix}
\Phi_{11} & 0 & 0 & 0 & \Phi_{13} & \Phi_{12} \\
0 & \Phi_{22} & 0 & \Phi_{23} & 0 & \Phi_{21} \\
0 & 0 & \Phi_{33} & \Phi_{32} & \Phi_{31} & 0 \\
0 & \frac{\Phi_{32}}{2} & \frac{\Phi_{23}}{2} & \frac{(\Phi_{33}+\Phi_{22})}{2} & \frac{\Phi_{21}}{2} & \frac{\Phi_{31}}{2} \\
\frac{\Phi_{31}}{2} & 0 & \frac{\Phi_{13}}{2} & \frac{\Phi_{12}}{2} & \frac{(\Phi_{11}+\Phi_{33})}{2} & \frac{\Phi_{32}}{2} \\
\frac{\Phi_{21}}{2} & \frac{\Phi_{12}}{2} & 0 & \frac{\Phi_{13}}{2} & \frac{\Phi_{23}}{2} & \frac{(\Phi_{11}+\Phi_{22})}{2}
\end{bmatrix} . \tag{3.90}
$$

Note that a similar symmetry rule applies for 6×6 matrix representations in cases B and C as in case A.

3.4.3 Matrix representation of damage effect tensors expressed in the principal coordinate system of asecond-rank damage tensor

Employing definitions given in Sect. 3.4.2, and assuming

$$
\mathbf{D} = \begin{Bmatrix} D_1 & 0 & 0 \\ 0 & D_2 & 0 \\ 0 & 0 & D_3 \end{Bmatrix}, \qquad D_{23} = D_{31} = D_{12} = 0, \tag{3.91}
$$

we obtain the following diagonal forms for $\mathbf{M}_1$, $\mathbf{M}_2$, $\mathbf{M}_3$ cases, respectively, (cf. Voyiadjis and Kattan, 1992; Voyiadjis and Park, 1996):

$$
[\mathbf{M}_1(D_1, D_2, D_3)] = \begin{bmatrix}
\frac{1}{1-D_1} & 0 & 0 & 0 & 0 & 0 \\
0 & \frac{1}{1-D_2} & 0 & 0 & 0 & 0 \\
0 & 0 & \frac{1}{1-D_3} & 0 & 0 & 0 \\
0 & 0 & 0 & \frac{1}{\sqrt{(1-D_2)(1-D_3)}} & 0 & 0 \\
0 & 0 & 0 & 0 & \frac{1}{\sqrt{(1-D_1)(1-D_3)}} & 0 \\
0 & 0 & 0 & 0 & 0 & \frac{1}{\sqrt{(1-D_1)(1-D_2)}}
\end{bmatrix}; \tag{3.92}
$$

$$
[\mathbf{M}_2(D_1, D_2, D_3)] = \begin{bmatrix}
\frac{1}{1-D_1} & 0 & 0 & 0 & 0 & 0 \\
0 & \frac{1}{1-D_2} & 0 & 0 & 0 & 0 \\
0 & 0 & \frac{1}{1-D_3} & 0 & 0 & 0 \\
0 & 0 & 0 & \frac{1}{1-\frac{D_2+D_3}{2}} & 0 & 0 \\
0 & 0 & 0 & 0 & \frac{1}{1-\frac{D_1+D_3}{2}} & 0 \\
0 & 0 & 0 & 0 & 0 & \frac{1}{1-\frac{D_1+D_2}{2}}
\end{bmatrix}; \tag{3.93}
$$

$$
[\mathbf{M}_3(D_1, D_2, D_3)] = \left[\begin{array}{cccccc}
\frac{1}{1-D_1} & 0 & 0 & 0 & 0 & 0 \\
0 & \frac{1}{1-D_2} & 0 & 0 & 0 & 0 \\
0 & 0 & \frac{1}{1-D_3} & 0 & 0 & 0 \\
0 & 0 & 0 & \frac{1}{2}\left(\frac{1}{1-D_2}+\frac{1}{1-D_3}\right) & 0 & 0 \\
0 & 0 & 0 & 0 & \frac{1}{2}\left(\frac{1}{1-D_1}+\frac{1}{1-D_3}\right) & 0 \\
0 & 0 & 0 & 0 & 0 & \frac{1}{2}\left(\frac{1}{1-D_1}+\frac{1}{1-D_2}\right)
\end{array} \right]. \tag{3.94}
$$

When Chaboche's notation is employed to express the hypothesis of energy equivalence (cf. Chaboche, Lesne and Maire, 1995),

$$
\widetilde{\boldsymbol{\sigma}} = \mathbf{M}_{\mathrm{Ch}}^{-1}(\mathrm{d}_\alpha) : \boldsymbol{\sigma} \qquad \text{and} \qquad \widetilde{\boldsymbol{\varepsilon}} = \mathbf{M}_{\mathrm{Ch}}(\mathrm{d}_\alpha) : \boldsymbol{\varepsilon}, \tag{3.95}
$$

the diagonal form in the principal damage coordinates d_α takes a representation equivalent to case A, with $\mathbf{M}_{\mathrm{Ch}} = \mathbf{M}_1^{-1}$ (cf. Qi and Bertram, 1997):

$$
[\mathbf{M}_{Ch}(d_1, d_2, d_3)] = \left[\begin{array}{cccccc}
1-d_1 & 0 & 0 & 0 & 0 & 0 \\
0 & 1-d_2 & 0 & 0 & 0 & 0 \\
0 & 0 & 1-d_3 & 0 & 0 & 0 \\
0 & 0 & 0 & \sqrt{(1-d_2)(1-d_3)} & 0 & 0 \\
0 & 0 & 0 & 0 & \sqrt{(1-d_3)(1-d_1)} & 0 \\
0 & 0 & 0 & 0 & 0 & \sqrt{(1-d_1)(1-d_2)}
\end{array} \right]. \tag{3.96}
$$

3.4.4 Example: Plane stress conditions

Under plane stress conditions, when vectorial representation is employed, a three-element stress vector is sufficient instead of a six-element one in a general three-dimensional stress state. Hence, the damage effect tensor M_{ijkl} can be represented by the 3×3 matrix, so that a plane transformation between the Cauchy stress tensor and the effective Cauchy stress tensor is as follows:

$$\begin{Bmatrix} \tilde{\sigma}_{11} \\ \tilde{\sigma}_{22} \\ \tilde{\sigma}_{12} \end{Bmatrix} = \begin{bmatrix} M_{11} & M_{12} & M_{13} \\ M_{21} & M_{22} & M_{23} \\ M_{31} & M_{32} & M_{33} \end{bmatrix} \begin{Bmatrix} \sigma_{11} \\ \sigma_{22} \\ \sigma_{12} \end{Bmatrix}. \tag{3.97}$$

Representation of a 3×3 matrix depends on the definition used for $[\mathbf{M}(\mathbf{D})]$ (cf. Sect. 3.4.2):

$$\mathbf{M}_1^2(\mathbf{D}) = \mathbf{P}^{-1}(\mathbf{D}), \tag{3.98}$$

$$[\mathbf{M}_1^2(\mathbf{D})] = \frac{1}{\nabla^2} \begin{bmatrix} (1-D_{22})^2 & D_{12}^2 & 2D_{12}(1-D_{22}) \\ D_{12}^2 & (1-D_{11})^2 & 2D_{12}(1-D_{11}) \\ D_{12}(1-D_{22}) & D_{12}(1-D_{11}) & D_{12}^2 + (1-D_{11})(1-D_{22}) \end{bmatrix}, \tag{3.99}$$

where: $\nabla^2 = [(1-D_{11})(1-D_{22}) - D_{12}^2]^2$;

$$\mathbf{M}_2(\mathbf{D}) = \left[\mathbf{I} - \widehat{\mathbf{D}}(\mathbf{D})\right]^{-1}, \tag{3.100}$$

$$[\mathbf{M}_2(\mathbf{D})] = \begin{bmatrix} \frac{1}{1-D_{11}}\left[1+\frac{D_{12}^2}{2\Delta(1-D_{11})}\right] & \frac{D_{12}^2}{2\Delta(1-D_{11})(1-D_{22})} & \frac{D_{12}}{\Delta(1-D_{11})} \\ \frac{D_{12}^2}{2\Delta(1-D_{11})(1-D_{22})} & \frac{1}{1-D_{22}}\left[1+\frac{D_{12}^2}{2\Delta(1-D_{22})}\right] & \frac{D_{12}}{\Delta(1-D_{22})} \\ \frac{D_{12}}{2\Delta(1-D_{11})} & \frac{D_{12}}{2\Delta(1-D_{22})} & \frac{1}{\Delta} \end{bmatrix}, \tag{3.101}$$

where $\Delta = 1 - \frac{1}{2}(D_{11} + D_{22}) - \frac{D_{12}^2}{2}\left(\frac{1}{1 - D_{11}} + \frac{1}{1 - D_{22}}\right)$;

$$\mathbf{M}_3(\mathbf{D}) = \widehat{\boldsymbol{\Phi}}(\mathbf{D}), \tag{3.102}$$

$$M_3(\mathbf{D}) = \frac{1}{\nabla}\begin{bmatrix} 1 - D_{22} & 0 & D_{12} \\ 0 & 1 - D_{11} & D_{12} \\ \dfrac{D_{12}}{2} & \dfrac{D_{12}}{2} & \dfrac{(1 - D_{11} + 1 - D_{22})}{2} \end{bmatrix}, \tag{3.103}$$

where $\nabla = (1 - D_{11})(1 - D_{22}) - D_{12}^2$.

Note that in this case the second-rank tensor $\boldsymbol{\Phi} = (\mathbf{1} - \mathbf{D})^{-1}$ has a plane 2×2 matrix representation, whereas the fourth-rank damage effect tensor $\widehat{\boldsymbol{\Phi}}(\boldsymbol{\Phi})$ has a 3×3 matrix form:

$$[\boldsymbol{\Phi}] = \left[(\mathbf{1} - \mathbf{D})^{-1}\right] = \begin{bmatrix} \dfrac{1 - D_{22}}{(1 - D_{11})(1 - D_{22}) - D_{12}^2} & \dfrac{D_{12}}{(1 - D_{11})(1 - D_{22}) - D_{12}^2} \\ \dfrac{D_{12}}{(1 - D_{11})(1 - D_{22}) - D_{12}^2} & \dfrac{1 - D_{11}}{(1 - D_{11})(1 - D_{22}) - D_{12}^2} \end{bmatrix} = \begin{bmatrix} \Phi_{11} & \Phi_{12} \\ \Phi_{21} & \Phi_{22} \end{bmatrix}, \tag{3.104}$$

$$\left[\widehat{\boldsymbol{\Phi}}(\boldsymbol{\Phi})\right] = \begin{bmatrix} \Phi_{11} & 0 & \Phi_{12} \\ 0 & \Phi_{22} & \Phi_{21} \\ \dfrac{\Phi_{21}}{2} & \dfrac{\Phi_{12}}{2} & \dfrac{(\Phi_{11} + \Phi_{22})}{2} \end{bmatrix}. \tag{3.105}$$

4

Three-dimensional anisotropic damage accumulation

4.1 Phenomenological models of creep-damage accumulation under nonproportional loadings

4.1.1 *Orthotropic damage growth in case of constant principal directions of the stress tensor*

Directional damage or, more precisely, damage anisotropy in creep conditions under nonproportional loading requires a modification of the simple scalar description of the damage growth rule (Chap. 2) and the creep-damage coupling in constitutive equations. The complexity of the description depends on the loading path or, more strictly, on the question whether the principal directions of the stress tensor are constant or rotate with respect to material particles, as examined, e.g., by Trąpczyński, Hayhurst, and Leckie, 1981 (Fig. 3.1). Chow and Lu (1992) developed and utilized a damage-coupled elasto-plastic model suitable for ductile fracture examination under both proportional and nonproportional loading conditions. It was based on a damage-perturbed updated Lagrangian formulation and an implicit concept of the objective derivative applied to the second-rank symmetric damage tensor. A similar problem was investigated by Lis (1992), who expressed damage rates in a rotating coordinate system coinciding with the principal directions of the stress tensor, and then accumulated them on a global sampling plane by an implicit concept of the objective derivative. In what follows, a concept of the damage induced creep anisotropy is developed using the second-rank damage tensor and the orthotropic damage growth rule applied to current principal stress directions. For simplicity, any effect of the damage anisotropy on the elastic stiffnesses is disregarded.

Consider first the simpler case when principal directions of the stress and damage tensors $\boldsymbol{\sigma}$, $\mathbf{D}$ coincide and do not change with time, such that the orthotropic theory of brittle damage coupled with the similarity of deviators of principal creep strain rates $\dot{\mathrm{e}}^{\mathrm{c}}$ and either the principal stress s (partly coupled) or the principal effective stress $\widetilde{\mathrm{s}}$ (fully coupled) are applicable (cf. Kachanov, 1986; Ganczarski and Skrzypek, 1994a). When formulated in the material axes of an orthotropic material, there is no coupling effect between normal stress and shear strain; therefore, in their principal directions the stress and damage tensors are:

$$[\boldsymbol{\sigma}] = \begin{bmatrix} \sigma_1 & 0 & 0 \\ 0 & \sigma_2 & 0 \\ 0 & 0 & \sigma_3 \end{bmatrix}, \quad [\mathbf{D}] = \begin{bmatrix} D_1 & 0 & 0 \\ 0 & D_2 & 0 \\ 0 & 0 & D_3 \end{bmatrix}, \quad \boldsymbol{\Psi} = \mathbf{1} - \mathbf{D} \tag{4.1}$$

and the orthotropic creep-damage growth rule (a direct extension of Kachanov's concept (2.26) to principal continuity ψ_i or damage D_i components holds, cf. Kachanov, 1986):

$$\dot{\psi}_i = \frac{\partial \psi_i}{\partial t} = -C_i \left\langle \frac{\sigma_i}{\psi_i} \right\rangle^{r_i} \tag{4.2}$$

or

$$\dot{D}_i = \frac{\partial D_i}{\partial t} = \left\langle \frac{\sigma_i}{A_i} \right\rangle^{r_i} (1 - D_i)^{-k_i}. \tag{4.3}$$

For the damage equivalent effective stress apply (3.9):

$$\tilde{\boldsymbol{\sigma}} = \frac{1}{2} \left[\boldsymbol{\sigma} : (\mathbf{1} - \mathbf{D})^{-1} + (\mathbf{1} - \mathbf{D})^{-1} : \boldsymbol{\sigma} \right]. \tag{4.4}$$

Hence, when principal stress and damage axes coincide and $D_{12} = D_{23} = D_{31} = 0$, $\sigma_{12} = \sigma_{23} = \sigma_{31} = 0, \tilde{\sigma}_{12} = \tilde{\sigma}_{23} = \tilde{\sigma}_{31} = 0$, the general matrix representation of the transformation (3.74) reduces to the form

$$\begin{Bmatrix} \tilde{\sigma}_1 \\ \tilde{\sigma}_2 \\ \tilde{\sigma}_3 \end{Bmatrix} = \begin{bmatrix} \dfrac{1}{1 - D_1} & 0 & 0 \\ 0 & \dfrac{1}{1 - D_2} & 0 \\ 0 & 0 & \dfrac{1}{1 - D_3} \end{bmatrix} \begin{Bmatrix} \sigma_1 \\ \sigma_2 \\ \sigma_3 \end{Bmatrix}. \tag{4.5}$$

Note that in the case considered, when $\boldsymbol{\sigma}$, $\tilde{\boldsymbol{\sigma}}$ and $\mathbf{D}$ are coaxial in their principal directions, all matrix representations of the damage effect tensor $\mathbf{M}(D_1, D_2, D_3)$, (3.92)–(3.94) coincide as well.

4.1.2 Orthotropic damage accumulation in case of variable principal directions of the stress tensor

Consider now a more general case, when principal directions $\alpha_i\,(1_\sigma, 2_\sigma, 3_\sigma,)$ of the stress tensor $\boldsymbol{\sigma}$ rotate through a small angle $\mathrm{d}\alpha_i$, in time t to $t+\mathrm{d}t$, to $\alpha_i'\,(1_{\sigma'}', 2_{\sigma'}', 3_{\sigma'}')$, for instance, if a specimen is subjected to a shear effect due to a single reverse torsion and steady tension, or a multiple reverse torsion and steady tension, etc., as shown in Fig. 3.1b, c (cf. Trąpczyński et al., 1981). After damage has occurred, the virgin isotropic material becomes orthotropic, and the principal directions $\beta_i\,(1_{\mathrm{D}}, 2_{\mathrm{D}}, 3_{\mathrm{D}})$ follow the principal stress axes rotation; however, by contrast to the previous case (Sect. 4.1.1),

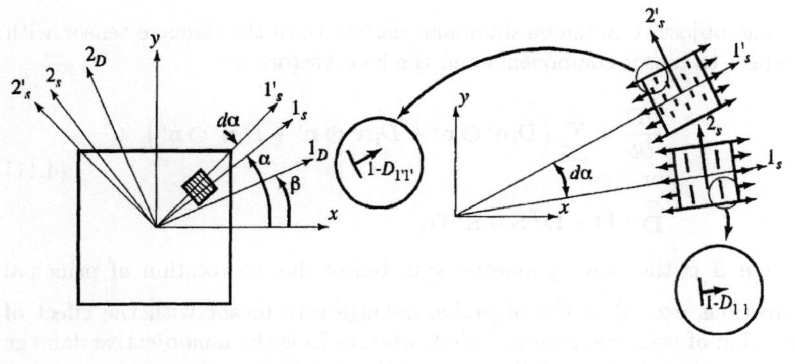

Fig. 4.1. Schematic representation of the accumulation of several orthotropic damage increments in variable principal directions

the stress and damage tensors $\boldsymbol{\sigma}$ and $\mathbf{D}$ are no longer coaxial in their principal axes, $\alpha_i \neq \beta_i$, as sketched in Fig. 4.1.

In other words, at current time t in current principal stress axes α_i $(1_\sigma, 2_\sigma, 3_\sigma,)$, the normative stress vector $\{{}^t\boldsymbol{\sigma}\}$ is expressed by its principal components

$$\{{}^t\boldsymbol{\sigma}\} = \{\sigma_{11}, \sigma_{22}, \sigma_{33}\}^{\mathrm{T}}, \quad \sigma_{12} = \sigma_{23} = \sigma_{31} = 0, \tag{4.6}$$

but the damage tensor $\{{}^t\mathbf{D}\}$ in the principal stress axes requires a full representation,

$$\{{}^t\mathbf{D}\} = \{D_{11}, D_{22}, D_{33}, D_{23}, D_{31}, D_{12}\}^{\mathrm{T}}, \tag{4.7}$$

whereas, for the damage rate tensor $\left\{{}^t\dot{\mathbf{D}}\right\}$ in the current principal stress space α_i, the representation in terms of its principal components becomes sufficient:

$$\left\{{}^t\dot{\mathbf{D}}\right\} = \left\{\dot{D}_{11}, \dot{D}_{22}, \dot{D}_{33}\right\}^{\mathrm{T}}, \quad \dot{D}_{12} = \dot{D}_{23} = \dot{D}_{31} = 0. \tag{4.8}$$

$\mathbf{D}$ denotes here the Murakami and Ohno (1981) second-rank damage tensor as represented through its principal values by (3.3)

$$\mathbf{D} = \sum_{i=1_D}^{3_D} D_i \mathbf{n}^i \otimes \mathbf{n}^i, \quad \mathbf{D} = \mathbf{1} - \boldsymbol{\Psi}, \tag{4.9}$$

whereas the nonobjective damage rate tensor $\dot{\mathbf{D}}$ is

$$\dot{\mathbf{D}} = \sum_{i=1_\sigma}^{3_\sigma} \dot{D}_i \mathbf{n}^i \otimes \mathbf{n}^i \tag{4.10}$$

because the principal axes of $\boldsymbol{\sigma}$ and $\dot{\mathbf{D}}$ coincide.

The objective Zaremba–Jaumann derivative of the damage tensor with respect to tensor components and the base vectors is

$$\frac{\partial \mathbf{D}}{\partial t} = \sum_{i=1}^{3} \left(\dot{D}_i \mathbf{n}^i \otimes \mathbf{n}^i + D_i \dot{\mathbf{n}}^i \otimes \mathbf{n}^i + D_i \mathbf{n}^i \otimes \dot{\mathbf{n}}^i \right)$$
or
$$\overset{\triangledown}{\mathbf{D}} = \dot{\mathbf{D}} - \mathbf{D}^T \mathbf{S} - \mathbf{S}^T \mathbf{D}, \tag{4.11}$$

where $\mathbf{S}$ is the skew-symmetric spin tensor due to rotation of principal directions $\mathrm{d}\alpha_i$, $\overset{\triangledown}{\mathbf{D}}$ is the objective damage rate tensor with the effect of rotation of principal axes included, whereas $\dot{\mathbf{D}}$ is the nonobjective damage rate in current principal directions of the stress tensor α_i.

When the nonobjective damage rate $\dot{\mathbf{D}}$ in current principal directions of the stress tensor α_i (effect of rotation of the base vector ignored) is assumed to be governed by the orthotropic damage growth rule, (4.2) and (4.3), and the skew-symmetric spin tensor representation in terms of $\mathrm{d}\alpha_i$ is used, we obtain:

$$\begin{aligned} \overset{\triangledown}{D}_{IJ} = \dot{D}_{IJ} - D_{IJ}^{\mathrm{T}} & \begin{bmatrix} 0 & \mathrm{d}\alpha_1 & -\mathrm{d}\alpha_2 \\ -\mathrm{d}\alpha_1 & 0 & \mathrm{d}\alpha_3 \\ \mathrm{d}\alpha_2 & -\mathrm{d}\alpha_3 & 0 \end{bmatrix} \\ & - \begin{bmatrix} 0 & -\mathrm{d}\alpha_1 & \mathrm{d}\alpha_2 \\ \mathrm{d}\alpha_1 & 0 & -\mathrm{d}\alpha_3 \\ -\mathrm{d}\alpha_2 & \mathrm{d}\alpha_3 & 0 \end{bmatrix} D_{IJ}, \end{aligned} \tag{4.12}$$

where

$$\dot{D}_{IJ} = C_{IJ} \left\langle \frac{\sigma_{IJ}}{1 - D_{IJ}} \right\rangle^{r_{IJ}}. \tag{4.13}$$

The new damage tensor $D_{I'J'}$ corresponding to the rotated basis $\alpha_i + \mathrm{d}\alpha_i$ is furnished next, to yield the damage accumulation in current principal stress directions

$$D_{I'J'}(t + \Delta t) = D_{IJ}(t) + \overset{\triangledown}{D}_{IJ}(t)\Delta t \tag{4.14}$$

and transformed then to the global coordinates (i, j)

$$D_{I'J'}(t + \Delta t) \xrightarrow{\text{transform}} D_{ij}(t + \Delta t). \tag{4.15}$$

Note that damage accumulation in the sampling space

$$D_{ij}(t + \Delta t) = D_{ij}(t) + \overset{\triangledown}{D}_{ij}(t)\Delta t \tag{4.16}$$

may also be used.

4.1.3 Creep-damage coupling formulations

I. Isotropic creep-damage scalar coupling (Skrzypek, 1993)

Assume the Mises-type flow rule (Penny and Marriott, 1995) (2.42),

$$\dot{\varepsilon}_{ij}^{c} = \frac{3}{2}\frac{\dot{\varepsilon}_{eq}^{c}}{\sigma_{eq}} s_{ij}, \tag{4.17}$$

the multiaxial time-hardening coupled with isotropic damage,

$$\dot{\varepsilon}_{eq}^{c} = \left(\frac{\sigma_{eq}}{1-D}\right)^{m} \dot{f}(t), \tag{4.18}$$

the Kachanov–Hayhurst isotropic damage growth (2.35) ($r = k$),

$$\dot{D} = C\left\langle \frac{\chi(\boldsymbol{\sigma})}{1-D}\right\rangle^{r}, \tag{4.19}$$

the Chaboche–Hayhurst invariant (scalar) damage equivalent stress (2.36),

$$\chi(\boldsymbol{\sigma}) = a\sigma_1 + 3b\sigma_H + c\sigma_{eq}, \tag{4.20}$$

where the following definitions hold:

$$d\varepsilon_{eq}^{c} = \sqrt{\frac{2}{3} d\varepsilon_{ij}^{c} d\varepsilon_{ij}^{c}}, \quad \dot{\varepsilon}_{eq}^{c} = \sqrt{\frac{2}{3}\dot{\varepsilon}_{ij}^{c}\dot{\varepsilon}_{ij}^{c}}, \quad \sigma_{eq} = \sqrt{\frac{3}{2} s_{ij} s_{ij}}. \tag{4.21}$$

In case of the plane stress and the creep incompressibility ($\sigma_3 = 0$) the damage coupled constitutive equations, (4.17) and (4.18) yield

$$\begin{gathered} \dot{\varepsilon}_{eq}^{c} = (2/\sqrt{3}) \left[(\dot{\varepsilon}_1^c)^2 + (\dot{\varepsilon}_2^c)^2 + \dot{\varepsilon}_1^c \dot{\varepsilon}_2^c\right]^{1/2}, \\ \sigma_{eq} = [\sigma_1^2 + \sigma_2^2 - \sigma_1\sigma_2]^{1/2}, \quad \sigma_H = (\sigma_1 + \sigma_2)/3 \end{gathered} \tag{4.22}$$

and

$$\begin{cases} \dot{\varepsilon}_1^c = \dfrac{\sigma_{eq}^{m-1}}{(1-D)^m}\left(\sigma_1 - \dfrac{\sigma_2}{2}\right)\dot{f}(t), \\ \dot{\varepsilon}_2^c = \dfrac{\sigma_{eq}^{m-1}}{(1-D)^m}\left(\sigma_2 - \dfrac{\sigma_1}{2}\right)\dot{f}(t), \quad \dot{\varepsilon}_3^c = -\dot{\varepsilon}_1^c - \dot{\varepsilon}_2^c, \end{cases} \tag{4.23}$$

since

$$[\boldsymbol{\sigma}] = \begin{bmatrix} \sigma_1 & 0 & 0 \\ 0 & \sigma_2 & 0 \\ 0 & 0 & 0 \end{bmatrix}, \quad [\mathbf{D}] = \begin{bmatrix} D & 0 & 0 \\ 0 & D & 0 \\ 0 & 0 & D \end{bmatrix},$$
$$[\dot{\boldsymbol{\varepsilon}}^{c}] = \begin{bmatrix} \dot{\varepsilon}_1^{c} & 0 & 0 \\ 0 & \dot{\varepsilon}_2^{c} & 0 \\ 0 & 0 & -\dot{\varepsilon}_1^{c} - \dot{\varepsilon}_2^{c} \end{bmatrix}. \tag{4.24}$$

II. Orthotropic creep-damage in case of constant principal directions (Ganczarski and Skrzypek, 1994a)

In the case of damage orthotropy, the creep-damage constitutive equations may be formulated in two different ways. First, the partly coupled approachconsists in a scalar coupling between the orthotropic damage growth rule and the isotropic creep flow rule, if the usual equivalent stress σ_{eq} in the time-hardening hypothesis is replaced by the effective equivalent stress $\widetilde{\sigma}_{\mathrm{eq}}$. This simplified approach is actually inconsistent, because damage orthotropy does not affects creep strain isotropy, and such an approach is justified only in the case of proportional loadings.

Table 4.1. Partly or fully coupled creep-orthotropic damage approach in case of constant principal directions

Partly coupled (Skrzypek, 1993)	Fully coupled (Ganczarski and Skrzypek, 1993)
isotropic flow rule $\dot{\varepsilon}_{ij}^{c} = \dfrac{3}{2}\dfrac{\dot{\varepsilon}_{\mathrm{eq}}^{c}}{\sigma_{\mathrm{eq}}} s_{ij}$	modified orthotropic flow rule $\dot{\varepsilon}_{ij}^{c} = \dfrac{3}{2}\dfrac{\dot{\varepsilon}_{\mathrm{eq}}^{c}}{\widetilde{\sigma}_{\mathrm{eq}}} \widetilde{s}_{ij}$
scalar coupling $\dot{\varepsilon}_{\mathrm{eq}}^{c} = (\widetilde{\sigma}_{\mathrm{eq}})^m \dot{\mathfrak{f}}(t)$	tensorial coupling $\dot{\varepsilon}_{\mathrm{eq}}^{c} = (\widetilde{\sigma}_{\mathrm{eq}})^m \dot{\mathfrak{f}}(t)$
orthotropic damage growth $\dot{D}_\nu = C_\nu \left\langle \dfrac{\sigma_\nu}{1 - D_\nu} \right\rangle^{r_\nu}$	orthotropic damage growth $\dot{D}_\nu = C_\nu \left\langle \dfrac{\sigma_\nu}{1 - D_\nu} \right\rangle^{r_\nu}$

$$\sigma_{\mathrm{eq}} = \sqrt{\frac{3}{2} s_{ij} s_{ij}}, \quad \widetilde{\sigma}_{\mathrm{eq}} = \sqrt{\frac{3}{2} \widetilde{s}_{ij} \widetilde{s}_{ij}}, \quad \dot{\varepsilon}_{\mathrm{eq}}^{c} = \sqrt{\frac{2}{3} \dot{\varepsilon}_{ij}^{c} \dot{\varepsilon}_{ij}^{c}}. \tag{4.25}$$

To avoid the above inconsistency, a second fully coupled approach is postulated where the modified orthotropic flow rule is used that assumes similarity of the creep strain rate $\dot{\boldsymbol{\varepsilon}}^{c}$ and the effective stress deviators $\widetilde{\mathbf{s}}$ instead of the usual stress deviator $\mathbf{s}$ in the previous formulation. This approach is consistent in the sense that, after the orthotropic damage has occurred, the virgin isotropic creep flow becomes orthotropic as well (cf. Table 4.1).

In the case of constant principal directions ($\alpha_i = \beta_i$), the stress $\boldsymbol{\sigma}$, the effective stress $\widetilde{\boldsymbol{\sigma}}$ and the damage tensor $\mathbf{D}$ are coaxial in their common principal directions; therefore, if the matrix representation is used, we have

$$[\boldsymbol{\sigma}] = \begin{bmatrix} \sigma_1 & 0 & 0 \\ 0 & \sigma_2 & 0 \\ 0 & 0 & \sigma_3 \end{bmatrix}, \quad [\mathbf{D}] = \begin{bmatrix} D_1 & 0 & 0 \\ 0 & D_2 & 0 \\ 0 & 0 & D_3 \end{bmatrix},$$
$$[\widetilde{\boldsymbol{\sigma}}] = \begin{bmatrix} \dfrac{\sigma_1}{1-D_1} & 0 & 0 \\ 0 & \dfrac{\sigma_2}{1-D_2} & 0 \\ 0 & 0 & \dfrac{\sigma_3}{1-D_3} \end{bmatrix}. \tag{4.26}$$

In a particular case when plane stress and creep incompressibility are assumed, the definitions (4.25) reduce to the form

$$\begin{aligned} \dot{\varepsilon}^{\mathrm{c}}_{\mathrm{eq}} &= (2/\sqrt{3}) \left[(\dot{\varepsilon}^{\mathrm{c}}_1)^2 + (\dot{\varepsilon}^{\mathrm{c}}_2)^2 + \dot{\varepsilon}^{\mathrm{c}}_1 \dot{\varepsilon}^{\mathrm{c}}_2 \right]^{1/2}, \\ \sigma_{\mathrm{eq}} &= \left[\sigma_1^2 + \sigma_2^2 - \sigma_1\sigma_2 \right]^{1/2}, \\ \widetilde{\sigma}_{\mathrm{eq}} &= \left[\left(\frac{\sigma_1}{1-D_1} \right)^2 + \left(\frac{\sigma_2}{1-D_2} \right)^2 - \frac{\sigma_1\sigma_2}{(1-D_1)(1-D_2)} \right]^{1/2}, \end{aligned} \tag{4.27}$$

whereas the plane stress creep-damage constitutive equations are obtained as follows:

$$\begin{aligned} \dot{\varepsilon}^{\mathrm{c}}_1 &= \frac{(\widetilde{\sigma}_{\mathrm{eq}})^m}{\sigma_{\mathrm{eq}}} \left(\sigma_1 - \frac{\sigma_2}{2} \right) \dot{\mathsf{f}}(t), \\ \dot{\varepsilon}^{\mathrm{c}}_2 &= \frac{(\widetilde{\sigma}_{\mathrm{eq}})^m}{\sigma_{\mathrm{eq}}} \left(\sigma_2 - \frac{\sigma_1}{2} \right) \dot{\mathsf{f}}(t), \quad \dot{\varepsilon}^{\mathrm{c}}_3 = -\dot{\varepsilon}^{\mathrm{c}}_1 - \dot{\varepsilon}^{\mathrm{c}}_2; \end{aligned} \tag{4.28}$$

$$\begin{aligned} \dot{\varepsilon}^{\mathrm{c}}_1 &= (\widetilde{\sigma}_{\mathrm{eq}})^{m-1} \left[\frac{\sigma_1}{1-D_1} - \frac{1}{2} \frac{\sigma_2}{(1-D_2)} \right] \dot{\mathsf{f}}(t), \\ \dot{\varepsilon}^{\mathrm{c}}_2 &= (\widetilde{\sigma}_{\mathrm{eq}})^{m-1} \left[\frac{\sigma_2}{1-D_2} - \frac{1}{2} \frac{\sigma_1}{(1-D_1)} \right] \dot{\mathsf{f}}(t), \quad \dot{\varepsilon}^{\mathrm{c}}_3 = -\dot{\varepsilon}^{\mathrm{c}}_1 - \dot{\varepsilon}^{\mathrm{c}}_2, \end{aligned} \tag{4.29}$$

when the partly coupled (scalar) or the fully coupled (tensor) approach is used.

III. Orthotropic creep-damage in case of variable principal directions

In the case of changing principal directions, the stress and damage tensors $\boldsymbol{\sigma}$ and $\mathbf{D}$ are not coaxial in their principal axes ($\alpha_i \neq \beta_i$). Therefore, either the partly or the fully coupled creep-damage approach may be used at current time t when the creep-damage constitutive equations and the

Table 4.2. Partly or fully coupled creep-damage approaches applied to current principal stress axes

Partly (scalar) coupled	Full (tensor) coupled
$\dot{\varepsilon}^{\mathrm{c}}_{IJ} = \frac{3}{2}\frac{\dot{\varepsilon}^{\mathrm{c}}_{\mathrm{eq}}}{\sigma_{\mathrm{eq}}} s_{IJ}$	$\dot{\varepsilon}^{\mathrm{c}}_{IJ} = \frac{3}{2}\frac{\dot{\varepsilon}^{\mathrm{c}}_{\mathrm{eq}}}{\widetilde{\sigma}_{\mathrm{eq}}} \widetilde{s}_{IJ}$
$\dot{\varepsilon}^{\mathrm{c}}_{\mathrm{eq}} = (\widetilde{\sigma}_{\mathrm{eq}})^m \dot{\mathrm{f}}(t)$	$\dot{\varepsilon}^{\mathrm{c}}_{\mathrm{eq}} = (\widetilde{\sigma}_{\mathrm{eq}})^m \dot{\mathrm{f}}(t)$
$\dot{D}_{IJ} = C_{IJ}\left\langle \frac{\sigma_{IJ}}{1-D_{IJ}} \right\rangle^{r_{IJ}}$	$\dot{D}_{IJ} = C_{IJ}\left\langle \frac{\sigma_{IJ}}{1-D_{IJ}} \right\rangle^{r_{IJ}}$

nonobjective damage rate $\dot{\mathbf{D}}$ are referred to the current principal stress axes $\alpha_i\,(I,J)$, Table 4.2.

The objective damage rate tensor $\overset{\triangledown}{\mathbf{D}}$ corresponding to the rotation of principal stress axes from α_i to $\alpha_i + \mathrm{d}\alpha_i$, with time changing from t to $t+\mathrm{d}t$, is obtained by the use of the Zaremba–Jaumann objective derivative (cf. Bathe, 1982)

$$\overset{\triangledown}{\mathbf{D}} = \dot{\mathbf{D}} - \mathbf{D}^{\mathrm{T}}\mathbf{S} - \mathbf{S}^{\mathrm{T}}\mathbf{D}. \tag{4.30}$$

When the transformation of the objective damage rate tensor $\overset{\triangledown}{\mathbf{D}}$ from the actual principal stress directions IJ to the global coordinates ij is performed,

$$\overset{\triangledown}{D}_{IJ} \longrightarrow \overset{\triangledown}{D}_{ij}, \tag{4.31}$$

the new (updated) damage tensor at time $t+\mathrm{d}t$, the components of which are represented in the global (sampling) space, is achieved:

$$D_{ij}(t+\Delta t) = D_{ij}(t) + \overset{\triangledown}{D}_{ij}(t)\Delta t. \tag{4.32}$$

In a particular case, when plane stress and creep incompressibility are assumed, the nonobjective damage rates are obtained from the orthotropic damage growth rule

$$\dot{D}_1 = C_1\left\langle \frac{\sigma_1}{1-D_1} \right\rangle^{r_1}, \qquad \dot{D}_2 = C_2\left\langle \frac{\sigma_2}{1-D_2} \right\rangle^{r_2}, \tag{4.33}$$

whereas the objective damage rate tensor components, associated with the plane rotation of principal stress axes by the angle $\mathrm{d}\alpha$, are

$$\begin{aligned} \begin{bmatrix} \overset{\triangledown}{D}_{1'1'} & \overset{\triangledown}{D}_{1'2'} \\ \overset{\triangledown}{D}_{2'1'} & \overset{\triangledown}{D}_{2'2'} \end{bmatrix} &= \begin{bmatrix} \dot{D}_{11} & 0 \\ 0 & \dot{D}_{22} \end{bmatrix} - \begin{bmatrix} D_{11} & D_{21} \\ D_{12} & D_{22} \end{bmatrix} \\ &\times \begin{bmatrix} 0 & \mathrm{d}\alpha \\ -\mathrm{d}\alpha & 0 \end{bmatrix} - \begin{bmatrix} 0 & -\mathrm{d}\alpha \\ \mathrm{d}\alpha & 0 \end{bmatrix} \begin{bmatrix} D_{11} & D_{12} \\ D_{21} & D_{22} \end{bmatrix}. \end{aligned} \tag{4.34}$$

4.2 Modeling of orthotropic time-dependent elastic-brittle damage in crystalline metallic solids

4.2.1 Basic equations of anisotropic elasticity coupled with damage

The directional nature of microcrack and void nucleation and growth, in initially homogeneous and isotropic solids, results in an anisotropic stress-strain law of elasticity, where a fourth-rank anisotropy tensor $\widetilde{\Lambda}_{ijkl}$ is time-dependent:

$$\varepsilon_{ij} = \widetilde{\Lambda}^{-1}_{ijkl}\sigma_{kl} \qquad \text{or} \qquad \boldsymbol{\varepsilon} = \widetilde{\boldsymbol{\Lambda}}^{-1}(\mathbf{D}^*) : \boldsymbol{\sigma}. \tag{4.35}$$

The representation the fourth-rank elasticity tensor of a damaged metallic material $\widetilde{\Lambda}_{ijkl}$ (derived by Litewka, 1985), when nonlinear forms with respect to damage are neglected, has the form

$$\begin{aligned}\widetilde{\Lambda}^{-1}_{ijkl} &= -\frac{\nu}{E}\delta_{ij}\delta_{kl} + \frac{1+\nu}{2E}(\delta_{ik}\delta_{jl} + \delta_{il}\delta_{jk}) \\ &+\frac{D_1^*}{4(1+D_1^*)E}(\delta_{ik}D^*_{jl} + \delta_{jl}D^*_{ik} + \delta_{il}D^*_{jk} + \delta_{jk}D^*_{il}).\end{aligned} \tag{4.36}$$

The equivalent equation of anisotropic elasticity coupled with damage is

$$\boldsymbol{\varepsilon} = -\frac{\nu}{E}(\mathrm{Tr}\boldsymbol{\sigma})\,\mathbf{1} + \frac{1+\nu}{E}\boldsymbol{\sigma} + \frac{D_1^*}{2(1+D_1^*)E}(\boldsymbol{\sigma} : \mathbf{D}^* + \mathbf{D}^* : \boldsymbol{\sigma}) \tag{4.37}$$

or

$$\varepsilon_{ij} = -\frac{\nu}{E}\sigma_{kk}I_{ij} + \frac{1+\nu}{E}\sigma_{ij} + \frac{D_1^*}{2(1+D_1^*)E}(\sigma_{ik}D^*_{kj} + D^*_{ik}\sigma_{kj}). \tag{4.38}$$

E and ν denote Young's modulus and Poisson's ratio of the virgin (undamaged) solid, whereas D_1^* is the dominant principal component of the modified damage tensor $\mathbf{D}^*$ the principal components of which are related to the classical Murakami and Ohno ones by $D_i^* = D_i/(1-D_i)$ (3.18). Note that for the undamaged material the above formulas reduce to the classical from the Hooke's law for the isotropic solid.

The elastic strain energy for the damaged solid takes the form:

$$\Phi^e(\boldsymbol{\sigma}, \mathbf{D}^*) = \frac{1-2\nu}{6E}\mathrm{Tr}^2\boldsymbol{\sigma} + \frac{1+\nu}{2E}\mathrm{Tr}(\boldsymbol{\sigma}'^2) + \frac{D_1^*}{2(1+D_1^*)E}\mathrm{Tr}(\boldsymbol{\sigma}^2 : \mathbf{D}^*), \tag{4.39}$$

where

$$\begin{aligned}&\mathrm{Tr}\boldsymbol{\sigma} = 3\sigma_{\mathrm{H}} = 3\sigma_{ii}, \quad \mathrm{Tr}(\boldsymbol{\sigma}'^2) = \frac{2}{3}\sigma_{\mathrm{eq}}^2 = s_{ij}s_{ij},\\ &\mathrm{Tr}(\boldsymbol{\sigma}^2 : \mathbf{D}^*) = \sigma_{ik}\sigma_{kl}D_{li}^*.\end{aligned} \tag{4.40}$$

When the definitions of hydrostatic and equivalent stresses are used, the above equation can be furnished as follows:

$$\begin{aligned}\Phi^{\mathrm{e}}(\boldsymbol{\sigma}, \mathbf{D}^*) = &\frac{\sigma_{\mathrm{eq}}^2}{2E}\left[\frac{2}{3}(1+\nu) + 3(1-2\nu)\left(\frac{\sigma_{\mathrm{H}}}{\sigma_{\mathrm{eq}}}\right)^2\right]\\ &+\frac{D_{\mathrm{I}}^*}{2(1+D_{\mathrm{I}}^*)E}(\sigma_{ik}\sigma_{kl}D_{li}^*).\end{aligned} \tag{4.41}$$

This time-dependent nonlinear function $\Phi^{\mathrm{e}}(\boldsymbol{\sigma}, \mathbf{D}^*)$ may be compared to the Chaboche concept of the strain energy density release rate Y (Sect. 2.3.1), but in this case the additive energy decomposition is used $\Phi^{\mathrm{e}}(\mathbf{D}^*) = \Phi^{\mathrm{e}}(\mathbf{0}) + \Phi^{\mathrm{d}}(\mathbf{D}^*)$, where the first term is responsible for the elastic energy of virgin (undamaged) material and the second is a nonlinear function of the damage evolution $\mathbf{D}^*(t)$. Recalling the notation of Sect. 2.3.1 the equivalent abbreviated form may also be used:

$$\begin{aligned}&\Phi^{\mathrm{e}}(\mathbf{D}^*) = \frac{\sigma_{\mathrm{eq}}^2}{2E}R_\nu + \Phi^{\mathrm{d}}(\mathbf{D}^*),\\ &R_\nu = \frac{2}{3}(1+\nu) + 3(1-2\nu)\left(\frac{\sigma_{\mathrm{H}}}{\sigma_{\mathrm{eq}}}\right)^2,\\ &\Phi^{\mathrm{d}}(\mathbf{D}^*) = \frac{D_{\mathrm{I}}^*}{2E(1+D_{\mathrm{I}}^*)}\mathrm{Tr}(\boldsymbol{\sigma}^2 : \mathbf{D}^*).\end{aligned} \tag{4.42}$$

4.2.2 Failure criterion and material identification

As a corresponding failure criterion the three-parameter damage affected isotropic scalar function of $\boldsymbol{\sigma}$ and $\mathbf{D}^*$ tensors is assumed:

$$F(\boldsymbol{\sigma}, \mathbf{D}^*) = C_1\mathrm{Tr}^2\boldsymbol{\sigma} + C_2\mathrm{Tr}\left(\boldsymbol{\sigma}'^2\right) + C_3\mathrm{Tr}\left(\boldsymbol{\sigma}^2 : \mathbf{D}^*\right) - \sigma_{\mathrm{u}}^2 = 0. \tag{4.43}$$

σ_{u} denotes the ultimate strength of the undamaged material (in general, temperature dependent), whereas constants C_1, C_2 and C_3 are to be obtained from the uniaxial tension direction (1), uniaxial tension direction (2), and biaxial tension (1+2) tests (cf. Litewka and Hult, 1989).

4.2.3 Damage evolution equation

In Litewka's theory the damage evolution rule is formulated by the use of the classical Murakami second-rank damage tensor $\mathbf{D}$ (3.3):

$$\dot{\mathbf{D}} = \mathcal{F}(\boldsymbol{\sigma}, \mathbf{D}). \tag{4.44}$$

Applying general tensor function representations (with nine tensor generators), the reduced two-terms form, which accounts for both the isotropic and the anisotropic damage, is proposed as sufficiently general to describe creep rupture in metals (Litewka, 1989):

$$\dot{\mathbf{D}} = B\,(\Phi^{\mathrm{e}})^m\,\mathbf{1} + C\,(\Phi^{\mathrm{e}})^n\,\boldsymbol{\sigma}^*. \tag{4.45}$$

$\boldsymbol{\sigma}^*$ is a modified stress tensor whose compressive principal components are replaced by zeros, whereas tensile ones are left unchanged. When the first isotropic term is omitted $B = 0$, and the exponent $n = 2$ is set, the simplified equation (4.45) takes the form (Litewka and Hult, 1989):

$$\begin{aligned}\dot{\mathbf{D}} = C\,(\Phi^{\mathrm{e}})^2\,\boldsymbol{\sigma}^* = C\Bigg[&\frac{1-2\nu}{6E}\mathrm{Tr}^2\boldsymbol{\sigma} + \frac{1+\nu}{2E}\mathrm{Tr}\left(\boldsymbol{\sigma}'^2\right)\\ &+\frac{D_1^*}{2E(1+D_1^*)}\mathrm{Tr}\left(\boldsymbol{\sigma}^2 : \mathbf{D}^*\right)\Bigg]^2\boldsymbol{\sigma}^*\end{aligned} \tag{4.46}$$

or, consistently applying Murakami's damage tensor $\mathbf{D}$,

$$\begin{aligned}&\dot{\mathbf{D}} = \frac{C}{2E}\left[\sigma_{\mathrm{eq}}^2 R_\nu + D_1\mathrm{Tr}(\boldsymbol{\sigma}^2 : \mathbf{D} : (\mathbf{1}-\mathbf{D})^{-1})\right]^2\boldsymbol{\sigma}^*,\\ &\sigma_i^* = \langle\sigma_i\rangle\,, \qquad D_1 = \max\{D_i\}, \qquad i = 1,2,3.\end{aligned} \tag{4.47}$$

Note that in Litewka's theory the damage evolution equation is generally not consistent in the thermodynamic sense.

4.2.4 Example: Plane stress, $\sigma_3 = 0$

When principal directions of the plane stress are used we have:

$$\begin{aligned}&\mathrm{Tr}\boldsymbol{\sigma} = \sigma_{kk} = \sigma_1 + \sigma_2,\\ &\mathrm{Tr}\left(\boldsymbol{\sigma}'^2\right) = \frac{2}{3}\sigma_{\mathrm{eq}}^2 = \frac{2}{3}(\sigma_1^2 - \sigma_1\sigma_2 + \sigma_2^2),\\ &\mathrm{Tr}[\boldsymbol{\sigma}^2 : \mathbf{D} : (\mathbf{1}-\mathbf{D})^{-1}] = \sigma_1^2\frac{D_1}{1-D_1} + \sigma_2^2\frac{D_2}{1-D_2},\\ &\Phi^{\mathrm{e}} = \frac{\sigma_1^2}{2E}\Bigg\{1 - 2\nu\left(\frac{\sigma_2}{\sigma_1}\right) + \left(\frac{\sigma_2}{\sigma_1}\right)^2\\ &\qquad + D_1\left[\frac{D_1}{1-D_1} + \left(\frac{\sigma_2}{\sigma_1}\right)^2\frac{D_2}{1-D_2}\right]\Bigg\}.\end{aligned} \tag{4.48}$$

Hence, the damage growth rule for plane stress is written as

$$\begin{aligned}\dot{\mathbf{D}} &= C\left(\Phi^{e}\right)^{2}\boldsymbol{\sigma}^{*} = \\ &= \frac{C}{4E^{2}}\left[A+BD_{1}\left(\frac{D_{1}}{1-D_{1}}+\left(\frac{\sigma_{2}}{\sigma_{1}}\right)^{2}\frac{D_{2}}{1-D_{2}}\right)\right]\sigma_{1}^{4}\boldsymbol{\sigma}^{*},\end{aligned} \tag{4.49}$$

where

$$\begin{aligned}A\left(\frac{\sigma_{2}}{\sigma_{1}}\right) &= 1-4\nu\frac{\sigma_{2}}{\sigma_{1}}+2(1+2\nu^{2})\left(\frac{\sigma_{2}}{\sigma_{1}}\right)^{2} \\ &\quad -4\nu\left(\frac{\sigma_{2}}{\sigma_{1}}\right)^{3}+\left(\frac{\sigma_{2}}{\sigma_{1}}\right)^{4}, \\ B\left(\frac{\sigma_{2}}{\sigma_{1}}\right) &= 2\left[1-2\nu\frac{\sigma_{2}}{\sigma_{1}}+2\left(\frac{\sigma_{2}}{\sigma_{1}}\right)^{2}\right].\end{aligned} \tag{4.50}$$

Constants C_1, C_2, and C_3 are obtained from the failure criterion (Sect. 4.2.2) as follows:

$$\begin{aligned}&C_{1}+\frac{2}{3}C_{2}+D_{1}^{*}C_{3}=\left(\sigma_{u}/\sigma_{1u}\right)^{2} && \text{direction } D_{1}^{*}, \\ &C_{1}+\frac{2}{3}C_{2}+D_{2}^{*}C_{3}=\left(\sigma_{u}/\sigma_{2u}\right)^{2} && \text{direction } D_{2}^{*}, \\ &4C_{1}+\frac{2}{3}C_{2}+\left(D_{1}^{*}+D_{2}^{*}\right)C_{3}=\left(\sigma_{u}/\sigma_{bu}\right)^{2} && \text{biaxial tension } (D_{1}^{*},D_{2}^{*}).\end{aligned} \tag{4.51}$$

Material constants σ_{1u}, σ_{2u}, and σ_{bu} are the ultimate strengths of the two uniaxial tests and the biaxial test for the damaged material, whose values are related to the ultimate tensile strength by (cf. Żuchowski, 1986):

$$\begin{aligned}\sigma_{1u}=\sigma_{bu}&=(1-D_{1})\sigma_{u}, \\ \sigma_{2u}&=(1-D_{2})\sigma_{u}.\end{aligned} \tag{4.52}$$

Note that, for a virgin (undamaged) material, $\sigma_{1u}=\sigma_{2u}=\sigma_{bu}=\sigma_{u}$ and parameter $C_1=0$, hence, with $D_1^*=D_2^*=0$, the general formula for the failure criterion reduces to the classical Huber–Mises–Hencky hypothesis $F(\boldsymbol{\sigma},\mathbf{0})=F^{\mathrm{HMH}}$. In other words, the proposed failure criterion is identified as the damage-influenced Mises-type failure criterion where it is assumed that the onset of material failure (first macrocrack initiation) is observed when the continuously shrunk failure surface (due to the damage growth) meets the stress vector actually applied at the point (Fig. 4.2).

When two cases for $m=\sigma_2/\sigma_1$ are considered, we get the damage evolution as:

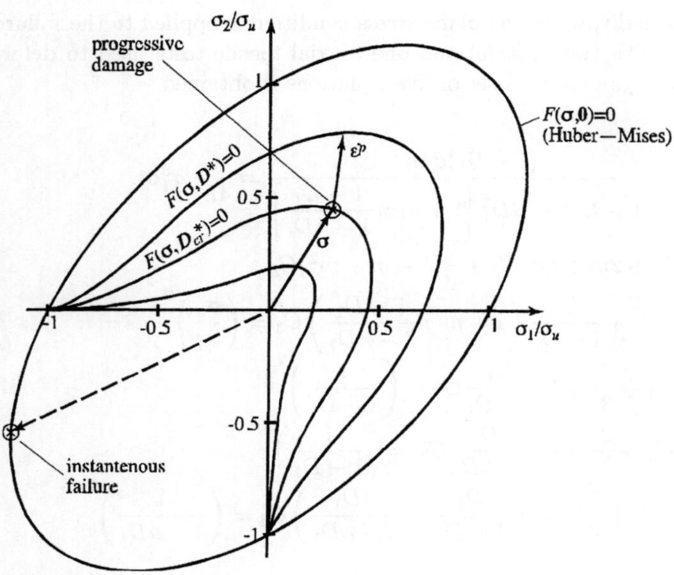

Fig. 4.2. Failure criterion (or isochronous rupture curves) for copper at 523 K (after Litewka and Hult, 1989)

$0 < m < 1$

$$\left\{ \begin{array}{c} \dot{D}_1 \\ \dot{D}_2 \end{array} \right\} = \left[A + BD_1 \left(\frac{D_1}{1-D_1} + m^2 \frac{D_2}{1-D_2} \right) \right] \frac{C}{4E^2} \sigma_1^4 \left\{ \begin{array}{c} \sigma_1 \\ \sigma_2 \end{array} \right\} \tag{4.53}$$

or $\dot{D}_2 = m\dot{D}_1,$

$m < 0$

$$\left\{ \begin{array}{c} \dot{D}_1 \\ \dot{D}_2 \end{array} \right\} = \left[A + BD_1 \left(\frac{D_1}{1-D_1} + m^2 \frac{D_2}{1-D_2} \right) \right] \frac{C}{4E^2} \sigma_1^4 \left\{ \begin{array}{c} \sigma_1 \\ 0 \end{array} \right\} \tag{4.54}$$

or $\dot{D}_2 = 0.$

Both cases may also be expressed by the unified damage growth formulas (Litewka and Hult, 1989)

$$\mathrm{d}D_1 = \left[A + B \frac{D_1^2}{1-D_1} \left(1 + m^2 n \frac{1-D_1}{1-nD_1} \right) \right] \frac{C}{4E^2} \sigma_1^5 \mathrm{d}t, \tag{4.55}$$
$$\mathrm{d}D_2 = n\mathrm{d}D_1,$$

where $n = m$ for $0 \leq m \leq 1$, and $n = 0$ for $m < 0$.

Eventually, when the plane stress condition is applied to the failure criterion, with two uniaxial and one biaxial tensile tests used to determine C_1, C_2, C_3, a reduced set of five equations is obtained:

$$\begin{aligned}
&\frac{(1-D_1)\mathrm{d}D_1}{A(1-D_1)+BD_1^2\left[1+m^2n\dfrac{1-D_1}{1-nD_1}\right]} = \frac{C}{4E^2}\sigma_1^5\mathrm{d}t,\\
&(1+2m+m^2)C_1+\frac{2}{3}(1-m+m^2)C_2\\
&\quad+\frac{D_1}{1-D_1}\left(1+m^2n\frac{1-D_1}{1-nD_1}\right)C_3=\left(\frac{\sigma_\mathrm{u}}{\sigma_1}\right)^2,\\
&C_1+\frac{2}{3}C_2+\frac{D_1}{1-D_1}C_3=\left(\frac{1}{1-D_1}\right)^2,\\
&C_1+\frac{2}{3}C_2+\frac{nD_1}{1-nD_1}C_3=\left(\frac{1}{1-D_1}\right)^2,\\
&4C_1+\frac{2}{3}C_2+\left(\frac{D_1}{1-D_1}+\frac{nD_1}{1-nD_1}\right)C_3=\left(\frac{1}{1-nD_1}\right)^2.
\end{aligned} \tag{4.56}$$

This system of equations that determine five unknown values D_1^{crit}, t_R, C_1, C_2, C_3 was numerically solved by the authors and compared with the experimental data by Johnson et al. (1956) and Murakami et al. (1986) for copper at 523 K at given σ_2/σ_1 ratios. The material constants used for theoretical predictions are as follows: $\nu = 0.35$, $\sigma_\mathrm{u} = 120$, MPa, $C/4E^2 = 2.49\times10^{-12}$ $[\mathrm{MPa}^5\mathrm{h}]^{-1}$.

Comments:

i. Although the results reported in this section exhibit that the theoretical isochronous curves obtained for the radially fixed stress ratios in principle remain in harmony with a certain number of experimental points, the practical application of this methods requires further improvement.

ii. In general, for a prescribed boundary problem, the damage induced redistribution of stresses results from the constitutive stress-strain-damage equation (Sect. 4.2.1). The elasticity equation coupled with damage is time-dependent and, hence, the constant stress ratios used to solve the basic system of equations should rather be replaced by integration along nonproportional paths.

iii. In addition, when the principal directions of the stress tensor change, the same happens with principal directions of the damage tensor, but the principal axes of both tensors do not coincide and, therefore, the objective derivative of the damage tensor should be applied to account for the rotation of principal stress axes (cf. Sect. 4.1.2).

iv. Note that under pure compression (third quarter), no progressive damage accumulation occurs and, therefore, when the applied stress vector meets the initial (undamaged) failure surface, instantaneous failure might occur at the point with no prior damage accumulation.

v. Litewka's model is, in general, inconsistent in the thermodynamic sense, since the coupled damage-constitutive equations are not consistently derived from the Helmholtz free energy function. Nevertheless, for metallic crystalline materials (like copper or some stainless steels) reasonable predictions may be obtained. The model is not applicable for elastic-brittle rock-like materials (like concrete) since damage evolution under compressive forces is not included.

4.3 Unified constitutive and damage theory of anisotropic elastic-brittle rock-like materials

4.3.1 Thermodynamically based equations of elastic-brittle damaged materials

The general thermodynamically based theory for constitutive and evolution equations of elastic-brittle damaged materials is due to Murakami and Kamiya (1997). It is based on the Helmholtz free energy as a function of the elastic strain tensor $\boldsymbol{\varepsilon}^{\mathrm{e}}$, the second-rank damage tensor $\mathbf{D}$, and another scalar damage variable β. By establishing a single dissipation potential, a unified description is possible instead of a separate formulation of constitutive and damage evolution equations.

The following representation of the Helmholtz free energy is postulated:

$$
\begin{aligned}
&\varrho\Psi(\boldsymbol{\varepsilon}^{\mathrm{e}},\mathbf{D},\beta) = \varrho\Psi^{\mathrm{e}}(\boldsymbol{\varepsilon}^{\mathrm{e}},\mathbf{D}) + \varrho\Psi^{\mathrm{d}}(\beta),\\
&\varrho\Psi^{\mathrm{e}}(\boldsymbol{\varepsilon}^{\mathrm{e}},\mathbf{D}) = \frac{1}{2}\lambda\mathrm{Tr}^2\boldsymbol{\varepsilon}^{\mathrm{e}} + \mu\mathrm{Tr}(\boldsymbol{\varepsilon}^{\mathrm{e}})^2 + \eta_1\mathrm{Tr}\mathbf{D}\mathrm{Tr}^2\boldsymbol{\varepsilon}^{\mathrm{e}}\\
&\qquad +\eta_2\mathrm{Tr}\mathbf{D}\mathrm{Tr}(\boldsymbol{\varepsilon}^{\mathrm{e}})^2 + \eta_3\mathrm{Tr}\boldsymbol{\varepsilon}^{\mathrm{e}}\mathrm{Tr}(\boldsymbol{\varepsilon}^{\mathrm{e}}:\mathbf{D}) + \eta_4\mathrm{Tr}[(\boldsymbol{\varepsilon}^{\mathrm{e}^*})^2:\mathbf{D}],\\
&\varrho\Psi^{\mathrm{d}}(\beta) = \frac{1}{2}K_d\beta^2,
\end{aligned}
\tag{4.57}
$$

where $\lambda = E\nu/(1+\nu)(1-2\nu)$ and $\mu = E/2(1+\nu)$ are Lamè constants for undamaged materials, η_1, η_2, η_3, η_4, and K_d are material constants, and $\boldsymbol{\varepsilon}^{\mathrm{e}^*}$ is a modified elastic strain tensor used to represent the unilateral damage response

$$\begin{aligned}
&\boldsymbol{\varepsilon}^{\mathrm{e}^*} = \langle \boldsymbol{\varepsilon}^{\mathrm{e}} \rangle - \zeta \langle -\boldsymbol{\varepsilon}^{\mathrm{e}} \rangle, \\
&[\langle \boldsymbol{\varepsilon}^{\mathrm{e}} \rangle] = \begin{bmatrix} \langle \varepsilon_1 \rangle & 0 & 0 \\ 0 & \langle \varepsilon_2 \rangle & 0 \\ 0 & 0 & \langle \varepsilon_3 \rangle \end{bmatrix}, \\
&[\langle -\boldsymbol{\varepsilon}^{\mathrm{e}} \rangle] = \begin{bmatrix} \langle -\varepsilon_1^{\mathrm{e}} \rangle & 0 & 0 \\ 0 & \langle -\varepsilon_2^{\mathrm{e}} \rangle & 0 \\ 0 & 0 & \langle -\varepsilon_3^{\mathrm{e}} \rangle \end{bmatrix}.
\end{aligned} \tag{4.58}$$

For the parameter $\zeta = 1$ the modified strain tensor $\boldsymbol{\varepsilon}^{\mathrm{e}^*}$ is identical to $\boldsymbol{\varepsilon}^{\mathrm{e}}$ and the unilateral damage opening/closure effect is not accounted for. For $\zeta = 0$, the strain tensor $\boldsymbol{\varepsilon}^{\mathrm{e}^*}$ is modified in such a way that negative principal strain components are replaced by zeros, whereas positive ones remain unchanged. The same rule was used in Sect. 4.2 when applied to the modified stress tensor $\boldsymbol{\sigma}^*$ in Litewka's model (cf. Litewka, 1985). In other words, in this limit case, cracking growth is stopped under compression. In general, neither of these two limit cases occurs, whereas ζ should be taken from the tension/compression test. In the case of high strength concrete application, the value $\zeta = 0.1$ was experimentally established (cf. Murakami and Kamiya, 1997). Applying (4.58) the following constitutive equations of anisotropic elasticity coupled with damage are furnished:

$$\begin{aligned}
\boldsymbol{\sigma} = \frac{\partial (\varrho \Psi^{\mathrm{e}})}{\partial \boldsymbol{\varepsilon}^{\mathrm{e}}} &= \widetilde{\boldsymbol{\Lambda}}(\mathbf{D}) : \boldsymbol{\varepsilon}^{\mathrm{e}} = [\lambda \mathrm{Tr} \boldsymbol{\varepsilon}^{\mathrm{e}} + 2\eta_1 \mathrm{Tr}\mathbf{D} \mathrm{Tr} \boldsymbol{\varepsilon}^{\mathrm{e}} \\
&+ \eta_3 \mathrm{Tr}(\boldsymbol{\varepsilon}^{\mathrm{e}} : \mathbf{D})] \mathbf{1} + 2(\mu + \eta_2 \mathrm{Tr}\mathbf{D}) \boldsymbol{\varepsilon}^{\mathrm{e}} + \eta_3 (\mathrm{Tr}\boldsymbol{\varepsilon}^{\mathrm{e}}) \mathbf{D} \\
&+ \eta_4 \frac{\partial \boldsymbol{\varepsilon}^{\mathrm{e}^*}}{\partial \boldsymbol{\varepsilon}^{\mathrm{e}}} \left(\boldsymbol{\varepsilon}^{\mathrm{e}^*} : \mathbf{D} + \mathbf{D} : \boldsymbol{\varepsilon}^{\mathrm{e}^*} \right),
\end{aligned} \tag{4.59}$$

whereas the thermodynamic damage conjugate forces of $\mathbf{D}$ and β are

$$\begin{aligned}
\mathbf{Y} &= -\frac{\partial (\varrho \Psi^{\mathrm{e}})}{\partial \mathbf{D}} = -\left[\eta_1 (\mathrm{Tr}\boldsymbol{\varepsilon}^{\mathrm{e}})^2 + \eta_2 \mathrm{Tr}(\boldsymbol{\varepsilon}^{\mathrm{e}})^2 \right] \mathbf{1} - \eta_3 (\mathrm{Tr}\boldsymbol{\varepsilon}^{\mathrm{e}}) \boldsymbol{\varepsilon}^{\mathrm{e}} - \eta_4 \boldsymbol{\varepsilon}^{\mathrm{e}^*} : \boldsymbol{\varepsilon}^{\mathrm{e}^*}, \\
B &= \frac{\partial (\varrho \Psi^{\mathrm{d}})}{\partial \beta} = K_d \beta.
\end{aligned} \tag{4.60}$$

$\widetilde{\boldsymbol{\Lambda}}(\mathbf{D})$ is a fourth-rank symmetric tensor, the secant stiffness, as a function of the second-rank damage tensor $\mathbf{D}$ (the damaged elastic stiffness). Thermodynamic conjugate force $\mathbf{Y}$, associated with $\mathbf{D}$, is known as the damage strain energy release rate and is the derivative of strain energy with respect to the damage variable $\mathcal{D}$ (the mechanical flux vector component). In case of the second rank-damage tensor $\mathbf{D}$, $\mathbf{Y}$ is the second-rank tensor as well. In case of the isotropic damage defined by the scalar D, Y is a scalar (see

Sect. 2.3.4, (2.135)). The damage criterion in the space $\{\mathbf{Y}, -B\}$ is also assumed in the form

$$
\begin{aligned}
&F(\mathbf{Y}, B) = Y_{\text{eq}} - (B_0 + B) = 0, \\
&Y_{\text{eq}} = \left(\tfrac{1}{2}\mathbf{Y} : \mathbf{L} : \mathbf{Y}\right)^{1/2}, \qquad L_{ijkl} = \tfrac{1}{2}\left(\delta_{ik}\delta_{jl} + \delta_{il}\delta_{jk}\right).
\end{aligned}
\tag{4.61}
$$

The evolution equations for damage are finally established as follows:

$$
\begin{aligned}
&\dot{\mathbf{D}} = -\dot{\lambda}_{\text{d}}\frac{\partial F}{\partial \mathbf{Y}}, \qquad \dot{\beta} = \dot{\lambda}_{\text{d}}\frac{\partial F}{\partial(-B)} = \dot{\lambda}_{\text{d}}, \\
&\dot{\lambda}_{\text{d}} = \frac{\partial F}{\partial \mathbf{Y}} : \dot{\mathbf{Y}} \Big/ \left(\frac{\partial B}{\partial \beta}\right) = \alpha \frac{\mathbf{L} : \mathbf{Y}}{2K_d \mathbf{Y}_{\text{eq}}} : \dot{\mathbf{Y}},
\end{aligned}
\tag{4.62}
$$

where $\alpha = 1$ if $F = 0$ and $\partial F/\partial \mathbf{Y} : \dot{\mathbf{Y}} > 0$ or $\alpha = 0$ if $F < 0$ and $\partial F/\partial \mathbf{Y} : \dot{\mathbf{Y}} \leq 0$.

4.3.2 Example: Application to high strength concrete

The above theory is applied to two cases, uniaxial compression and uniaxial tension tests of a high strength concrete. The modified elastic strain tensor takes the form

$$
\left[\boldsymbol{\varepsilon}^{\text{e}^*}\right]^{\text{com}} = \begin{bmatrix} \zeta\varepsilon^{\text{e}}_{11} & 0 & 0 \\ 0 & \varepsilon^{\text{e}}_{22} & 0 \\ 0 & 0 & \varepsilon^{\text{e}}_{33} \end{bmatrix}
$$

or

$$
\left[\boldsymbol{\varepsilon}^{\text{e}^*}\right]^{\text{ten}} = \begin{bmatrix} \varepsilon^{\text{e}}_{11} & 0 & 0 \\ 0 & \zeta\varepsilon^{\text{e}}_{22} & 0 \\ 0 & 0 & \zeta\varepsilon^{\text{e}}_{33} \end{bmatrix},
\tag{4.63}
$$

for the uniaxial compression or the uniaxial tension, respectively. Additionally, the following holds:

$$
\begin{aligned}
&\text{Tr}\boldsymbol{\varepsilon}^{\text{e}} = \varepsilon^{\text{e}}_{11} + \varepsilon^{\text{e}}_{22} + \varepsilon^{\text{e}}_{33}, \\
&\text{Tr}\left(\boldsymbol{\varepsilon}^{\text{e}}\right)^2 = \left(\varepsilon^{\text{e}}_{11}\right)^2 + \left(\varepsilon^{\text{e}}_{22}\right)^2 + \left(\varepsilon^{\text{e}}_{33}\right)^2, \\
&\left(\text{Tr}\boldsymbol{\varepsilon}^{\text{e}}\right)^2 = \left(\varepsilon^{\text{e}}_{11} + \varepsilon^{\text{e}}_{22} + \varepsilon^{\text{e}}_{33}\right)^2, \\
&\text{Tr}\left(\boldsymbol{\varepsilon}^{\text{e}} : \mathbf{D}\right) = \varepsilon^{\text{e}}_{11}D_{11} + \varepsilon^{\text{e}}_{22}D_{22} + \varepsilon^{\text{e}}_{33}D_{33}.
\end{aligned}
\tag{4.64}
$$

The high strength concrete identification, based on the uniaxial compression test, yields (cf. Murakami and Kamiya, 1997)

$$
\begin{aligned}
&E_0 = 21.4 \text{ GPa}, && \nu_0 = 0.2, && \eta_1 = -400 \text{ MPa}, \\
&\eta_2 = -900 \text{ MPa}, && \eta_3 = 100 \text{ MPa}, && \eta_4 = -23500 \text{ MPa}, \\
&\zeta = 0.1, && K_d = 0.04, && B_0 = 2.6 \times 10^{-3} \text{ MPa}.
\end{aligned}
\tag{4.65}
$$

When the matrix representation is used, the following damage coupled constitutive equations for the uniaxial compression or the uniaxial tension are furnished:

$$
\left\{ \begin{matrix} \sigma_{11} \\ \sigma_{22} = 0 \\ \sigma_{33} = 0 \end{matrix} \right\} = \left[\begin{matrix} \Lambda_{1111} & \Lambda_{1122} & \Lambda_{1133} \\ \Lambda_{2211} & \Lambda_{2222} & \Lambda_{2233} \\ \Lambda_{3311} & \Lambda_{3322} & \Lambda_{3333} \end{matrix} \right] \left\{ \begin{matrix} \varepsilon^{\mathrm{e}}_{11} \\ \varepsilon^{\mathrm{e}}_{22} \\ \varepsilon^{\mathrm{e}}_{33} \end{matrix} \right\}, \tag{4.66}
$$

where the components of symmetric 3×3 matrices are

$$
\begin{aligned}
\Lambda_{1111} &= \lambda + 2\mu + 2(\eta_1 + \eta_2)\,\mathrm{Tr}\mathbf{D} + 2\left(\eta_3 + \eta_4\zeta^2\right) D_{11}, \\
\Lambda_{2222} &= \lambda + 2\mu + 2(\eta_1 + \eta_2)\,\mathrm{Tr}\mathbf{D} + 2(\eta_3 + \eta_4) D_{22}, \\
\Lambda_{3333} &= \lambda + 2\mu + 2(\eta_1 + \eta_2)\,\mathrm{Tr}\mathbf{D} + 2(\eta_3 + \eta_4) D_{33}, \\
\Lambda_{1122} &= \Lambda_{2211} = \lambda + 2\eta_1 \mathrm{Tr}\mathbf{D} + \eta_3 (D_{11} + D_{22}), \\
\Lambda_{1133} &= \Lambda_{3311} = \lambda + 2\eta_1 \mathrm{Tr}\mathbf{D} + \eta_3 (D_{11} + D_{33}), \\
\Lambda_{2233} &= \Lambda_{3322} = \lambda + 2\eta_1 \mathrm{Tr}\mathbf{D} + \eta_3 (D_{22} + D_{33})
\end{aligned}
\tag{4.67}
$$

or

$$
\begin{aligned}
\Lambda_{1111} &= \lambda + 2\mu + 2(\eta_1 + \eta_2)\,\mathrm{Tr}\mathbf{D} + 2(\eta_3 + \eta_4) D_{11}, \\
\Lambda_{2222} &= \lambda + 2\mu + 2(\eta_1 + \eta_2)\,\mathrm{Tr}\mathbf{D} + 2\left(\eta_3 + \eta_4\zeta^2\right) D_{22}, \\
\Lambda_{3333} &= \lambda + 2\mu + 2(\eta_1 + \eta_2)\,\mathrm{Tr}\mathbf{D} + 2\left(\eta_3 + \eta_4\zeta^2\right) D_{33}, \\
\Lambda_{1122} &= \Lambda_{2211} = \lambda + 2\eta_1 \mathrm{Tr}\mathbf{D} + \eta_3 (D_{11} + D_{22}), \\
\Lambda_{1133} &= \Lambda_{3311} = \lambda + 2\eta_1 \mathrm{Tr}\mathbf{D} + \eta_3 (D_{11} + D_{33}), \\
\Lambda_{2233} &= \Lambda_{3322} = \lambda + 2\eta_1 \mathrm{Tr}\mathbf{D} + \eta_3 (D_{22} + D_{33}),
\end{aligned}
\tag{4.68}
$$

for the uniaxial compression or tension, respectively. For both cases, the damage rates (anisotropic damage evolution) may be expressed as follows:

$$\left\{\begin{array}{c}\dot{D}_{11}\\ \dot{D}_{22}\\ \dot{D}_{33}\end{array}\right\}=\frac{\alpha\left(Y_{11}\dot{Y}_{11}+Y_{22}\dot{Y}_{22}+Y_{33}\dot{Y}_{33}\right)}{2K_d\left(Y_{11}^2+Y_{22}^2+Y_{33}^2\right)}\left\{\begin{array}{c}Y_{11}\\ Y_{22}\\ Y_{33}\end{array}\right\}. \tag{4.69}$$

However, the components of the damage conjugate force **Y** are different for both cases, and are given as follows:

$$\begin{aligned}
Y_{11}&=-\eta_1\mathrm{Tr}^2\boldsymbol{\varepsilon}^{\mathrm{e}}-\eta_2\mathrm{Tr}\left(\boldsymbol{\varepsilon}^{\mathrm{e}}\right)^2-\eta_3\left(\mathrm{Tr}\boldsymbol{\varepsilon}^{\mathrm{e}}\right)\varepsilon_{11}^{\mathrm{e}}-\eta_4\left(\varepsilon_{11}^{\mathrm{e}^*}\right)^2,\\
Y_{22}&=-\eta_1\mathrm{Tr}^2\boldsymbol{\varepsilon}^{\mathrm{e}}-\eta_2\mathrm{Tr}\left(\boldsymbol{\varepsilon}^{\mathrm{e}}\right)^2-\eta_3\left(\mathrm{Tr}\boldsymbol{\varepsilon}^{\mathrm{e}}\right)\varepsilon_{22}^{\mathrm{e}}-\eta_4\left(\varepsilon_{22}^{\mathrm{e}^*}\right)^2,\\
Y_{33}&=-\eta_1\mathrm{Tr}^2\boldsymbol{\varepsilon}^{\mathrm{e}}-\eta_2\mathrm{Tr}\left(\boldsymbol{\varepsilon}^{\mathrm{e}}\right)^2-\eta_3\left(\mathrm{Tr}\boldsymbol{\varepsilon}^{\mathrm{e}}\right)\varepsilon_{33}^{\mathrm{e}}-\eta_4\left(\varepsilon_{33}^{\mathrm{e}^*}\right)^2,
\end{aligned} \tag{4.70}$$

where the appropriate components of the modified strain tensor $\boldsymbol{\varepsilon}^{\mathrm{e}^*}$ (4.63) for the uniaxial compression or tension are used.

4.3.3 Unilateral elastic-damage response of concrete under uniaxial compression versus uniaxial tension

Strong unilateral behaviour of high strength concrete, when subjected to uniaxial compression or uniaxial tension, is observed. Stress versus strain diagrams illustrate this phenomenon (Fig. 4.3).

The development of damage components under a uniaxial compression shows that, in spite of uniaxial stress state, microcracks in concrete both perpendicular (D_{11}) and parallel ($D_{22}=D_{33}$) to the loading direction propagate with strain increase and transverse components are dominant. The critical value of damage under compression is $D_{\mathrm{cr}}^{\mathrm{com}}\approx 0.4$, whereas the critical stress is $\sigma_{\mathrm{cr}}^{\mathrm{com}}=-52$ MPa.

In contrast to the previous phenomenon, the development of damage in concrete under uniaxial tension is much more anisotropic, with microcracks perpendicular to the tension direction (D_{11}) dominant and transverse components negligible. At critical stress level $\sigma_{\mathrm{cr}}^{\mathrm{ten}}=12.3$ MPa the critical damage components are $D_{11\mathrm{cr}}=0.13$, $D_{22\mathrm{cr}}=D_{33\mathrm{cr}}=0.01$.

In both cases a significant Young's modulus drop with damage is predicted; however, under uniaxial compression it is accompanied by an increase of Poisson's ratio, whereas under uniaxial tension no essential change in Poisson's ratio occurs. The above discussed unilateral response of elastic-brittle material like concrete is connected with the effect of cracks opening under tension and closuring under compression. Hence, the damage development is strongly anisotropic, with two specific cracking orientations, different for tension and compression. They might be identified as the transverse planar isotropy produced by $\sigma_1>0$, $\sigma_2=\sigma_3=0$ (uniaxial tension), and the cylindrical transverse isotropy produced by $\sigma_1<0$, $\sigma_2=\sigma_3=0$ (uniaxial compression) (cf. Chaboche, 1993). These are illustrated in Fig. 3.10.

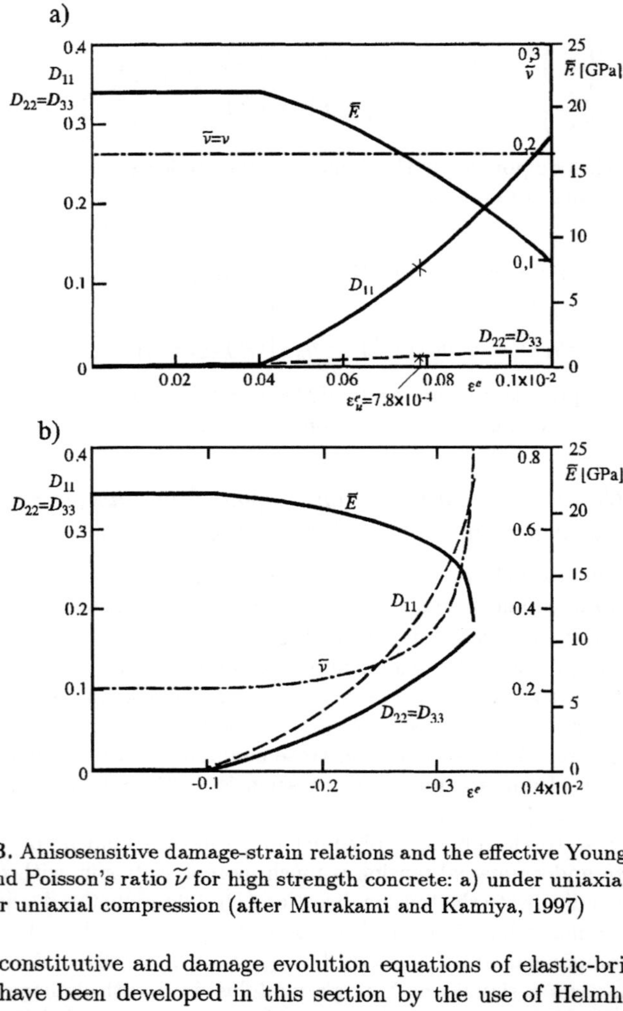

Fig. 4.3. Anisosensitive damage-strain relations and the effective Young's modulus $\widetilde{E}$ and Poisson's ratio $\widetilde{\nu}$ for high strength concrete: a) under uniaxial tension, b) under uniaxial compression (after Murakami and Kamiya, 1997)

The constitutive and damage evolution equations of elastic-brittle materials have been developed in this section by the use of Helmholtz free energy where damage conjugate forces are expressed as a function of the elastic strain tensor, (4.60)–(4.62). However, it is more convenient to define damage conjugate forces as functions of the stress tensor by the use of the Gibbs thermodynamic potential. Such an approach was recently applied by Hayakawa and Murakami (1998) to elastic-plastic-brittle materials for which the Gibbs potential accounts for complementary energy due to elastic deformation, the potential related to plastic deformation, and the damage potential related to microvoid nucleation.

4.4 Matrix representation of fourth-rank elasticity tensors for damaged materials

4.4.1 Representation of elasticity tensors expressed in the principal axes of damage tensor

In unified approach, reported in the previous sections, a fourth-rank symmetric tensor of secant stiffness $\widetilde{\mathbf{\Lambda}}$ is defined by applying the Helmholtz free energy of damaged material

$$\boldsymbol{\sigma} = \widetilde{\mathbf{\Lambda}}(\mathbf{D}) : \boldsymbol{\varepsilon}^{\mathrm{e}} \quad \text{or} \quad \sigma_{ij} = \widetilde{\Lambda}_{ijkl}\varepsilon^{\mathrm{e}}_{kl}. \tag{4.71}$$

If the stiffness tensor $\widetilde{\mathbf{\Lambda}}$ is expressed in terms of principal components of the second-rank damage tensor D_α , the constitutive equation of elasticity coupled with damage takes the following matrix representation (Murakami and Kamiya, 1997):

$$\{\boldsymbol{\sigma}\} = \left[\widetilde{\mathbf{\Lambda}}_{\mathrm{MK}}(D_\alpha)\right]\{\boldsymbol{\varepsilon}^{\mathrm{e}}\}, \tag{4.72}$$

$$\begin{Bmatrix} \sigma_{11} \\ \sigma_{22} \\ \sigma_{33} \\ \sigma_{23} \\ \sigma_{31} \\ \sigma_{12} \end{Bmatrix} = \left[\begin{matrix} \lambda + 2\mu + 2(\eta_1+\eta_2)\,\mathrm{Tr}\mathbf{D} + 2(\eta_3+\eta_4)\,D_{11} \\ \lambda + 2\eta_1\mathrm{Tr}\mathbf{D} + \eta_3(D_{11}+D_{22}) \\ \lambda + 2\eta_1\mathrm{Tr}\mathbf{D} + \eta_3(D_{11}+D_{33}) \\ 0 \\ 0 \\ 0 \end{matrix}\right.$$

$$\begin{matrix} \lambda + 2\eta_1\mathrm{Tr}\mathbf{D} + \eta_3(D_{11}+D_{22}) \\ \lambda + 2\mu + 2(\eta_1+\eta_2)\,\mathrm{Tr}\mathbf{D} + 2(\eta_3+\eta_4)\,D_{22} \\ \lambda + 2\eta_1\mathrm{Tr}\mathbf{D} + \eta_3(D_{22}+D_{33}) \\ 0 \\ 0 \\ 0 \end{matrix}$$

$$\begin{matrix} \lambda + 2\eta_1\mathrm{Tr}\mathbf{D} + \eta_3(D_{11}+D_{33}) \\ \lambda + 2\eta_1\mathrm{Tr}\mathbf{D} + \eta_3(D_{33}+D_{22}) \\ \lambda + 2\mu + 2(\eta_1+\eta_2)\,\mathrm{Tr}\mathbf{D} + 2(\eta_3+\eta_4)\,D_{33} \\ 0 \\ 0 \\ 0 \end{matrix}$$

$$\begin{matrix} 0 \\ 0 \\ 0 \\ 2\mu + 2\eta_2 \mathrm{Tr}\mathbf{D} + \eta_4 \left(D_{33} + D_{22}\right) \\ 0 \\ 0 \end{matrix}$$

$$\begin{matrix} 0 \\ 0 \\ 0 \\ 0 \\ 2\mu + 2\eta_2 \mathrm{Tr}\mathbf{D} + \eta_4 \left(D_{11} + D_{33}\right) \\ 0 \end{matrix} \tag{4.73}$$

$$\left. \begin{matrix} 0 \\ 0 \\ 0 \\ 0 \\ 0 \\ 2\mu + 2\eta_2 \mathrm{Tr}\mathbf{D} + \eta_4 \left(D_{11} + D_{22}\right) \end{matrix} \right] \begin{Bmatrix} \varepsilon_{11} \\ \varepsilon_{22} \\ \varepsilon_{33} \\ \varepsilon_{23} \\ \varepsilon_{31} \\ \varepsilon_{12} \end{Bmatrix}.$$

λ and μ are Lamè coefficients of the isotropic undamaged solid: $\lambda = E\nu / \left(1+\nu\right)\left(1-2\nu\right)$, $\mu = E/2\left(1+\nu\right)$. For simplicity, the unilateral damage effect has been excluded ($\zeta = 1$). The secant stiffness tensor $\widetilde{\mathbf{\Lambda}}$ may also be defined in a different fashion when the fourth-rank damage effect tensors $\mathbf{M}\left(\mathbf{D}\right)$ are used (cf. Sect. 3.4.2). If, for example, Chaboche's concept is adopted to define the effective stresses and the effective strains,

$$\widetilde{\boldsymbol{\sigma}} = \mathbf{M}_{\mathrm{Ch}}^{-1}\left(d_\alpha\right) : \boldsymbol{\sigma}, \qquad \boldsymbol{\varepsilon}^{\mathrm{e}} = \mathbf{M}_{\mathrm{Ch}}\left(d_\alpha\right) : \boldsymbol{\varepsilon}^{\mathrm{e}} \tag{4.74}$$

and

$$\widetilde{\boldsymbol{\sigma}} = \mathbf{\Lambda} : \widetilde{\boldsymbol{\varepsilon}}^{\mathrm{e}}, \qquad \boldsymbol{\sigma} = \widetilde{\mathbf{\Lambda}} : \boldsymbol{\varepsilon}^{\mathrm{e}}, \tag{4.75}$$

the elasticity tensor modified by damage $\widetilde{\mathbf{\Lambda}}\left(d_\alpha\right)$ is furnished as follows (cf. Chaboche, Lesne, and Moire, 1995):

$$\boldsymbol{\sigma} = \left[\mathbf{M}_{\mathrm{Ch}}\left(d_\alpha\right) : \mathbf{\Lambda} : \mathbf{M}_{\mathrm{Ch}}\left(d_\alpha\right)\right] : \boldsymbol{\varepsilon}^{\mathrm{e}} = \widetilde{\mathbf{\Lambda}}\left(d_\alpha\right) : \boldsymbol{\varepsilon}^{\mathrm{e}}. \tag{4.76}$$

Recalling the matrix representation for $\mathbf{M}_{\mathrm{Ch}}\left(d_1, d_2, d_3\right)$, represented by a diagonal matrix of principal damage components d_1, d_2, d_3 (3.96):

$$\mathbf{M}_{\mathrm{Ch}}\left(d_\alpha\right)=\begin{bmatrix} 1-d_1 & 0 & 0 & 0 & 0 & 0 \\ 0 & 1-d_2 & 0 & 0 & 0 & 0 \\ 0 & 0 & 1-d_3 & 0 & 0 & 0 \\ 0 & 0 & 0 & \sqrt{\left(1-d_2\right)\left(1-d_3\right)} & 0 & 0 \\ 0 & 0 & 0 & 0 & \sqrt{\left(1-d_3\right)\left(1-d_1\right)} & 0 \\ 0 & 0 & 0 & 0 & 0 & \sqrt{\left(1-d_1\right)\left(1-d_2\right)} \end{bmatrix} \tag{4.77}$$

and the well-known matrix representation of the elasticity tensor for undamaged isotropic materials

$$\mathbf{\Lambda}=\begin{bmatrix} \lambda+2\mu & \lambda & \lambda & 0 & 0 & 0 \\ \lambda & \lambda+2\mu & \lambda & 0 & 0 & 0 \\ \lambda & \lambda & \lambda+2\mu & 0 & 0 & 0 \\ 0 & 0 & 0 & 2\mu & 0 & 0 \\ 0 & 0 & 0 & 0 & 2\mu & 0 \\ 0 & 0 & 0 & 0 & 0 & 2\mu \end{bmatrix} \tag{4.78}$$

we obtain the following matrix representation of the elasticity tensor modified by damage

$$\widetilde{\mathbf{\Lambda}}_{\mathrm{Ch}}\left(d_\alpha\right)=\begin{bmatrix} (\lambda+2\mu)\left(1-d_1\right)^2 & \lambda\left(1-d_1\right)\left(1-d_2\right) & \lambda\left(1-d_1\right)\left(1-d_3\right) & 0 & \cdots \\ \lambda\left(1-d_1\right)\left(1-d_2\right) & (\lambda+2\mu)\left(1-d_2\right)^2 & \lambda\left(1-d_2\right)\left(1-d_3\right) & 0 & \cdots \\ \lambda\left(1-d_1\right)\left(1-d_3\right) & \lambda\left(1-d_2\right)\left(1-d_3\right) & (\lambda+2\mu)\left(1-d_3\right)^2 & 0 & \cdots \\ 0 & 0 & 0 & 2\mu\left(1-d_2\right)\left(1-d_3\right) & \cdots \\ 0 & 0 & 0 & 0 & \cdots \\ 0 & 0 & 0 & 0 & \cdots \end{bmatrix}$$

$$\begin{array}{cc} 0 & 0 \\ 0 & 0 \\ 0 & 0 \\ 0 & 0 \\ 2\mu\,(1-d_3)\,(1-d_1) & 0 \\ 0 & 2\mu\,(1-d_1)\,(1-d_2) \end{array}\Bigg] . \tag{4.79}$$

The appropriate constitutive equation for the elastic-brittle damaged material is established as

$$\begin{aligned} \{\boldsymbol{\sigma}\} &= \left[\widetilde{\boldsymbol{\Lambda}}_{\mathrm{Ch}}\,(d_\alpha)\right]\{\boldsymbol{\varepsilon}^{\mathrm{e}}\}\,, \\ \{\boldsymbol{\sigma}\}^{\mathrm{T}} &= \{\sigma_{11}, \sigma_{22}, \sigma_{33}, \sigma_{23}, \sigma_{31}, \sigma_{12}\}\,, \\ \{\boldsymbol{\varepsilon}\}^{\mathrm{T}} &= \{\varepsilon_{11}, \varepsilon_{22}, \varepsilon_{33}, \varepsilon_{23}, \varepsilon_{31}, \varepsilon_{12}\}\,. \end{aligned} \tag{4.80}$$

In case of the inverse formulation (cf. Sect. 3.4), the elasticity equation coupled with damage is expressed as follows:

$$\boldsymbol{\varepsilon}^{\mathrm{e}} = \widetilde{\boldsymbol{\Lambda}}^{-1} : \boldsymbol{\sigma} \qquad \text{or} \qquad \varepsilon^{\mathrm{e}}_{ij} = \widetilde{\Lambda}^{-1}_{ijkl}\sigma_{kl}. \tag{4.81}$$

The inverse elastic matrix (compliance) $\widetilde{\boldsymbol{\Lambda}}^{-1}\,(\mathbf{D})$ may be defined in a similar fashion when one of the matrix representations of the damage effect tensor (cf. Sect. 3.4.3) is used. Hence, if the following definitions are employed

$$\begin{aligned} \widetilde{\boldsymbol{\sigma}} &= \mathbf{M}\,(d_\alpha) : \boldsymbol{\sigma}, \quad & \widetilde{\boldsymbol{\varepsilon}}^{\mathrm{e}} &= \mathbf{M}^{-1}\,(d_\alpha) : \boldsymbol{\varepsilon}^{\mathrm{e}}, \\ \widetilde{\boldsymbol{\varepsilon}}^{\mathrm{e}} &= \boldsymbol{\Lambda}^{-1} : \widetilde{\boldsymbol{\sigma}}, & \boldsymbol{\varepsilon}^{\mathrm{e}} &= \widetilde{\boldsymbol{\Lambda}}^{-1} : \boldsymbol{\sigma}, \end{aligned} \tag{4.82}$$

the elasticity equation is furnished as follows:

$$\boldsymbol{\varepsilon}^{\mathrm{e}} = \left[\mathbf{M}_{\mathrm{Ch}}\,(d_\alpha) : \boldsymbol{\Lambda}^{-1} : \mathbf{M}_{\mathrm{Ch}}\,(d_\alpha)\right] : \boldsymbol{\sigma} = \widetilde{\boldsymbol{\Lambda}}^{-1}\,(d_\alpha) : \boldsymbol{\sigma}. \tag{4.83}$$

When the definitions of $\mathbf{M}_1$, $\mathbf{M}_2$ or $\mathbf{M}_3$ (cf. Chen and Chow, 1995, Sect. 3.4.3) are used in their matrix form in terms of the principal damage components, $D_1 = D_{11}$, $D_2 = D_{22}$, $D_3 = D_{33}$, $D_{23} = D_{31} = D_{12} = 0$, the following symmetric compliance matrices modified by damage are obtained:

$$\widetilde{\mathbf{\Lambda}}_1^{-1} = \frac{1}{E}\begin{bmatrix} \frac{1}{(1-D_1)^2} & -\frac{\nu}{(1-D_1)(1-D_2)} & -\frac{\nu}{(1-D_1)(1-D_3)} & 0 & 0 & 0 \\ -\frac{\nu}{(1-D_2)(1-D_1)} & \frac{1}{(1-D_2)^2} & -\frac{\nu}{(1-D_2)(1-D_3)} & 0 & 0 & 0 \\ -\frac{\nu}{(1-D_3)(1-D_1)} & -\frac{\nu}{(1-D_3)(1-D_2)} & \frac{1}{(1-D_3)^2} & 0 & 0 & 0 \\ 0 & 0 & 0 & \frac{1+\nu}{(1-D_2)(1-D_3)} & 0 & 0 \\ 0 & 0 & 0 & 0 & \frac{1+\nu}{(1-D_1)(1-D_3)} & 0 \\ 0 & 0 & 0 & 0 & 0 & \frac{1+\nu}{(1-D_1)(1-D_2)} \end{bmatrix}; \tag{4.84}$$

$$\widetilde{\mathbf{\Lambda}}_2^{-1} = \frac{1}{E}\begin{bmatrix} \frac{1}{(1-D_1)^2} & -\frac{\nu}{(1-D_1)(1-D_2)} & -\frac{\nu}{(1-D_1)(1-D_3)} & 0 & 0 & 0 \\ -\frac{\nu}{(1-D_2)(1-D_1)} & \frac{1}{(1-D_2)^2} & -\frac{\nu}{(1-D_2)(1-D_3)} & 0 & 0 & 0 \\ -\frac{\nu}{(1-D_3)(1-D_1)} & -\frac{\nu}{(1-D_3)(1-D_2)} & \frac{1}{(1-D_3)^2} & 0 & 0 & 0 \\ 0 & 0 & 0 & \frac{1+\nu}{\left(1-\frac{D_2+D_3}{2}\right)^2} & 0 & 0 \\ 0 & 0 & 0 & 0 & \frac{1+\nu}{\left(1-\frac{D_3+D_1}{2}\right)^2} & 0 \\ 0 & 0 & 0 & 0 & 0 & \frac{1+\nu}{\left(1-\frac{D_1+D_2}{2}\right)^2} \end{bmatrix}; \tag{4.85}$$

$$
\widetilde{\mathbf{\Lambda}}_3^{-1} = \frac{1}{E}
\begin{bmatrix}
\frac{1}{(1-D_1)^2} & -\frac{\nu}{(1-D_1)(1-D_2)} & -\frac{\nu}{(1-D_1)(1-D_3)} & 0 & 0 & 0 \\
-\frac{\nu}{(1-D_2)(1-D_1)} & \frac{1}{(1-D_2)^2} & -\frac{\nu}{(1-D_2)(1-D_3)} & 0 & 0 & 0 \\
-\frac{\nu}{(1-D_3)(1-D_1)} & -\frac{\nu}{(1-D_3)(1-D_2)} & \frac{1}{(1-D_3)^2} & 0 & 0 & 0 \\
0 & 0 & 0 & \frac{1+\nu}{4}\left(\frac{1}{1-D_2}+\frac{1}{1-D_3}\right)^2 & 0 & 0 \\
0 & 0 & 0 & 0 & \frac{1+\nu}{4}\left(\frac{1}{1-D_1}+\frac{1}{1-D_3}\right)^2 & 0 \\
0 & 0 & 0 & 0 & 0 & \frac{1+\nu}{4}\left(\frac{1}{1-D_1}+\frac{1}{1-D_2}\right)^2
\end{bmatrix} . \tag{4.86}
$$

For comparison, the appropriate elasticity matrix that follows from the Litewka model (cf. Sect. 4.2) is also quoted:

$$
\widetilde{\mathbf{\Lambda}}_L^{-1} = \frac{1}{E} \times
$$

$$
\times
\begin{bmatrix}
1+\frac{D_1^{*2}}{1+D_1^*} & -\nu & -\nu & 0 \\
-\nu & 1+\frac{D_1^* D_2^*}{1+D_1^*} & -\nu & 0 \\
-\nu & -\nu & 1+\frac{D_1^* D_3^*}{1+D_1^*} & 0 \\
0 & 0 & 0 & (1+\nu)\left[1+\frac{D_1^*(D_2^*+D_3^*)}{2(1+\nu)(1+D_1^*)}\right] \\
0 & 0 & 0 & 0 \\
0 & 0 & 0 & 0
\end{bmatrix}
$$

$$
\begin{matrix}
0 & 0 \\
0 & 0 \\
0 & 0 \\
0 & 0 \\
(1+\nu)\left[1+\frac{D_1^*(D_1^*+D_3^*)}{2(1+\nu)\left(1+D_1^*\right)}\right] & 0 \\
0 & (1+\nu)\left[1+\frac{D_1^*(D_1^*+D_2^*)}{2(1+\nu)\left(1+D_1^*\right)}\right]
\end{matrix}
\Bigg] . \qquad (4.87)
$$

In the last case, only the diagonal components of the elastic matrix are affected by damage when represented by the principal components D_1^*, D_2^*, D_3^* of the modified damage tensor $D_\alpha^* \in< 0, \infty)$, which is a limitation of the model.

4.4.2 Representation of elasticity tensors in a general case of damage anisotropy

In a more general case, when the damage tensor $\mathbf{D}$ (or $\mathbf{D}^*$), the stress tensor $\boldsymbol{\sigma}$ and the strain tensor $\boldsymbol{\varepsilon}^e$ are not coaxial in their principal axes, a complete representation with non-zero off-diagonal components must be used (damage induced anisotropy). We shall discuss this effect for two models: Litewka's model (Sect. 4.2) and Murakami and Kamiya's unified model (Sect. 4.3).

In general, when principal axes of stress and strain tensors rotate due to the stress and strain redistribution following damage evolution in solids, the principal axes of the second-rank damage tensor also rotate. However, the stress, strain, and damage tensors are no longer coaxial in their principal component coordinate axes. Hence, when the modified damage tensor is expressed in terms of six components D_{11}^*, D_{22}^*, D_{33}^*, $D_{23}^* = D_{32}^*$, $D_{31}^* = D_{13}^*$, $D_{12}^* = D_{21}^*$, for Litewka's model a generalized form of the constitutive equations of elasticity coupled with damage may be furnished:

$$
\{\varepsilon\} = \left[\widetilde{\boldsymbol{\Lambda}}_L\left(\mathbf{D}^*\right)\right]\{\sigma\} \qquad (4.88)
$$

and

$$\left\{\begin{array}{c}\varepsilon_{11}\\ \varepsilon_{22}\\ \varepsilon_{33}\\ \varepsilon_{23}\\ \varepsilon_{13}\\ \varepsilon_{12}\end{array}\right\} = E^{-1}\left[\begin{array}{cccccc}
1+\frac{D_1^*}{1+D_1^*}D_{11}^* & -\nu & -\nu & 0 & \frac{D_1^*}{1+D_1^*}D_{13}^* & \frac{D_1^*}{1+D_1^*}D_{12}^* \\
-\nu & 1+\frac{D_1^*}{1+D_1^*}D_{22}^* & -\nu & \frac{D_1^*}{1+D_1^*}D_{23}^* & 0 & \frac{D_1^*}{1+D_1^*}D_{12}^* \\
-\nu & -\nu & 1+\frac{D_1^*}{1+D_1^*}D_{33}^* & \frac{D_1^*}{1+D_1^*}D_{23}^* & \frac{D_1^*}{1+D_1^*}D_{13}^* & 0 \\
0 & \frac{D_1^*}{2(1+D_1^*)}D_{23}^* & \frac{D_1^*}{2(1+D_1^*)}D_{23}^* & 1+\nu+\frac{D_1^*}{2(1+D_1^*)}\left(D_{22}^*+D_{33}^*\right) & \frac{D_1^*}{2(1+D_1^*)}D_{12}^* & \frac{D_1^*}{2(1+D_1^*)}D_{13}^* \\
\frac{D_1^*}{2(1+D_1^*)}D_{13}^* & 0 & \frac{D_1^*}{2(1+D_1^*)}D_{13}^* & \frac{D_1^*}{2(1+D_1^*)}D_{12}^* & 1+\nu+\frac{D_1^*}{2(1+D_1^*)}\left(D_{11}^*+D_{33}^*\right) & \frac{D_1^*}{2(1+D_1^*)}D_{23}^* \\
\frac{D_1^*}{2(1+D_1^*)}D_{12}^* & \frac{D_1^*}{2(1+D_1^*)}D_{12}^* & 0 & \frac{D_1^*}{2(1+D_1^*)}D_{13}^* & \frac{D_1^*}{2(1+D_1^*)}D_{23}^* & 1+\nu+\frac{D_1^*}{2(1+D_1^*)}\left(D_{11}^*+D_{22}^*\right)
\end{array}\right]\left\{\begin{array}{c}\sigma_{11}\\ \sigma_{22}\\ \sigma_{33}\\ \sigma_{23}\\ \sigma_{13}\\ \sigma_{12}\end{array}\right\}. \tag{4.89}$$

In a similar fashion, for the Murakami–Kamiya model, the general elasticity equation coupled with damage for the initially isotropic material (damage induced anisotropy) is obtained:

$$\{\boldsymbol{\sigma}\} = \left[\widetilde{\boldsymbol{\Lambda}}_{\mathrm{MK}}\left(\mathbf{D}^*\right)\right]\{\boldsymbol{\varepsilon}\} \tag{4.90}$$

and

$$\left\{\begin{array}{c}\sigma_{11}\\ \sigma_{22}\\ \sigma_{33}\\ \sigma_{23}\\ \sigma_{13}\\ \sigma_{12}\end{array}\right\} = \left[\begin{array}{c}
\lambda+2\mu+2\left(\eta_1+\eta_2\right)\mathrm{Tr}\mathbf{D}+2\left(\eta_3+\eta_4\right)D_{11} \\
\lambda+2\eta_1\mathrm{Tr}\mathbf{D}+\eta_3\left(D_{11}+D_{22}\right) \\
\lambda+2\eta_1\mathrm{Tr}\mathbf{D}+\eta_3\left(D_{11}+D_{33}\right) \\
\eta_3 D_{23} \\
\left(\eta_3+\eta_4\right)D_{13} \\
\left(\eta_3+\eta_4\right)D_{12}
\end{array}\right.$$

$$
\begin{matrix}
\lambda + 2\eta_1 \mathrm{Tr}\mathbf{D} + \eta_3 \left(D_{11} + D_{22}\right) \\
\lambda + 2\mu + 2\left(\eta_1 + \eta_2\right) \mathrm{Tr}\mathbf{D} + 2\left(\eta_3 + \eta_4\right) D_{22} \\
\lambda + 2\eta_1 \mathrm{Tr}\mathbf{D} + \eta_3 \left(D_{22} + D_{33}\right) \\
\left(\eta_3 + \eta_4\right) D_{23} \\
\eta_3 D_{13} \\
\left(\eta_3 + \eta_4\right) D_{12}
\end{matrix}
$$

$$
\begin{matrix}
\lambda + 2\eta_1 \mathrm{Tr}\mathbf{D} + \eta_3 \left(D_{11} + D_{33}\right) \\
\lambda + 2\eta_1 \mathrm{Tr}\mathbf{D} + \eta_3 \left(D_{22} + D_{33}\right) \\
\lambda + 2\mu + 2\left(\eta_1 + \eta_2\right) \mathrm{Tr}\mathbf{D} + 2\left(\eta_3 + \eta_4\right) D_{33} \\
\left(\eta_3 + \eta_4\right) D_{23} \\
\left(\eta_3 + \eta_4\right) D_{13} \\
\eta_3 D_{12}
\end{matrix}
$$

$$
\begin{matrix}
2\eta_3 D_{23} & 2\left(\eta_3 + \eta_4\right) D_{13} \\
2\left(\eta_3 + \eta_4\right) D_{23} & 2\eta_3 D_{13} \\
2\left(\eta_3 + \eta_4\right) D_{23} & 2\left(\eta_3 + \eta_4\right) D_{13} \\
2\left(\mu + \eta_2 \mathrm{Tr}\mathbf{D}\right) + \eta_4 \left(D_{22} + D_{33}\right) & \eta_4 D_{12} \\
\eta_4 D_{12} & 2\left(\mu + \eta_2 \mathrm{Tr}\mathbf{D}\right) + \eta_4 \left(D_{11} + D_{33}\right) \\
\eta_4 D_{13} & \eta_4 D_{23}
\end{matrix}
$$

$$
\left.\begin{matrix}
2\left(\eta_3 + \eta_4\right) D_{12} \\
2\left(\eta_3 + \eta_4\right) D_{12} \\
2\eta_3 D_{12} \\
\eta_4 D_{13} \\
\eta_4 D_{13} \\
2\left(\mu + \eta_2 \mathrm{Tr}\mathbf{D}\right) + \eta_4 \left(D_{11} + D_{22}\right)
\end{matrix}\right]
\left\{\begin{matrix}
\varepsilon_{11} \\ \varepsilon_{22} \\ \varepsilon_{33} \\ \varepsilon_{23} \\ \varepsilon_{13} \\ \varepsilon_{12}
\end{matrix}\right\}. \tag{4.91}
$$

Note that in Litewka's model the current dominant principal value of the damage tensor D_1 plays the essential role in the damage affected terms of the elasticity matrix $[\tilde{\mathbf{\Lambda}}^{-1}]$.

4.4.3 Constitutive and damage evolution equations by use of the Gibbs thermodynamic potential

In Sects.4.3.1 – 4.3.3, the constitutive and damage evolution equations of elastic-plastic-brittle materials were developed by the use of the Helmholtz free energy, where the damage conjugate forces were expressed as a function of elastic strain tensor, (4.60)–(4.62). However, the experimental validation of this theory is difficult for elastic-plastic-damage materials. For this reason it is more convenient to define the damage conjugate forces as functions of the stress tensor by using the Gibbs thermodynamic potential Γ that consists of the complementary energy Γ^{e} due to the elastic deformation, the potential related to the plastic deformation Γ^{p}, and the damage potential related to the free surface energy due to the microcavities nucleation Γ^{d}:

$$\Gamma(\boldsymbol{\sigma}, r, \mathbf{D}, \beta) = \Gamma^{\mathrm{e}}(\boldsymbol{\sigma}, \mathbf{D}) + \Gamma^{\mathrm{p}}(r) + \Gamma^{\mathrm{d}}(\beta). \tag{4.92}$$

The elastic complementary energy $\Gamma^{\mathrm{e}}(\boldsymbol{\sigma}, \mathbf{D})$ is assumed to be quadratic in $\boldsymbol{\sigma}$ and linear in $\mathbf{D}$, hence

$$\begin{aligned}\Gamma^{\mathrm{e}}(\boldsymbol{\sigma}, \mathbf{D}) = &-\frac{\nu}{2E}(\mathrm{Tr}\boldsymbol{\sigma})^2 + \frac{1+\nu}{2E}\mathrm{Tr}\boldsymbol{\sigma}^2 + \vartheta_1 \mathrm{Tr}\mathbf{D}(\mathrm{Tr}\boldsymbol{\sigma})^2 \\ &+\vartheta_2 \mathrm{Tr}\mathbf{D}\mathrm{Tr}\boldsymbol{\sigma}^{*2} + \vartheta_3 \mathrm{Tr}\boldsymbol{\sigma}(\mathrm{Tr}\boldsymbol{\sigma}\mathbf{D}) + \vartheta_4 \mathrm{Tr}\left(\boldsymbol{\sigma}^{*2}\mathbf{D}\right)\end{aligned} \tag{4.93}$$

where ϑ_1, ϑ_2, ϑ_3 and ϑ_4 are material constants and $\boldsymbol{\sigma}^*$ is the modified stress tensor responsible for the opening/closure effect defined in an analogous way as the modified elastic strain tensors (cf. (4.58)):

$$\boldsymbol{\sigma}^* = \langle\boldsymbol{\sigma}\rangle - \zeta\langle-\boldsymbol{\sigma}\rangle. \tag{4.94}$$

For plastic and damage terms $\Gamma^{\mathrm{p}}(r)$ and $\Gamma^{\mathrm{d}}(\beta)$ the following formulas are used:

$$\begin{aligned}\Gamma^{\mathrm{p}}(r) &= R_\infty\left[r + \frac{1}{b}\exp(-br)\right], \\ \Gamma^{\mathrm{d}}(\beta) &= \frac{1}{2}K_{\mathrm{d}}\beta^2,\end{aligned} \tag{4.95}$$

where R_∞, b and K_{d} are material constants. Eventually, the elastic-damage constitutive equation is furnished as

$$\begin{aligned}\boldsymbol{\varepsilon}^{\mathrm{e}} = \frac{\partial\Gamma^{\mathrm{e}}}{\partial\boldsymbol{\sigma}} = &-\frac{\nu}{E}(\mathrm{Tr}\boldsymbol{\sigma})\mathbf{1} + \frac{1+\nu}{E}\boldsymbol{\sigma} + 2\vartheta_1(\mathrm{Tr}\mathbf{D}\mathrm{Tr}\boldsymbol{\sigma})\mathbf{1} \\ &+2\vartheta_2(\mathrm{Tr}\mathbf{D})\boldsymbol{\sigma}^* : \frac{\partial\boldsymbol{\sigma}^*}{\partial\boldsymbol{\sigma}} \\ &+\vartheta_3\left[\mathrm{Tr}(\boldsymbol{\sigma}\mathbf{D})\mathbf{1} + (\mathrm{Tr}\boldsymbol{\sigma})\mathbf{D}\right] + \vartheta_4(\boldsymbol{\sigma}^*\mathbf{D} + \mathbf{D}\boldsymbol{\sigma}^*) : \frac{\partial\boldsymbol{\sigma}^*}{\partial\boldsymbol{\sigma}}.\end{aligned} \tag{4.96}$$

and the forces conjugate to internal variables $\mathbf{D}$, r and β are

$$\begin{aligned}\mathbf{Y} &= \frac{\partial\Gamma^{\mathrm{e}}}{\partial\mathbf{D}} = \left[\vartheta_1(\mathrm{Tr}\boldsymbol{\sigma})^2 + \vartheta_2\mathrm{Tr}\boldsymbol{\sigma}^{*2}\right]\mathbf{1} + \vartheta_3(\mathrm{Tr}\boldsymbol{\sigma})\boldsymbol{\sigma} + \vartheta_4\boldsymbol{\sigma}^{*2}, \\ R &= \frac{\partial\Gamma^{\mathrm{p}}}{\partial r} = R_\infty\left[1 - \exp(-br)\right], \\ B &= \frac{\partial\Gamma^{\mathrm{e}}}{\partial\beta} = K_{\mathrm{d}}\beta.\end{aligned} \tag{4.97}$$

Assuming also the Mises-type yield condition of the damaged materials in the form

$$F^{\mathrm{p}}(\boldsymbol{\sigma}, R, \mathbf{D}) = \widetilde{\sigma}_{\mathrm{eq}} - (\sigma_y + R) = 0 \tag{4.98}$$

the constitutive equations for plastic strain rate $\dot{\varepsilon}^{\mathrm{p}}_{ij}$ and the rate of isotropic hardening $\dot{r}$ hold:

$$\begin{aligned} \dot{\varepsilon}^{\mathrm{p}}_{ij} &= \dot{\Lambda}^{\mathrm{p}} \frac{\partial F^{\mathrm{p}}}{\partial \sigma'_{ij}} = \frac{3}{2} \dot{\Lambda}^{\mathrm{p}} \frac{M_{ijkl} \sigma'_{kl}}{\widetilde{\sigma}_{\mathrm{eq}}} \\ \dot{r} &= \dot{\Lambda}^{\mathrm{p}} \frac{\partial F^{\mathrm{p}}}{\partial (-R)} = \dot{\Lambda}^{\mathrm{p}}, \end{aligned} \tag{4.99}$$

where $\widehat{\mathbf{M}}(\mathbf{D})$ is a fourth-rank damage effect tensor

$$\widehat{\mathbf{M}}(\mathbf{D}) = \frac{1}{2}\left(\delta_{ik}\delta_{jl} + \delta_{il}\delta_{jk}\right) + \frac{1}{2} c^{\mathrm{p}} \left(\delta_{ik} D_{jl} + D_{ik}\delta_{jl} + \delta_{il} D_{jk} + D_{il}\delta_{jk}\right) \tag{4.100}$$

and the effective Mises-type equivalent stress $\widetilde{\sigma}_{\mathrm{eq}}$ is

$$\widetilde{\sigma}_{\mathrm{eq}} = \left[(3/2)\, \boldsymbol{\sigma}' : \widehat{\mathbf{M}}(\mathbf{D}) : \boldsymbol{\sigma}'\right]^{1/2}. \tag{4.101}$$

In order to establish evolution equations of damage $\mathbf{D}$ and β, the damage dissipation potential is assumed in the form

$$F^{\mathrm{d}}(\mathbf{Y}, B, \mathbf{D}, r) = Y_{\mathrm{eq}} + c^{\mathrm{r}} r \mathrm{Tr}\mathbf{D} \mathrm{Tr}\mathbf{Y} - (B_0 + B) = 0 \tag{4.102}$$

that extends (4.61) by the additional damage-plasticity term corresponding to isotropic hardening r. The fourth-rank tensor $\widetilde{\mathbf{L}}(\mathbf{D})$ is given by the formula analogous to (4.100) with the material constant c^{p} replaced by the new constant c^{d}. Hence, the evolution equations are furnished as follows:

$$\begin{aligned} \dot{\mathbf{D}} &= \dot{\Lambda}^{\mathrm{d}} \frac{\partial F^{\mathrm{d}}}{\partial \mathbf{Y}} = \dot{\Lambda}^{\mathrm{d}} \left[\frac{\widetilde{\mathbf{L}}:\mathbf{Y}}{2Y_{\mathrm{eq}}} + c^{\mathrm{r}} r \left(\mathrm{Tr}\mathbf{D}\right) \mathbf{1}\right], \\ \dot{\beta} &= \dot{\Lambda}^{\mathrm{d}} \frac{\partial F^{\mathrm{d}}}{\partial (-B)} = \dot{\Lambda}^{\mathrm{d}}. \end{aligned} \tag{4.103}$$

The plasticity and damage multipliers $\dot{\Lambda}^{\mathrm{p}}$ and $\dot{\Lambda}^{\mathrm{d}}$ must be derived from the consistency conditions for the plastic yield surface (4.98) and damage surface (4.102) (cf. Hayakawa et al., 1998):

$$\begin{aligned} \dot{\Lambda}^{\mathrm{p}} &= \frac{\frac{\partial F^{\mathrm{p}}}{\partial \sigma} : \dot{\boldsymbol{\sigma}}}{\frac{\partial R}{\partial r}} + \frac{\left(\frac{\partial F^{\mathrm{d}}}{\partial \mathbf{D}} : \frac{\partial F^{\mathrm{d}}}{\partial \mathbf{Y}}\right)\left(\frac{\partial F^{\mathrm{d}}}{\partial \mathbf{Y}} : \dot{\mathbf{Y}}\right)}{\left(\frac{\partial R}{\partial r}\right)\left(\frac{\partial B}{\partial \beta} - \frac{\partial F^{\mathrm{d}}}{\partial \mathbf{D}} : \frac{\partial F^{\mathrm{d}}}{\partial \mathbf{Y}}\right)}, \\ \dot{\Lambda}^{\mathrm{d}} &= \frac{\frac{\partial F^{\mathrm{d}}}{\partial \mathbf{Y}} : \dot{\mathbf{Y}}}{\frac{\partial B}{\partial \beta} - \frac{\partial F^{\mathrm{d}}}{\partial \mathbf{D}} : \frac{\partial F^{\mathrm{d}}}{\partial \mathbf{Y}}}. \end{aligned} \tag{4.104}$$

The material constants determined for the cast iron FCD400 are (Hayakawa et al., 1998):

$$
\begin{array}{ll}
E = 169 \text{ MPa}, & \nu = 0.285, \\
\zeta = 0.89, & \vartheta_1 = -3.95 \times 10^{-7} \text{ MPa}^{-1}, \\
\vartheta_2 = 4.00 \times 10^{-6} \text{ MPa}^{-1}, & \vartheta_3 = -4.00 \times 10^{-7} \text{ MPa}^{-1}, \\
\vartheta_4 = 2.50 \times 10^{-6} \text{ MPa}^{-1}, & b = 15, \\
R_0 = 293.0 \text{ MPa}, & R_\infty = 250.0 \text{ MPa}, \\
K_\mathrm{d} = 1.3, & B_0 = 0.273, \\
c^\mathrm{p} = 1.0, & c^\mathrm{d} = -15.0, \\
c^\mathrm{r} = 50.0. &
\end{array}
$$

5

Coupled thermo-damage and damage-fracture fields

5.1 Damage effect on heat transfer in solids under thermo-mechanical loadings

5.1.1 *Concepts of a thermo-creep-damage coupling*

The creep process and the associated material deterioration are temperature sensitive phenomena. A classical approach consists in accounting for the effect of temperature on the material functions in the constitutive and the evolution equations of a damaged solid (cf. Ganczarski and Skrzypek, 1991), whereas the temperature field remains steady state.

In a general case, when thermo-mechanical loadings are applied to the structure, in addition to the constitutive and evolution state equations with the appropriate mechanical boundary conditions, the heat transfer equation must simultaneously be solved to yield a transient temperature field which satisfies the thermal boundary conditions. Material nonhomogeneity, which results from the deterioration process in a solid, influences both the mechanical moduli represented by elasticity tensors $\widetilde{\mathbf{\Lambda}}(\mathbf{x},t)$ or $\widetilde{\mathbf{\Lambda}}^{-1}(\mathbf{x},t)$, stiffness or compliance, and the thermal properties $\widetilde{\mathbf{L}}(\mathbf{x},t)$ or $\widetilde{\mathbf{\Gamma}}(\mathbf{x},t)$, where $\widetilde{\mathbf{\Lambda}}$ or $\widetilde{\mathbf{\Lambda}}^{-1}$ and $\widetilde{\mathbf{L}}$, $\widetilde{\mathbf{\Gamma}}$ are fourth-rank elasticity tensors and second-rank thermal conductivity and emissivity symmetric tensors, respectively, all defined at a given material particle $\mathbf{x}$. In fact, the tensor nature of thermal conductivity is a question of debate. Carslow and Jeager (1959) and Fung (1965) introduced a symmetric, positive definite matrix L_{ij} of thermal conductivity moduli, whereas Nowacki (1970) defined the thermal conductivity as $L_{ij} = L_{ij}/T^2$ and postulated considering it as a symmetric tensor, when the temperature change is limited to be small enough when compared to the natural state, such that L_{ij} can be assumed as constant. For porous media Kaviany (1995) introduced the thermal diffusivity tensor $\mathbf{L} = \mathbf{L}_{\text{ef}}/\varrho c_p + f\mathbf{L}^{\text{d}}$, where $\mathbf{L}_{\text{ef}}$ is the effective thermal conductivity tensor, $\mathbf{L}^{\text{d}}$ is the thermal dispersion tensor, and f denotes porosity. The author assumed that $\mathbf{L}$ is a positive-definite, symmetric tensor, the off-diagonal elements of which vanish in case of isotropic media. Recently Saanouni, Forster, and Ben Hatira (1994) when formulating the general constitutive law of the coupled isotropic damage-elasto-(visco)plasticity, also introduced a symmetric second-rank tensor of thermal conductivity $\widetilde{\mathbf{k}}$ (cf. Sect. 2.3.4).

In what follows, the tensor nature of $\widetilde{L}_{ij}$ and $\widetilde{\Gamma}_{ij}$ matrices is postulated, in particular, when second-rank damage tensor D_{ij} affects thermal properties of a solid resulting in an anisotropic thermal conductivity and radiation, though in a virgin material the thermal isotropy holds (cf. Ganczarski and Skrzypek, 1995, 1997; Skrzypek and Ganczarski, 1998b).

Tanigawa (1988) formulated a coupled thermo-elastic problem for time-independent, nonhomogeneous but isotropic structural materials. If a body is isotropic and nonhomogeneous, the steady state heat conduction equation, without the internal sources, has the form:

$$\frac{\partial}{\partial x}\left[\lambda_0\left(x,y,z\right)\frac{\partial T}{\partial x}\right]+\frac{\partial}{\partial y}\left[\lambda_0\left(x,y,z\right)\frac{\partial T}{\partial y}\right]+\frac{\partial}{\partial z}\left[\lambda_0\left(x,y,z\right)\frac{\partial T}{\partial z}\right]=0, \tag{5.1}$$

where $T=T(x,y,z)$.

The above equation needs to be extended when thermally nonhomogeneous solid suffers from a creep-damage process, hence, the thermal conductivity function of a virgin solid $\lambda_0(x,y,z)$ is replaced by a new time-dependent, generally anisotropic tensorial function $\widetilde{\mathbf{L}}\left(x,y,z,t\right)$ that characterizes the thermal properties of a partly damaged solid. The material nonhomogeneity is no longer time-independent, following damage evolution. Hence, in the simplest case, when the isotropic damage is assumed as governed by a single scalar variable $D\left(\mathbf{x},t\right)$, a more general form instead of (5.1) is required:

$$\begin{aligned}&\frac{\partial}{\partial x}\left\{\widetilde{\lambda}\left[\mathbf{x},D\left(\mathbf{x},t\right)\right]\frac{\partial T\left(\mathbf{x}\right)}{\partial x}\right\}+\frac{\partial}{\partial y}\left\{\widetilde{\lambda}\left[\mathbf{x},D\left(\mathbf{x},t\right)\right]\frac{\partial T\left(\mathbf{x}\right)}{\partial y}\right\}\\&+\frac{\partial}{\partial z}\left\{\widetilde{\lambda}\left[\mathbf{x},D\left(\mathbf{x},t\right)\right]\frac{\partial T\left(\mathbf{x}\right)}{\partial z}\right\}=0,\end{aligned} \tag{5.2}$$

whereas for nonsteady states, with internal heat sources $\dot{q}_v$, the extended equation (5.1) takes a form:

$$\begin{aligned}&\frac{\partial}{\partial x}\left\{\widetilde{\lambda}\left[\mathbf{x},D\left(\mathbf{x},t\right)\right]\frac{\partial T\left(\mathbf{x},t\right)}{\partial x}\right\}+\frac{\partial}{\partial y}\left\{\widetilde{\lambda}\left[\mathbf{x},D\left(\mathbf{x},t\right)\right]\frac{\partial T\left(\mathbf{x},t\right)}{\partial y}\right\}\\&+\frac{\partial}{\partial z}\left\{\widetilde{\lambda}\left[\mathbf{x},D\left(\mathbf{x},t\right)\right]\frac{\partial T\left(\mathbf{x},t\right)}{\partial z}\right\}+\frac{\partial q_v}{\partial t}=c_v\varrho\frac{\partial T\left(\mathbf{x},t\right)}{\partial t}\end{aligned} \tag{5.3}$$

or

$$\operatorname{div}\left\{\widetilde{\lambda}\left[\mathbf{x},D\left(\mathbf{x},t\right)\right]\mathbf{grad}T\right\}+\dot{q}_v=c_v\varrho\dot{T} \tag{5.4}$$

in a more general case.

The effect of damage on thermal properties is described here by the single scalar variable $\widetilde{\lambda}\left[\mathbf{x}, D\left(\mathbf{x},t\right)\right]$. The mass density and the specific heat, ϱ, c_v, are assumed to be time-independent constants.

The aim of this section is to specify time-dependent functions $\widetilde{\lambda}\left[\mathbf{x}, D\left(\mathbf{x}, t\right)\right]$ that introduce coupling between the heat conductivity and the isotropic damage evolution. Three models of the scalar thermo-damage coupling are proposed (cf. Skrzypek and Ganczarski, 1998b).

A. Direct extension of the equation of thermal conductivity for damaged solids

The simplest model is based on the assumption of linear heat conductivity drop with damage (cf. Ganczarski and Skrzypek, 1995):

$$\widetilde{\lambda}\left[\mathbf{x}, D\left(\mathbf{x},t\right)\right] = \lambda_0\left(\mathbf{x}\right)\left[1 - D\left(\mathbf{x},t\right)\right] \tag{5.5}$$

where $\lambda_0\left(\mathbf{x}\right)$ denotes generally nonhomogeneous distribution of the thermal conductivity in a virgin (undamaged) solid, whereas the scalar variable D defines the current damage level (e.g., governed by Hayhurst and Chaboche's rule, (2.35)–(2.36). In this model, when material is locally completely damaged, $D\left(\mathbf{x},t\right) \equiv 1$, the thermal conductivity coefficient drops at this point to zero $\widetilde{\lambda}(D = 1) = 0$ and, hence, local heat conductivity through the completely damaged surface element must also drop to zero.

In other words, the fully damaged RVE is assumed to be free from any kind of stress and unable to support heat conduction. This property was also used by Saanouni et al. (1994); however, it is not obvious when a more general heat transfer model is applied unless the mechanisms other than conductivity are excluded. Note also that, when in the Saanouni et al. (1994) approach the energy based equivalence principle was used instead of the linear conductivity drop (5.5), the other formula is derived from the state potential, namely $\widetilde{\mathbf{k}} = (1 - D)^{1/2}\mathbf{k}$, so that the following isotropic model may also be proposed:

$$\widetilde{\lambda}\left[\mathbf{x}, D\left(\mathbf{x},t\right)\right] = \lambda_0\left(\mathbf{x}\right)\left[1 - D\left(\mathbf{x},t\right)\right]^{1/2}. \tag{5.6}$$

B. Concept of a combined evolution of thermal conductivity and radiation through partly damaged solid

Further extension of the Model A accounts for an additional heat flow term through the damaged surface element portion, by application of the Stefan–Boltzmann radiation law. Hence, when both conduction and radiation mechanisms of heat transfer are admitted, the following extension of (5.3) was proposed by Ganczarski and Skrzypek (1995):

$$
\begin{aligned}
&\frac{\partial}{\partial x}\left\{\lambda_0(\mathbf{x})\left[1-D(\mathbf{x},t)\right]\frac{\partial T(\mathbf{x},t)}{\partial x}-\sigma\epsilon_0(\mathbf{x},t)\,D(\mathbf{x},t)\,T^4\right\}\\
&+\frac{\partial}{\partial y}\left\{\lambda_0(\mathbf{x})\left[1-D(\mathbf{x},t)\right]\frac{\partial T(\mathbf{x},t)}{\partial y}-\sigma\epsilon_0(\mathbf{x},t)\,D(\mathbf{x},t)\,T^4\right\}\\
&+\frac{\partial}{\partial z}\left\{\lambda_0(\mathbf{x})\left[1-D(\mathbf{x},t)\right]\frac{\partial T(\mathbf{x},t)}{\partial z}-\sigma\epsilon_0(\mathbf{x},t)\,D(\mathbf{x},t)\,T^4\right\}\\
&+\frac{\partial q_v}{\partial t}=c_v\varrho\frac{\partial T(\mathbf{x},t)}{\partial t}.
\end{aligned}
\tag{5.7}
$$

In the Model B under consideration a combined conductivity/radiation mechanism allows for a heat flux even though the damage at a point reaches level 1 (due to radiation across the microcracks). However, as will be shown further, the model exhibits an essential inconsistency. The form of terms associated with radiation suggests, namely, that there exists heat exchange caused by a redistribution of damage only, even though the temperature remains constant. To omit this inconsistency, it is necessary to use the second law of thermodynamics and to cutoff inadmissible temperature distributions (cf. Sect. 7.4.3)

C. Concept of the equivalent (reduced) coefficient of thermal conductivity for a combined conductivity/radiation heat flux through partly damaged solid

Another way that consists in accounting for a combined heat exchange, when the conductivity is assumed to be a dominant phenomenon, was presented by Ganczarski and Skrzypek (1998b). A combined heat flux is characterized by the substitutive coefficient of thermal conductivity modified in order to take into account a simultaneous influence of the conductivity through the RVE at the point $\mathbf{x}$, $\widetilde{\lambda}$ and the radiation from $\mathbf{x}$ to $\mathbf{x}+\mathrm{d}\mathbf{x}$. The equivalent coefficient of thermal conductivity λ^{eq} is expressed, therefore, by the equation:

$$
\lambda^{\mathrm{eq}}\left[\mathbf{x},D(\mathbf{x},t),T(\mathbf{x},t)\right]=\widetilde{\lambda}\left[\mathbf{x},D(\mathbf{x},t)\right]+\mathrm{d}\widetilde{\lambda}^{\mathrm{rad}}\left[\mathrm{d}\mathbf{x},D(\mathbf{x},t),T(\mathbf{x},t)\right].
\tag{5.8}
$$

Consequently, the equation of heat transfer (5.3) may be extended to the following form:

$$
\frac{\partial}{\partial x}\left\{\lambda^{\mathrm{eq}}\left[\mathbf{x},D(\mathbf{x},t),T(\mathbf{x},t)\right]\frac{\partial T(\mathbf{x},t)}{\partial x}\right\}
$$

$$
\begin{aligned}
&+\frac{\partial}{\partial y}\left\{\lambda^{\mathrm{eq}}\left[\mathbf{x}, D(\mathbf{x}, t), T(\mathbf{x}, t)\right] \frac{\partial T(\mathbf{x}, t)}{\partial y}\right\} \\
&+\frac{\partial}{\partial z}\left\{\lambda^{\mathrm{eq}}\left[\mathbf{x}, D(\mathbf{x}, t), T(\mathbf{x}, t)\right] \frac{\partial T(\mathbf{x}, t)}{\partial z}\right\} \\
&+\frac{\partial q_v}{\partial t}=c_v \varrho \frac{\partial T(\mathbf{x}, t)}{\partial t} .
\end{aligned} \tag{5.9}
$$

The equivalent (substitutive) coefficient of thermal conductivity λ^{eq} is obtained by equating the heat flux due to conductivity and radiation through the partly damaged cross section which the heat flux due to the corresponding conductivity through the fictitious pseudo-undamaged cross section. The specific formulas for $\mathrm{d}\widetilde{\lambda}^{\mathrm{rad}}$ will be discussed in the following section.

Conclusion: In Model C a combined conductivity and radiation mechanism through undamaged (solid) and damaged (voided) material, respectively, is reduced to the equivalent conductivity through the fictive pseudo-undamaged material, when a substitutive coefficient of thermal conductivity λ^{eq} is introduced to the Fourier conductivity law for a partly damaged solid (5.9) instead of linearly decreasing with damage coefficient $\widetilde{\lambda}$ used in Model A. Note that in Model C, in the case when the material damage parameter locally reaches level $D = 1$ (macrocrack initiation), the equivalent coefficient $\lambda^{\mathrm{eq}}(D = 1)$ remains nonzero and, hence, the residual fictive heat conductivity through the pseudo-undamaged surface element, equivalent to the heat radiation through the completely damaged real element, remains nonzero as well. On the other hand, Model C, in contrast to Model B, is free from an inadmissible heat exchange phenomenon caused by the damage redistribution when the temperature gradient drops to zero.

5.1.2 *Uniaxial (1D) heat transfer through isotropic damaged solids*

Consider a uniaxial representative volume element $\mathrm{d}x\mathrm{d}A_0$ as a rod which undergoes brittle damage at elevated temperature, Fig. 5.1.

The actual state of damage in the element is determined by the damage variable D interlinked with the continuity variable ψ as $D + \psi = 1$, or $\mathrm{d}D + \mathrm{d}\psi = 0$. Hence, we can easily interpret the products $D\mathrm{d}A_0$ and $\psi\mathrm{d}A_0$ as the damaged and the undamaged portions of the elementary cross section area $\mathrm{d}A_0$, respectively. Due to the dual nature of a partly damaged cross section, the total heat flow rate needs to be decomposed into two parts: the classical Fourier conductivity through the undamaged portion of cross section q^{cond} and the Stefan–Boltzmann radiation through the damaged portion of cross section q^{rad}:

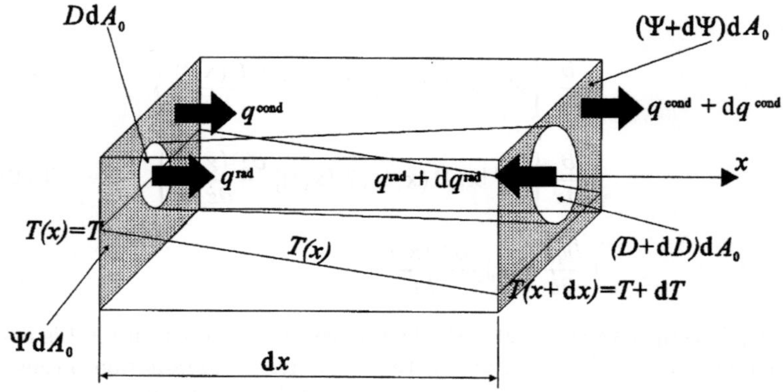

Fig. 5.1. Schematics of heat transfer through partly damaged uniaxial RVE $\mathrm{d}x\mathrm{d}A_0$

$$q^{\mathrm{cond}} = -\lambda_0 \frac{\mathrm{d}T}{\mathrm{d}x}, \qquad q^{\mathrm{rad}} = \sigma\epsilon_0 \left[T^4(x) - T^4(x+\mathrm{d}x)\right], \tag{5.10}$$

where λ_0, ϵ_0, and σ denote thermal conductivity, emissivity of the gray body in a virgin state, and the Stefan–Boltzmann constant.

Consider heat flux through the element taking into account infinitesimal changes of damage and continuity on $\mathrm{d}x$:

$$\begin{aligned} q^{\mathrm{cond}}\psi + q^{\mathrm{rad}}D - \left(q^{\mathrm{cond}} + \frac{\partial q^{\mathrm{cond}}}{\partial x}\mathrm{d}x\right)\left(\psi + \frac{\partial \psi}{\partial x}\mathrm{d}x\right) \\ - \left(q^{\mathrm{rad}} + \frac{\partial q^{\mathrm{rad}}}{\partial x}\mathrm{d}x\right)\left(D + \frac{\partial D}{\partial x}\mathrm{d}x\right) + \frac{\partial q_v}{\partial t}\mathrm{d}x = c_v\varrho\frac{\partial T}{\partial t}\mathrm{d}x. \end{aligned} \tag{5.11}$$

Neglecting second-order terms and substituting (5.10) for q^{cond} and q^{rad}, the modified equation of uniaxial heat transfer through partly damaged body in the form (Model B),

$$\frac{\partial}{\partial x}\left[\lambda_0(1-D)\frac{\partial T}{\partial x} - \sigma\epsilon_0 D T^4\right] + \frac{\partial q_v}{\partial t} = c_v\varrho\frac{\partial T}{\partial t}, \tag{5.12}$$

is eventually obtained, where q_v is intensity of the inner heat source (if it exists). In the case of a pure thermal conductivity (Model A), setting emissivity to zero, $\epsilon_0 = 0$, the equation (5.12) reduces to the following form:

$$\frac{\partial}{\partial x}\left(\tilde{\lambda}\frac{\partial T}{\partial x}\right) + \frac{\partial q_v}{\partial t} = c_v\varrho\frac{\partial T}{\partial t}, \qquad \tilde{\lambda} = \lambda_0(1-D). \tag{5.13}$$

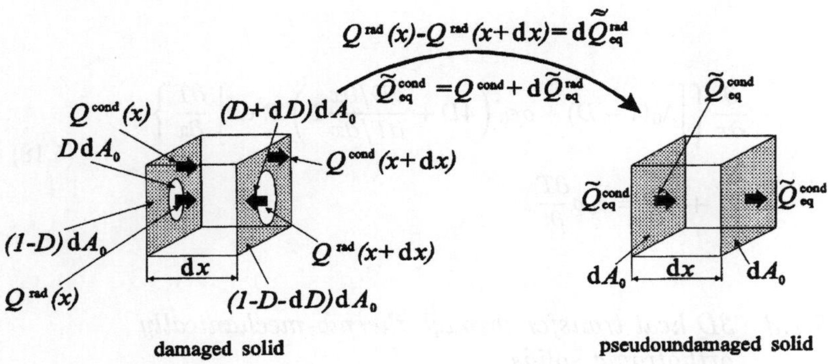

Fig. 5.2. One-dimensional concept of the equivalent coefficient of thermal conductivity

The concept of an equivalent coefficient of thermal conductivity λ^{eq} (Model C) requires, first of all, comparison of the heat flux by radiation through the partly damaged cross section and the additional heat flux by conduction through the fictive pseudo-undamaged cross section (Fig. 5.2):

$$\sigma\epsilon_0\left[DT^4-(D+\mathrm{d}D)(T+\mathrm{d}T)^4\right]\mathrm{d}A_0=-\mathrm{d}\widetilde{\lambda}^{\mathrm{rad}}\mathrm{d}A_0\frac{\partial T}{\partial x}. \tag{5.14}$$

Next, expanding temperature $T(x+\mathrm{d}x)$ and damage parameter $D(x+\mathrm{d}x)$ in Taylor series for x and introducing these into (5.14) we have:

$$\begin{aligned}&\sigma\epsilon_0\left[DT^4-\left(D+\frac{\partial D}{\partial x}\mathrm{d}x+\dots\right)\right.\\&\qquad\left.\times\left(T^4+4T^3\frac{\partial T}{\partial x}\mathrm{d}x+\dots\right)\right]=-\mathrm{d}\widetilde{\lambda}^{\mathrm{rad}}\frac{\partial T}{\partial x}.\end{aligned} \tag{5.15}$$

When higher-order terms in (5.15) are neglected, the additional substitutive coefficient of thermal conductivity in pseudo-undamaged material responsible for the radiation in damaged material $\mathrm{d}\widetilde{\lambda}^{\mathrm{rad}}$ is expressed by the formula:

$$\mathrm{d}\widetilde{\lambda}^{\mathrm{rad}}=\sigma\epsilon_0\left(4DT^3+\frac{\partial D/\partial x}{\partial T/\partial x}T^4\right)\mathrm{d}x. \tag{5.16}$$

Therefore, the equation of uniaxial heat transfer takes the form:

$$\frac{\partial}{\partial x}\left(\lambda^{\mathrm{eq}}\frac{\partial T}{\partial x}\right)+\frac{\partial q_v}{\partial t}=c_v\varrho\frac{\partial T}{\partial t},\qquad\lambda^{\mathrm{eq}}=\widetilde{\lambda}+\mathrm{d}\widetilde{\lambda}^{\mathrm{rad}} \tag{5.17}$$

or, when the explicit formulas for $\widetilde{\lambda}$ and $\mathrm{d}\widetilde{\lambda}^{\mathrm{rad}}$ are used, we obtain:

$$\frac{\partial}{\partial x}\left\{\left[\lambda_0(1-D)+\sigma\epsilon_0\left(4D+\frac{\partial D/\partial x}{\partial T/\partial x}T\right)T^3\mathrm{d}x\right]\frac{\partial T}{\partial x}\right\}$$
$$+\frac{\partial q_v}{\partial t}=c_v\varrho\frac{\partial T}{\partial t}. \tag{5.18}$$

5.1.3 3D heat transfer through thermo-mechanically orthotropic solids

Let us extend equations (5.12), (5.13), and (5.18) to the most general case of thermo-mechanical orthotropy. The anisotropic nature of damage requires the symmetric second-order tensors of damage and continuity to be used instead of corresponding scalar variables (cf. Murakami and Ohno, 1981). Consider an infinitesimal tetrahedron defined by $\mathrm{d}x$, $\mathrm{d}y$, $\mathrm{d}z$ and the inclined plane of the unit normal vector $\mathbf{n}=(n_x,n_y,n_z)$, Fig. 5.3 (cf. Skrzypek and Ganczarski, 1998b).

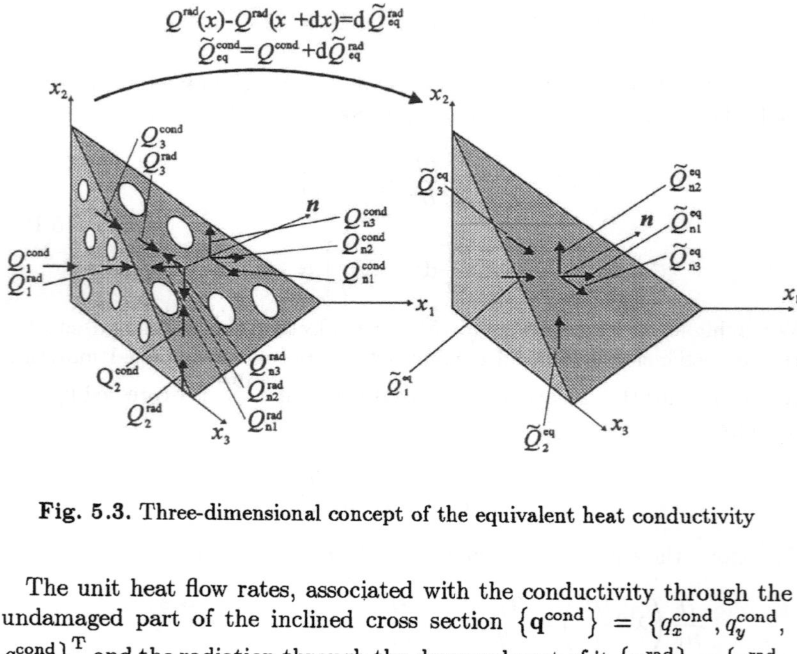

Fig. 5.3. Three-dimensional concept of the equivalent heat conductivity

The unit heat flow rates, associated with the conductivity through the undamaged part of the inclined cross section $\{\mathbf{q}^{\mathrm{cond}}\}=\{q_x^{\mathrm{cond}}, q_y^{\mathrm{cond}}, q_z^{\mathrm{cond}}\}^{\mathrm{T}}$ and the radiation through the damaged part of it $\{\mathbf{q}^{\mathrm{rad}}\}=\{q_x^{\mathrm{rad}}, q_y^{\mathrm{rad}}, q_z^{\mathrm{rad}}\}^{\mathrm{T}}$, are as follows:

$$
\left\{\begin{array}{c} q_x^{\text{cond}} \\ q_y^{\text{cond}} \\ q_z^{\text{cond}} \end{array}\right\} = -\underbrace{\lambda_0 \begin{bmatrix} \Psi_{xx} & \Psi_{yx} & \Psi_{zx} \\ & \Psi_{yy} & \Psi_{zy} \\ & & \Psi_{zz} \end{bmatrix}}_{\widetilde{L}_{ij}} \left\{\begin{array}{c} \partial T/\partial x \\ \partial T/\partial y \\ \partial T/\partial z \end{array}\right\},
$$

$$
\left\{\begin{array}{c} q_x^{\text{rad}} \\ q_y^{\text{rad}} \\ q_z^{\text{rad}} \end{array}\right\} = \underbrace{\sigma\epsilon_0 \begin{bmatrix} D_{xx} & D_{yx} & D_{zx} \\ & D_{yy} & D_{zy} \\ & & D_{zz} \end{bmatrix}}_{\widetilde{\Gamma}_{ij}} \left\{\begin{array}{c} n_x \\ n_y \\ n_z \end{array}\right\} T^4, \tag{5.19}
$$

where terms associated with off-diagonal components of the corresponding (3×3) matrices play role of the diffusional conductivity/radiation portions, respectively, due to the transverse temperature gradients. When the above decomposition of the unit heat flow rates is introduced, the tensor of thermal conductivity $\widetilde{L}_{ij}$ and the tensor of radiation $\widetilde{\Gamma}_{ij}$ are defined as follows:

$$
\widetilde{L}_{ij} = \lambda_0(I_{ij} - D_{ij}), \qquad \widetilde{\Gamma}_{ij} = \sigma\epsilon_0 D_{ij}. \tag{5.20}
$$

Both tensors defined above $\widetilde{\mathbf{L}}$, $\widetilde{\mathbf{\Gamma}}$ are coaxial with the damage tensor $\mathbf{D}$ in their principal axes, therefore, there exists a locally orthogonal frame coinciding with directions of damage orthotropy such that (5.20) can be written as:

$$
\widetilde{L}_\nu = \lambda_0(1 - D_\nu), \qquad \widetilde{\Gamma}_\nu = \sigma\epsilon D_\nu, \qquad \nu = 1, 2, 3. \tag{5.21}
$$

Consequently, the heat flux rates expressed in terms of damage tensor eigenvalues take the form:

$$
\left\{\begin{array}{c} q_1^{\text{cond}} \\ q_2^{\text{cond}} \\ q_3^{\text{cond}} \end{array}\right\} = -\lambda_0 \begin{bmatrix} 1 - D_1 & 0 & 0 \\ 0 & 1 - D_2 & 0 \\ 0 & 0 & 1 - D_3 \end{bmatrix} \left\{\begin{array}{c} \partial T/\partial x_1 \\ \partial T/\partial x_2 \\ \partial T/\partial x_3 \end{array}\right\},
$$

$$
\left\{\begin{array}{c} q_1^{\text{rad}} \\ q_2^{\text{rad}} \\ q_3^{\text{rad}} \end{array}\right\} = \sigma\epsilon_0 \begin{bmatrix} D_1 & 0 & 0 \\ 0 & D_2 & 0 \\ 0 & 0 & D_3 \end{bmatrix} \left\{\begin{array}{c} n_1 \\ n_2 \\ n_3 \end{array}\right\} T^4. \tag{5.22}
$$

When (5.21) and (5.22) are introduced into the heat flux equation we arrive at:

$$\frac{\partial}{\partial x_1}\left[\lambda_0(1-D_1)\frac{\partial T}{\partial x_1}-\sigma\epsilon_0 D_1 T^4\right]$$
$$+\frac{\partial}{\partial x_2}\left[\lambda_0(1-D_2)\frac{\partial T}{\partial x_2}-\sigma\epsilon_0 D_2 T^4\right] \tag{5.23}$$
$$+\frac{\partial}{\partial x_3}\left[\lambda_0(1-D_3)\frac{\partial T}{\partial x_3}-\sigma\epsilon_0 D_3 T^4\right]+\frac{\partial q_v}{\partial t}=c_v\varrho\frac{\partial T}{\partial t},$$

where the problem of a heat transfer in an anisotropic solid is reduced to the equivalent problem of a heat transfer in a local thermo-mechanically orthotropic solid.

A particular case of pure conductivity (3D extention of Model A) yields the equation:

$$\frac{\partial}{\partial x_1}\left[\lambda_0(1-D_1)\frac{\partial T}{\partial x_I}\right]+\frac{\partial}{\partial x_2}\left[\lambda_0(1-D_2)\frac{\partial T}{\partial x_2}\right]$$
$$+\frac{\partial}{\partial x_3}\left[\lambda_0(1-D_3)\frac{\partial T}{\partial x_3}\right]+\frac{\partial q_v}{\partial t}=c_v\varrho\frac{\partial T}{\partial t}. \tag{5.24}$$

An extension of the equivalent thermal conductivity to the case of orthotropic damage evolution (extension of Model C) consists in introducing the substitutive conductivity diagonal tensor $\mathrm{d}\widetilde{\mathbf{L}}_v^{\mathrm{rad}}$ that corresponds to the equivalent conductivity through the fictive, pseudo-undamaged material. Equating the heat flux by radiation through the partly damaged (real) cross section and the heat flux by conduction through the fictive pseudo-undamaged cross section along each of the coordinate axes, we can write for axis x_1 (Fig. 5.3):

$$\sigma\epsilon_0\left[D_1(x_1,x_2,x_3)n_1T^4(x_1,x_2,x_3)\right.$$
$$\left.-D_1(x_1+\mathrm{d}x_1,x_2,x_3)n_1T^4(x_1+\mathrm{d}x_1,x_2,x_3)\right]=-\mathrm{d}\widetilde{L}_1^{rad}\frac{\partial T}{\partial x_1},$$

etc.

(5.25)

When the procedure of expansion of temperature and damage in Taylor series is applied and the higher-order terms are neglected, we find the formulas:

$$\mathrm{d}\widetilde{L}_1^{\mathrm{rad}}=\sigma\epsilon_0\left(4D_1T^3+\frac{\partial D_1/\partial x_1}{\partial T/\partial x_1}T^4\right)\mathrm{d}x_1,$$
$$\mathrm{d}\widetilde{L}_2^{\mathrm{rad}}=\sigma\epsilon_0\left(4D_2T^3+\frac{\partial D_2/\partial x_2}{\partial T/\partial x_2}T^4\right)\mathrm{d}x_2, \tag{5.26}$$
$$\mathrm{d}\widetilde{L}_3^{\mathrm{rad}}=\sigma\epsilon_0\left(4D_3T^3+\frac{\partial D_3/\partial x_3}{\partial T/\partial x_3}T^4\right)\mathrm{d}x_3,$$

which, when substituted into the equation of heat transfer, yield the following 3D equivalent heat flux equation in terms of three components of the diagonal substitutive conductivity tensor $\widetilde{\mathbf{L}}_\nu^{\mathrm{eq}}$:

$$\begin{aligned}&\frac{\partial}{\partial x_1}\left(\widetilde{L}_1^{\mathrm{eq}}\frac{\partial T}{\partial x_1}\right)+\frac{\partial}{\partial x_2}\left(\widetilde{L}_2^{\mathrm{eq}}\frac{\partial T}{\partial x_2}\right)\\&+\frac{\partial}{\partial x_3}\left(\widetilde{L}_3^{\mathrm{eq}}\frac{\partial T}{\partial x_3}\right)+\frac{\partial q_v}{\partial t}=c_v\varrho\frac{\partial T}{\partial t},\end{aligned}\tag{5.27}$$

where

$$\widetilde{\mathbf{L}}_\nu^{\mathrm{eq}}=\widetilde{\mathbf{L}}_\nu+\mathrm{d}\widetilde{\mathbf{L}}_\nu^{\mathrm{rad}}.\tag{5.28}$$

Note that in the general case, when the damaged solid is anisotropic, complete thermal conductivity and radiation tensors L_{ij} and Γ_{ij} must be used instead of their diagonalized representations, hence, the additional terms connected with diffusion due to the transverse temperature gradients would appear when thermodynamical balance along the coordinate axes $\mathbf{x}$ is considered. On the other hand, when principal directions of damage change following a rotation of principal directions of stress, the combined thermo-damage equations must be considered at current principal damage directions to yield current heat flux orthotropy, though in a reference space (invariant) a general heat anisotropy (with diffusion included) occurs.

5.1.4 General anisotropic thermo-creep-damage coupling for initially isotropic material

Consider a general anisotropic case when the complete representation of the second-rank conductivity and radiation tensors $\widetilde{\mathbf{L}}(\mathbf{D})$ and $\widetilde{\mathbf{\Gamma}}(\mathbf{D})$ in the x, y, z frame, defined either by the second-rank damage tensor $\mathbf{D}$ or by the continuity tensor $\mathbf{\Psi}=\mathbf{1}-\mathbf{D}$, is as follows:

$$\widetilde{\mathbf{L}}(\mathbf{D})=\lambda_0(\mathbf{1}-\mathbf{D}),\qquad \widetilde{\mathbf{\Gamma}}(\mathbf{D})=\sigma\epsilon_0\mathbf{D}\tag{5.29}$$

and

$$\begin{aligned}\left[\widetilde{\mathbf{L}}(\mathbf{D})\right]&=\lambda_0\begin{bmatrix}1-D_{xx} & -D_{xy} & -D_{xz}\\ & 1-D_{yy} & -D_{yz}\\ & & 1-D_{zz}\end{bmatrix},\\ \left[\widetilde{\mathbf{\Gamma}}(\mathbf{D})\right]&=\sigma\epsilon_0\begin{bmatrix}D_{xx} & D_{xy} & D_{xz}\\ & D_{yy} & D_{yz}\\ & & D_{zz}\end{bmatrix}.\end{aligned}\tag{5.30}$$

The virgin material was assumed to be isotropic, with thermal properties characterized by the coefficients of thermal conductivity and emissivity of a gray body λ_0 and ϵ_0 and the Stefan–Boltzmann constant σ. Introducing the vector of temperature gradient $\mathbf{grad}T$ and the normal vector $\mathbf{n}$,

$$\mathbf{grad}T = \left\{\frac{\partial T}{\partial x}, \frac{\partial T}{\partial y}, \frac{\partial T}{\partial z}\right\}, \qquad \mathbf{n} = \{n_x, n_y, n_z\} \tag{5.31}$$

and defining the operations

$$\begin{aligned}
\mathbf{\Psi grad}T = \quad & \Big\{\Psi_{xx}\tfrac{\partial T}{\partial x} + \Psi_{xy}\tfrac{\partial T}{\partial y} + \Psi_{xz}\tfrac{\partial T}{\partial z}, \\
& \Psi_{yx}\tfrac{\partial T}{\partial x} + \Psi_{yy}\tfrac{\partial T}{\partial y} + \Psi_{yz}\tfrac{\partial T}{\partial z}, \\
& \Psi_{zx}\tfrac{\partial T}{\partial x} + \Psi_{zy}\tfrac{\partial T}{\partial y} + \Psi_{zz}\tfrac{\partial T}{\partial z}, \Big\}, \\
\mathbf{Dn} = \quad & \{D_{xx}n_x + D_{xy}n_y + D_{xz}n_z, \\
& D_{yx}n_x + D_{yy}n_y + D_{yz}n_z, \\
& D_{zx}n_x + D_{zy}n_y + D_{zz}n_z, \},
\end{aligned} \tag{5.32}$$

the general representation of the anisotropic damage coupled heat transfer equation is furnished in one of the equivalent forms as follows (Model B):

$$\begin{aligned}
&\text{div}\left(\lambda_0 \mathbf{\Psi grad}T - \sigma\epsilon_0 \mathbf{Dn}T^4\right) + \dot{q}_v = c_v \varrho \dot{T}, \\
&\text{div}\left[\lambda_0 \left(\mathbf{1} - \mathbf{D}\right) \mathbf{grad}T - \sigma\epsilon_0 \mathbf{Dn}T^4\right] + \dot{q}_v = c_v \varrho \dot{T}
\end{aligned} \tag{5.33}$$

or,

$$\frac{\partial}{\partial x_i}\left[\lambda_0 \left(I_{ij} - D_{ij}\right)\frac{\partial T}{\partial x_j} - \sigma\epsilon_0 D_{ij} n_j T^4\right] + \dot{q}_v = c_v \varrho \dot{T}, \tag{5.34}$$

where the absolute or the indices notation was applied. Eventually, when the explicit representation is used, (5.34) may be written as:

$$\begin{aligned}
&\frac{\partial}{\partial x}\left[\lambda_0 \left(\Psi_{xx}\frac{\partial T}{\partial x} + \Psi_{xy}\frac{\partial T}{\partial y} + \Psi_{xz}\frac{\partial T}{\partial z}\right)\right. \\
&\quad \left. -\sigma\epsilon_0 \left(D_{xx}n_x + D_{xy}n_y + D_{xz}n_z\right) T^4\right] \\
&\quad +\frac{\partial}{\partial y}\left[\lambda_0 \left(\Psi_{yx}\frac{\partial T}{\partial x} + \Psi_{yy}\frac{\partial T}{\partial y} + \Psi_{yz}\frac{\partial T}{\partial z}\right)\right. \\
&\quad \left. -\sigma\epsilon_0 \left(D_{yx}n_x + D_{yy}n_y + D_{yz}n_z\right) T^4\right] \\
&\quad +\frac{\partial}{\partial z}\left[\lambda_0 \left(\Psi_{zx}\frac{\partial T}{\partial x} + \Psi_{zy}\frac{\partial T}{\partial y} + \Psi_{zz}\frac{\partial T}{\partial z}\right)\right. \\
&\quad \left. -\sigma\epsilon_0 \left(D_{zx}n_x + D_{zy}n_y + D_{zz}n_z\right) T^4\right] \\
&\quad +\dot{q}_v = c_v \varrho \dot{T}.
\end{aligned} \tag{5.35}$$

In case of thermo-mechanically orthotropic solids, the off-diagonal components of (5.30) disappear for the x_1, x_2, x_3 axes; hence, the orthotropic heat flux equation (5.22) is recovered.

5.1.5 Coupled constitutive thermo-creep-damage equations

I. Isotropic (scalar) thermo-damage coupling (cf. Ganczarski and Skrzypek, 1995)

In order to solve the coupled thermo-creep-isotropic damage problem, the heat transfer equation (5.3) or (5.7) or (5.9) must be combined with constitutive creep and damage evolution equations (cf. Sect. 4.1.3(I)). Hence, the coupled thermo-creep-damage state equations are (Model A and Model B)

$$\left.\begin{aligned}
&\frac{\partial}{\partial x}\left\{\widetilde{\lambda}\left[\mathbf{x}, D(\mathbf{x},t)\right]\frac{\partial T}{\partial x} - \sigma\epsilon_0 D(\mathbf{x},t)\,T^4\right\} \\
&\quad +\frac{\partial}{\partial y}\left\{\widetilde{\lambda}\left[\mathbf{x}, D(\mathbf{x},t)\right]\frac{\partial T}{\partial y} - \sigma\epsilon_0 D(\mathbf{x},t)\,T^4\right\} \\
&\quad +\frac{\partial}{\partial z}\left\{\widetilde{\lambda}\left[\mathbf{x}, D(\mathbf{x},t)\right]\frac{\partial T}{\partial z} - \sigma\epsilon_0 D(\mathbf{x},t)\,T^4\right\} \\
&\quad +\dot{q}_v = c_v\widetilde{\varrho}\left[\mathbf{x}, D(\mathbf{x},t)\right]\dot{T}, \\
&\left.\begin{aligned}
\widetilde{\lambda}\left[\mathbf{x}, D(\mathbf{x},t)\right] &= \lambda_0(\mathbf{x})\left[1 - D(\mathbf{x},t)\right], \\
\widetilde{\varrho}\left[\mathbf{x}, D(\mathbf{x},t)\right] &= \varrho_0\left[1 - D(\mathbf{x},t)\right]^{3/2},
\end{aligned}\right\} \\
&\dot{D}(\mathbf{x},t) = C(T)\left\langle\frac{\chi(\boldsymbol{\sigma})}{1 - D(\mathbf{x},t)}\right\rangle^{r(T)}, \\
&\dot{\varepsilon}_{ij} = \frac{3}{2}\frac{\dot{\varepsilon}^{\mathrm{c}}_{\mathrm{eq}}}{\sigma_{\mathrm{eq}}}s_{ij}, \qquad \dot{\varepsilon}^{\mathrm{c}}_{\mathrm{eq}} = \left(\frac{\sigma_{\mathrm{eq}}}{1 - D(\mathbf{x},t)}\right)^{m(T)}\dot{\mathrm{f}}(t).
\end{aligned}\right\} \tag{5.36}$$

For the sake of generality, an initially nonhomogeneous isotropic material was used where both the coefficients of thermal conductivity $\widetilde{\lambda}$ and the mass density $\widetilde{\varrho}$ change with damage, whereas the Kachanov–Hayhurst damage growth rule is coupled with the Mises-type creep flow rule and the multiaxial time-hardening hypothesis, and $C(T)$, $r(T)$, $m(T)$ are temperature dependent material constants. When Model C is used (cf. Sect. 5.1.1) Eq. (5.9) should be substituted for the first of Eqs. (5.36), where the substitutive thermal conductivity is given by the formula

$$\lambda^{\mathrm{eq}} = \lambda_0(\mathbf{x})(1 - D) + \sigma\epsilon_0\left(4D + \frac{\partial D/\partial \mathbf{x}}{\partial T/\partial \mathbf{x}}T\right)T^3\mathrm{d}\mathbf{x}. \tag{5.37}$$

II. Orthotropic thermo-creep-damage coupling in constant principal directions (cf. Ganczarski and Skrzypek, 1997)

In the case that there is an orthogonal frame x_1, x_2, x_3 of thermo-mechanical orthotropy, such that the heat flux in the one of directions of orthotropy, say x_1, is affected by the temperature gradient in the same direction, $\partial T/\partial x_1$, but is not influenced by the other two, $\partial T/\partial x_2$ and $\partial T/\partial x_3$, we arrive at:

$$\left.\begin{aligned}
&\frac{\partial}{\partial x_1}\left\{\lambda_1\left[\mathbf{x}, D_1(\mathbf{x},t)\right]\frac{\partial T}{\partial x_1} - \sigma\epsilon_0 D_1(\mathbf{x},t)\,T^4\right\}\\
&\quad+\frac{\partial}{\partial x_2}\left\{\lambda\left[\mathbf{x}, D_2(\mathbf{x},t)\right]\frac{\partial T}{\partial x_2} - \sigma\epsilon_0 D_2(\mathbf{x},t)\,T^4\right\}\\
&\quad+\frac{\partial}{\partial x_3}\left\{\lambda\left[\mathbf{x}, D_3(\mathbf{x},t)\right]\frac{\partial T}{\partial x_3} - \sigma\epsilon_0 D_3(\mathbf{x},t)\,T^4\right\}\\
&\quad+\dot{q}_v = c_v\varrho_0\dot{T},\\
&\lambda_\nu\left[\mathbf{x}, D_\nu(\mathbf{x},t)\right] = \lambda_0(\mathbf{x})\left[1 - D_\nu(\mathbf{x},t)\right],\\
&\dot{D}_\nu(\mathbf{x},t) = C_\nu(T)\left\langle\frac{\boldsymbol{\sigma}_\nu(\mathbf{x})}{1 - D_\nu(\mathbf{x},t)}\right\rangle^{r_\nu(T)},\\
&\dot{\varepsilon}_{ij} = \frac{3}{2}\frac{\dot{\varepsilon}^{\mathrm{c}}_{\mathrm{eq}}}{\widetilde{\sigma}_{\mathrm{eq}}}\widetilde{s}_{ij}, \qquad \dot{\varepsilon}^{\mathrm{c}}_{\mathrm{eq}} = \widetilde{\sigma}^{m(T)}_{\mathrm{eq}}\dot{\mathsf{f}}(t).
\end{aligned}\right\} \tag{5.38}$$

The fully coupled creep-damage approach in (5.38) has been used as it is more consistent (cf. Sect. 4.1.3(II)).

5.2 The local approach to fracture using the CDM approach

5.2.1 *Effective elastic moduli of cracked solids*

A transition between the atomic, the micro, or the mesoscale and the fourth-rank elasticity tensors $\widetilde{\boldsymbol{\Lambda}}$ or $\widetilde{\boldsymbol{\Lambda}}^{-1}$ for stiffness or compliance, and the second-rank thermal properties tensors $\widetilde{\mathbf{L}}$ and $\widetilde{\mathbf{L}}^{-1}$ for conductivity and emissivity, etc., requires a proper selection of the representative volume element (RVE). The RVE maps a finite volume of linear size λ_{RVE} of the piecewise-discontinuous and heterogeneous solid, the state of damage in which is determined by the topology, size, orientation, and number of microcracks, microvoids, microslips, etc., on a material point of the pseudo-undamaged quasicontinuum. This effective quasicontinuum method, also called the CDM method, is based on the assumptions that (cf. Krajcinovic, 1995):

- each defect within the RVE is subjected to the same stress field derived from the external tractions applied at the boundary of the element, and
- the effect of other defects within the RVE on the observed defect is measured through the change of the effective thermo-mechanical properties (cf. Sect. 1.1.2).

In other words, the exact spatial correlation of the defects within the RVE has a negligible influence on the effective properties defined in the element. The minimum linear size of λ_{RVE} of the RVE must be large enough to include a sufficient number of damage entities to provide a statistically homogeneous representation of the microstructure or the mesostructure. At the same time, however, the size λ_{RVE} must be small enough for the stress field to be considered as homogeneous within the RVE. The existence of the RVE, that allows the heterogeneous and discontinuous material to be considered as statistically homogeneous within the element, is the condition for a local approach (LA) in which there are no scale parameters involved. Hence, the effective moduli of a damaged solid depend on the average distribution of sizes, orientations, and spatial positions of defects within the RVE. Spatially averaged damage variables are, generally, a sufficiently good approximation for the stiffness and thermal flux characterization. Damage evolution, on the other hand, depends more on the extreme values of the defect distribution, for instance, the largest defect size, the minimum neighbor distance, etc., such that the effect of damage patterning on the local driving forces should also be incorporated to the higher-order macroscopic damage descriptors (cf. Lacy et al., 1997).

The effect of crack systems on the effective moduli of linearly elastic isotropic solids was critically reviewed by M. Kachanov (1992). For non-interacting cracks in the isotropic matrix material the effective moduli can be determined exactly for a random, arbitrary crack distribution, particularly at low crack densities. In the approximation of non-interacting cracks, each of them is regarded as isolated and free of any influence from other cracks. Hence, the compliance is linear in crack density. The second-rank or the fourth-rank crack density tensors $\boldsymbol{\alpha}$ or $\widetilde{\boldsymbol{\alpha}}$ provide an adequate description of a crack array in the 2D or the 3D cases, respectively. If crack distribution is nonrandom, interactions can be strong even at small density. For interacting cracks, the determination of the effective moduli requires considering a problem of direct interaction for each crack configuration including their exact orientation, position, and size, and then a subsequent averaging over them. Particular approximate models (self-consistent, differential, generalized self-consistent, Mori–Tanaka, scheme and others) are generally based on the analysis of one isolated crack placed into a matrix with the effective moduli, such that the influence of interaction on a considered crack is accounted for by the reduced stiffness of the surrounding material (cf. Budiansky and O'Connel, 1976; Hashin, 1988; Sayers and Kachanov, 1991; Mori and Tanaka, 1973; Christiansen and Lo, 1979). Usually, the interaction produces a softening effect on the effective moduli with the exception of the Mori–Tanaka approach for which the predicted moduli coincide with those obtained for noninteracting cracks (cf. Kachanov, 1992).

5.2.2 Crack growth by a local approach – general features

The situation becomes much more complicated when a coupled damage-fracture mechanism is determined by the nucleation (pre-critical) and the growth (post-critical) of a single macrocrack (or the macrocracks pattern), the geometry of which is explicitly determined in the fracture (cracking) process. Generally speaking, crack propagation through the solid with a heterogeneous microstructure may be arrested and the continuum damage accumulation prior to the macrofracture may occur. As a consequence, a strong interaction between the macrocrack(s) and the damage field in fracturing process is observed and, hence, the nonlocal approach (NLA) should be used rather than the local one, mainly due to stress concentration at a crack tip. Nevertheless, an approximate fracture analysis by applying the local approach (LA) to the stress, strain, and damage fields at a crack tip may also be recommended for its simplicity for the creep damage-fracture analysis (cf. Murakami, Kawai and Rong, 1988; Liu, Murakami and Kanagawa, 1994; Murakami and Liu, 1995) and the elastic-brittle damage-fracture analysis (cf. Skrzypek, Kuna-Ciskał, and Ganczarski, 1998c).

The local approach to fracture (LAF) based on continuum damage mechanics (CDM), when a free surface is produced on the macrocrack, is usually combined with the finite element method (FEM), hence, the crucial question is the mesh dependence and its regularization. This problem is examined in the paper by Murakami and Liu (1995), where the FEM was applied to the creep-fracture analysis by the use of the elastic-creep material model with isotropic damage, the one-parameter Kachanov–Rabotnov–Sdobyrev model (Sect. 2.2.2(II)) and the linear scalar Young's modulus drop with damage, all implemented in the UMAT of ABAQUS FEM code. By the use of this simple model, an assembly of fractured elements is considered as a crack when the stress in the element is released after the scalar damage variable in the element has reached the critical value D_{cr}, and a free surface is created. In this approach, the crack width is governed by the size of the finite element, and the crack cannot develop in the direction of its width. As a consequence, heavy mesh-dependence of both the crack length growth rate and the stress and damage concentration, particularly in the region around the crack tip, is observed. Possibilities for regularization were examined in the frame of the local approach by the use of a nonlocal damage variable (averaged over a neighborhood of the crack tip), a simplified stress limitation (ideal plasticity), and a modification of the damage evolution law (cf. Sect. 5.2.3).

Recently, the local approach to fracture was applied to a coupled elastic-brittle damage-fracture analysis by Skrzypek, Kuna-Ciskał, and Ganczarski (1998). In contrast to the material model used by Murakami and Liu (1995), the damage anisotropy was accounted for in this paper by the application of the anisotropic elasticity coupled with damage. The damage evolution law by Litewka and Hult (1989) was generalized to the case of a rotation of principal stress and damage axes (due to the shear effect included) and the extended time-dependent elastic-brittle constitutive model as originated by Litewka (1985, 1989) was combined with the failure criterion in the form of an isotropic scalar function of stress and damage tensors (cf. Sect. 4.2). All these constitutive models were implemented in the UMAT of ABAQUS. Interaction of the two mechanisms, releasing the kinematic boundary conditions on the fixed-edge element face and/or fully removing the element that has failed, governs an unstable process of structural fracturing which leads to the complete fragmentation of the structure.

This combined macrocrack penetration through the volume of an elastic-brittle-damaged solid consists in a mixed-controlled mechanism. First, it is observed mostly as tensile stress controlled crack length growth if the anticipated crack is formed along the a priori known structure of the fixed edge (if any) after the failure criterion has been satisfied in a neighboring (damaged) element and, as the consequence, the appropriate kinematic boundary conditions are released to allow for the crack opening on the free surface produced. The element disconnected from the rigid edge is left in the FE mesh to be able to carry the shear stress, although the tensile stress in the direction normal to the crack has been released. Second, it is recognized as a combined tension/shear controlled crack branching mechanism that allows the crack to deviate from the primary direction along the fixed edge into the interior. The neighboring element which has been caused to fail is then fully removed from the mesh and, in this way, a secondary crack of the width of the element is formed. In both cases, a significant cumulative continuum damage field, prior to the failure prediction in the element as governed by the failure criterion legislated, is observed to develop in particular in the primary and/or the secondary crack tip surroundings. In addition, in a region where the compressive stresses predominate, little or no continuum damage prior to fracture is observed, hence, the element which is led to failure when the stress vector meets the initial failure surface is instantaneously crushed and removed from the mesh (cf. Fig. 5.4).

The possibility to examine the complex crack patterns in a structure, with changing crack directions and crack branching allowed, is a benefit of the local approach to fracture when applied to anisotropic elastic-brittle-damage structures. It was observed by the authors that the failure criterion for elastic-brittle-damaged solids, as proposed by Litewka and Hult, 1989, introduces naturally a stress and damage limitation such that no artificial methods for stress and damage field regularization are required. On the other hand, the overall crack pattern and the complete structure fracture prediction t_F were found not to be as strongly mesh-dependent as in the Murakami and Liu approach (cf. Skrzypek et al., 1998c).

5.2.3 Local approach of elastic-creep fracture versus elastic-brittle fracture

A quantitative comparison of the local approach to fracture by the FEM when applied to two material models, the coupled elastic-creep damaged solid (Murakami, Kawai and Rhong, 1988; Murakami and Liu , 1995) and the elastic-brittle orthotropic damaged solid (Litewka, 1985; Skrzypek, Kuna-Ciskał, and Ganczarski, 1998c), both implemented using FE codes, is presented in what follows.

I. Elastic-creep damage model of fractured material

A discussion of a local approach to the analysis of crack growth in a particular creep-orthotropic damaged solid when the stress and the damage tensors are coinciding in their principal axes (no rotation allowed) is due to Murakami et al. (1988). The damage evolution equation, under simplifying assumptions that the damage rate is described by the net area reduction on the planes perpendicular to the direction $\mathbf{n}^{(1)}$ of the maximum principal stress σ_1 combined with the isotropic area reduction and that its magnitude is governed by the Hayhurst-type isochronous rupture function (cf. (2.36)), was postulated by the authors in the following fashion:

$$\begin{aligned}\dot{\mathbf{D}} = B &\left[\xi\sigma_1 + \zeta\sigma_{eq} + \frac{1-\xi-\zeta}{3}\mathrm{Tr}\left(\boldsymbol{\sigma}\right)\right]^k \\ &\times \left[\mathrm{Tr}\left\{\left(\mathbf{1}-\mathbf{D}\right)^{-1}\left(\mathbf{n}^{(1)}\otimes\mathbf{n}^{(1)}\right)\right\}\right]^l \\ &\times \left[\left(1-\eta\right)\mathbf{1} + \eta\mathbf{n}^{(1)}\otimes\mathbf{n}^{(1)}\right].\end{aligned} \tag{5.39}$$

Let us mention that, for particular cases $\eta = 0$ and $\eta = 1$, (5.39) reduces to purely isotropic damage evolution and purely orthotropic microcrack growth in planes perpendicular to the maximum tensile stress, respectively, whereas for $0 < \eta < 1$ a mixed isotropic/maximum principal stress controlled damage growth mechanism occurs. A combined the McVetty and the Mises-type creep flow rules together with the strain hardening hypothesis was selected as the isotropic constitutive law:

$$\begin{aligned}\dot{\boldsymbol{\varepsilon}}^{\mathrm{c}} &= \frac{3}{2}\left[A_1\sigma_{\mathrm{eq}}^{n_1-1}\alpha\exp\left(-\alpha\bar{t}\right)\boldsymbol{\sigma}' + A_2\widetilde{\boldsymbol{\sigma}}_{\mathrm{eq}}^{n_2-1}\widetilde{\boldsymbol{\sigma}}'\right] \\ \varepsilon_{\mathrm{eq}}^{\mathrm{c}}\left(t\right) &= A_1\sigma_{\mathrm{eq}}^{n_1}\left(t\right)\left[1-\exp\left(-\alpha\bar{t}\right)\right] + A_2\widetilde{\boldsymbol{\sigma}}_{\mathrm{eq}}^{n_2}\left(t\right)\bar{t},\end{aligned} \tag{5.40}$$

where:

$$\sigma_1 = \max \sigma_i, \qquad \sigma_{\mathrm{eq}} = \left[(3/2)\,\boldsymbol{\sigma}' : \boldsymbol{\sigma}'\right]^{1/2},$$

$$\dot{\varepsilon}^{\mathrm{c}}_{\mathrm{eq}} = \left[(3/2)\,\dot{\boldsymbol{\varepsilon}}^{\mathrm{c}} : \dot{\boldsymbol{\varepsilon}}^{\mathrm{c}}\right]^{1/2},$$

$$\tilde{\boldsymbol{\sigma}} = (1/2)\left[\boldsymbol{\sigma} : (\mathbf{1} - \mathbf{D})^{-1} + (\mathbf{1} - \mathbf{D})^{-1} : \boldsymbol{\sigma}\right], \tag{5.41}$$

$$\tilde{\sigma}_{\mathrm{eq}} = \left[(3/2)\,\tilde{\boldsymbol{\sigma}}' : \tilde{\boldsymbol{\sigma}}'\right]^{1/2}, \qquad \boldsymbol{\sigma}' = \boldsymbol{\sigma} - (1/3)\,(\mathrm{Tr}\boldsymbol{\sigma})\,\mathbf{1},$$

$$\tilde{\boldsymbol{\sigma}}' = \tilde{\boldsymbol{\sigma}} - (1/3)\,(\mathrm{Tr}\tilde{\boldsymbol{\sigma}})\,\mathbf{1},$$

whereas B, k, l, ξ, ζ, η and A_1, A_2, n_1, n_2, α are material constants and a fictitious time $\bar{t}$ is to be eliminated from (5.40).

The above described coupled constitutive and damage evolution equations were implemented on a FACOM M–382 system and applied to crack growth analysis in a square plate of copper at 250°C with a width to thickness ratio $b/c = 30$ and an initial crack length to thickness ratio $a_0/c = 10$, subjected to a biaxial proportional or nonproportional loading, when the material constants of (5.39)–(5.41) were as follows:

$$\begin{aligned}
&B = 4.46 \times 10^{-13}\left[\mathrm{MPa}^{-k}\mathrm{h}^{-l}\right], && l = 5.0,\\
&k = 5.55, \quad \xi = 1.0, && \zeta = 0.0,\\
&A_1 = 2.40 \times 10^{-7}\left[\mathrm{MPa}^{-n_1}\right], && n_1 = 2.60,\\
&A_2 = 3.00 \times 10^{-16}\left[\mathrm{MPa}^{-n_2}\mathrm{h}^{-1}\right], && n_2 = 7.10,\\
&E = 66.240\,[\mathrm{MPa}], && \alpha = 0.05\left[\mathrm{h}^{-1}\right],
\end{aligned} \tag{5.42}$$

whereas the parameter of anisotropy η was taken as $\eta = 0.0$ or $\eta = 0.5$ or $\eta = 1.0$ for the purely orthotropic or the mixed or the purely isotropic case, respectively. The failure of the element was defined as the state in which the maximum principal damage value of the damage tensor D_1 (damage orthotropy) or the scalar damage variable D attained the critical value $D_{\mathrm{cr}} = 0.99$, at which point the rigidity of the failed element was released to zero, whereas the assembly of fully cracked elements was considered as a part of the global crack.

The effect of damage orthotropy on the creep-crack pattern and the final time to failure prediction was examined under the biaxial nonproportional loadings, as shown in Fig. 5.4.

The results obtained for crack initiation t_I are approximately equal when the isotropic, $t_I^{\mathrm{is}} = 26$ h, and the orthotropic, $t_{II}^{\mathrm{or}} = 27$ h, models are

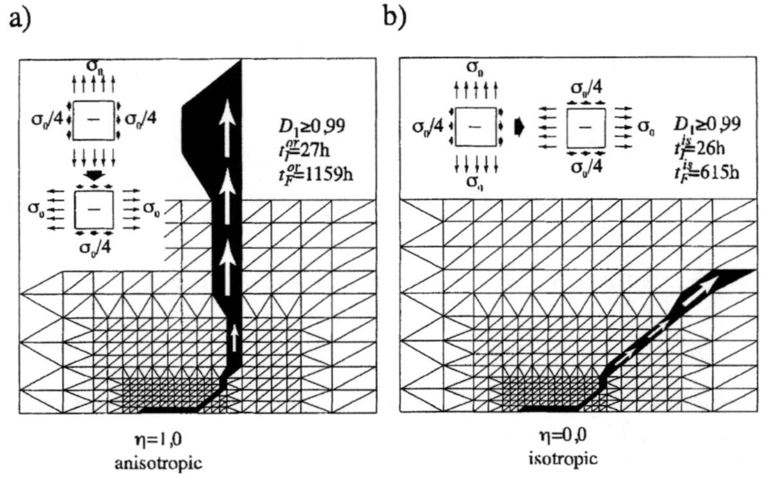

Fig. 5.4. Effect of creep-damage anisotropy on the crack-pattern in a fractured plate under cross-changing loading: a) anisotropic damage model $\eta = 1.0$, b) isotropic damage model $\eta = 0.0$, (after Murakami, Kawai and Rhong, 1988)

used. However, the anisotropy improves the time to failure prediction from $t_{\mathrm{F}}^{\mathrm{is}} = 615$ h for the purely isotropic model, to $t_{F}^{\mathrm{or}} = 1159$ h in the case of the purely orthotropic mechanism. Additionally, creep-crack patterns differ from each other in that when the purely isotropic damage evolution law is applied ($\eta = 0.0$) the main crack direction is, roughly speaking, insensitive to the maximum tensile stress direction change, whereas in the purely orthotropic one ($\eta = 1.0$) after the external load change the main crack also gradually deviates from the primary direction to finally reach the direction perpendicular to the maximum principal tension for the second phase of loading, Fig. 5.4.

II. Regularization methods in local approach to creep-fracture (cf. Murakami and Liu, 1995)

A simple elastic-creep-isotropic damage material model that ignores the damage anisotropy is employed by the authors in the form

$$
\begin{aligned}
&\boldsymbol{\sigma} = \widetilde{\boldsymbol{\Lambda}}(D) : (\boldsymbol{\varepsilon} - \boldsymbol{\varepsilon}^{\mathrm{c}}), && \widetilde{\boldsymbol{\Lambda}}(D) = \boldsymbol{\Lambda}(1-D), \\
&\dot{\boldsymbol{\varepsilon}}^{\mathrm{c}} = \frac{3}{2} A \left(\frac{\sigma_{\mathrm{eq}}}{1-D} \right)^{n} \frac{\boldsymbol{\sigma}'}{\sigma_{\mathrm{eq}}}, && \dot{D} = \frac{B}{q+1} \frac{[\chi(\boldsymbol{\sigma})]^{p}}{(1-D)^{q}}, \\
&\chi(\boldsymbol{\sigma}) = \alpha \sigma_{\mathrm{eq}} + (1-\alpha)\sigma_1,
\end{aligned}
\tag{5.43}
$$

where Λ is the elasticity tensor for isotropic materials, D is the scalar damage variable ($D_{cr} = 0.99$) and A, B, n, p, q are material constants.

The effect of mesh-dependence in the local approach to creep fracture and possible regularization methods for it are studied by Murakami and Liu (1995). Mesh dependence of crack growth was studied on an axisymmetric thick-walled tube uniformly pulled along the periphery (Fig. 5.5).

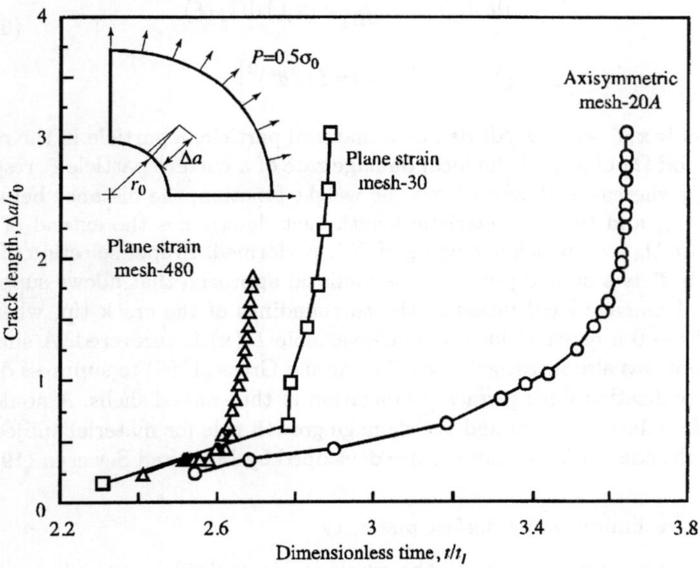

Fig. 5.5. Effect of FE mesh on the crack length growth versus dimensionless time t/t_I in an axisymmetric thick-walled tube under uniform exterior tension (after Murakami and Liu, 1995)

A discussion by the authors may be summarized as follows.

In the Murakami and Liu model using the local approach to fracture, the assembly of fractured elements E_f is considered as a crack. The crack region Ω can be determined by the local fracture criterion

$$\Omega = \{E_f; D(E_f) = D_{cr}\}. \tag{5.44}$$

In other words, the stress in an element that undergoes fracture is totally released when the damage level in an element reaches its critical value D_{cr}; hence, the stress on the free surface of the crack must vanish so that the crack cannot develop in the transverse direction to the crack length and the crack width is governed by the element size.

Three ways to suppress the mesh-dependence of the local approach to fracture were examined by Murakami and Liu (1995).

A. Nonlocal damage variable $\overline{D}$

Regularization of the local variation of the damage field $D(\mathbf{x})$ is achieved by averaging a nonlocal damage variable $\overline{D}(\mathbf{x},\Omega_\mathrm{d})$ over the neighborhood $\Omega_\mathrm{d}(\xi)$ of $\mathbf{x}$:

$$\frac{\mathrm{d}\overline{D}(\mathbf{x},\Omega_\mathrm{d})}{\mathrm{d}t} = \frac{\int_{\Omega_\mathrm{d}} \frac{\mathrm{d}D(\boldsymbol{\xi})}{\mathrm{d}t} \phi(\mathbf{x},\xi)\,\mathrm{d}\Omega_\mathrm{d}(\boldsymbol{\xi})}{\int_{\Omega_\mathrm{d}} \phi(\mathbf{x},\xi)\,\mathrm{d}\Omega_\mathrm{d}(\boldsymbol{\xi})},$$
$$\phi(\mathbf{x},\xi) = \exp\left[-\left(d(\mathbf{x},\xi)/d^*\right)^2\right]. \tag{5.45}$$

Symbols $\mathbf{x}$, ξ, and $\mathrm{d}D/\mathrm{d}t$ denote a material particle, a particle in the neigborhood Ω_d of $\mathbf{x}$, and the local damage rate of a current particle ξ, respectively, whereas ϕ, d, and d^* are the weight function, the distance between $\mathbf{x}$ and ξ, and the characteristic length that determines the extend of the domain Ω_d over which averaging of D is performed. Proper selection of the length d^* is a crucial point of this nonlocal approach that allows suppression of damage localization in the surroundings of the crack tip, whereas for $d^* \to 0$ a classical local damage variable $D(\mathbf{x})$ is recovered. A similar concept was also investigated by Baasar and Gross (1998) to suppress damage localization during crack propagation in thin-walled shells. A nonlocal brittle failure criterion and the damage growth rule for material subjected to multiaxial variable loadings are developed by Mróz and Seweryn (1998).

B. Stress limitation by perfect plasticity

The stress concentration at the crack tip in metallic materials may be limited by incorporating a perfect plasticity criterion into the model by assuming a modified stress $\boldsymbol{\sigma}^*$ for damage evaluation as follows:

$$\sigma_{ij} = \begin{cases} \sigma_{ij}, & \sigma_\mathrm{eq} \le \sigma_0, \\ k\sigma_{ij}, & \sigma_\mathrm{eq} > \sigma_0, \end{cases}$$
$$k\sqrt{(3/2)\,\sigma'_{ij}\sigma'_{ij}} - \sigma_0 = 0, \tag{5.46}$$

where σ_eq is the Huber–Mises–Hencky equivalent stress and a factor k is determined from the yield criterion.

C. Modification of the damage evolution law and reduction of the critical damage value D_cr

When the damage parameter in the classical damage evolution law (5.43) approaches the critical level $D_\mathrm{cr} = 1$, a strong stress sensitivity of the damage evolution is observed. To suppress this effect, a modified exponential form of the damage growth rule was proposed by the authors:

$$\frac{\mathrm{d}D}{\mathrm{d}t} = \frac{B}{q'} [\chi(\boldsymbol{\sigma})]^p \exp\left(q'D\right), \tag{5.47}$$

where B, p, q' are modified material constants that should be determined by comparing the modified (5.47) and the classical (5.43) damage evolution for constant stress. Additionally, limitation of the critical damage to the level $D_{\mathrm{cr}} = 0.7$ was applied for better FEM convergence.

All above regularization methods were tested by the authors to yield a significant reduction of the mesh dependence in the local approach to creep–fracture analysis by FEM.

III. Local approach to elastic-brittle fracture by the extended Litewka model (cf. Skrzypek, Kuna-Ciskał, and Ganczarski, 1988c)

When the modified CDM Litewka model of anisotropic elastic-brittle damage in metallic material (Sect. 4.2) is used to fracture analysis for structures, a problem of mesh-dependence is met as well (cf. Fig.5.6).

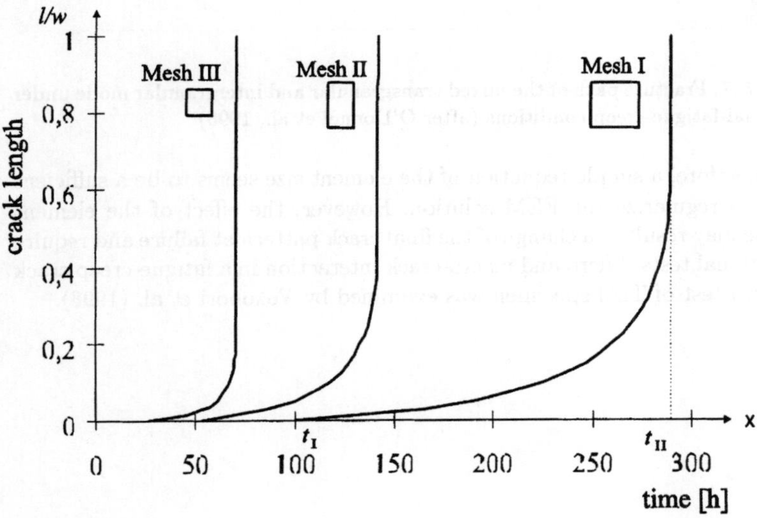

Fig. 5.6. Effect of FE mesh on the crack length growth in a 2D structure made of anisotropic elastic-brittle damaged steel (after Skrzypek, Kuna-Ciskał, and Ganczarski, 1998c)

In the material model considered, both the stress and the damage concentration at the crack tip are limited due to the isotropic scalar failure criterion function (4.43) used to define a crack opening. Additionally, when the failure criterion is satisfied in a fractured element the kinematic boundary conditions are released at the node and free surface (crack) may be produced without removing the element from the mesh. Consequently, the crack width is less dependent on the element size and the crack can be developed in the direction transverse to the primary crack length (crack pattern branching). Hence, the model is capable of predicting complex crack patterns like those observed by O'Donnel et al. (1998) in 316L stainless steel under thermal fatigue-creep loading conditions, Fig. 5.7.

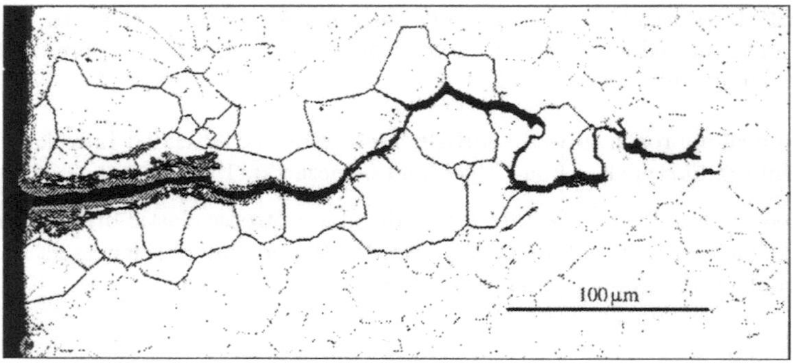

Fig. 5.7. Fracture path of the mixed transgranular and intergranular mode under thermal-fatigue-creep conditions (after O'Donnel et al., 1998)

Therefore, a simple reduction of the element size seems to be a sufficient way to regularize the FEM solution. However, the effect of the element shape may result in a change of the final crack pattern at failure and require additional tests. Micro and macro-crack interaction in a fatigue-creep crack growth test of TiAl specimen was examined by Yokobori et al. (1998).

Part II

Analysis of damage and failure of simple structures

6

Creep damage and failure of axisymmetric structures

6.1 Analysis of creep failure process in structures

When the damage evolution in a structure is considered the following stages of the failure advance may be distinguished (Table 6.1).

Table 6.1. Stages of failure advance

Isotropic material damage at a point D–scalar parameter	time	Orthotropic material damage at a point D_1, D_2, D_3–principal damage components
	0	
1. Damage incubation period : $D(\mathbf{x},t) \equiv 0$ $(\mathbf{x} \in V, t < t_0)$		$D_\nu(\mathbf{x},t) \equiv 0$ $(\nu = 1,2,3; \mathbf{x} \in V, t < t_0)$
2. Time of initial damage	t_0	microcrack initiation (if $t_0 = 0$: no incubation)
3.Damage growth period : $0 < D(\mathbf{x},t) < D_{\mathrm{crit}}$ $(t_0 < t < t_I)$		$0 < \sup_{(1,2,3)} D_\nu(\mathbf{x},t) < D^{(\nu)}_{\mathrm{crit}}$ $(\nu = 1,2,3; \mathbf{x} \in V; t_0 < t < t_I)$
4. Time of initiation of failure	t_I	first macrocrack initiation $(t_I = t_{\mathrm{R}})$
5. Fracture propagation period : $D(\mathbf{x},t) \equiv D_{\mathrm{crit}}$ (failed zone : $\mathbf{x} \in V_{\mathrm{f}}$) $0 < D(\mathbf{x},t) < D_{\mathrm{crit}}$ (unfailed zone : $\mathbf{x} \in V_{\mathrm{unf}}$)		$\sup_{(1,2,3)} D_\nu(\mathbf{x},t) \equiv D^{(\nu)}_{\mathrm{crit}}$ $(\mathbf{x} \in V_{\mathrm{f}})$ $0 < \sup_{(1,2,3)} D_\nu(\mathbf{x},t) < D^{(\nu)}_{\mathrm{crit}}$ $(\mathbf{x} \in V_{\mathrm{unf}})$
6. Time of structural failure	t_{II}	failure mechanism of a structure $(t_{II} = t_{\mathrm{F}})$
	time	

6.2 Example: Transient creep and creep failure of a thick-walled pressurized tube

6.2.1 Rotationally symmetric plane-strain problem for isotropic material

When a transient primary and secondary creep state analysis of a thick-walled tube is considered, the effect of stress redistribution on the creep

behavior has to be taken into account (cf. Penny and Marriott, 1995). This analysis was extended by incorporating the isotropic damage growth period in the tertiary creep phase, $t \leq t_I$, by Boyle and Spence (1983), whereas the creep-fracture process of an axisymmetric thick-walled tube by the local approach, when CDM based FEM was used for the crack growth analysis, $t_I \leq t \leq t_{II}$, is due to Murakami and Liu (1995). In what follows we confine ourselves to the damage growth period in a tube governed by the Mises-type rule, the multiaxial time-hardening hypothesis, and the Kachanov-type isotropic damage law. Hence, following Boyle and Spence (1983) the governing equations for the rotationally symmetric plane-strain problems when small total strains are split into the elastic and anelastic parts can be written as:
Geometric equations

$$\begin{aligned}
\varepsilon_r &= \frac{\mathrm{d}u}{\mathrm{d}r} = \frac{\sigma_r - \nu(\sigma_\vartheta + \sigma_z)}{E} + \varepsilon_r^{\mathrm{c}}, \\
\varepsilon_\vartheta &= \frac{u}{r} = \frac{\sigma_\vartheta - \nu(\sigma_z + \sigma_r)}{E} + \varepsilon_\vartheta^{\mathrm{c}}, \\
\varepsilon_z &= \varepsilon_0 = \frac{\sigma_z - \nu(\sigma_r + \sigma_\vartheta)}{E} + \varepsilon_z^{\mathrm{c}} = \mathrm{const},
\end{aligned} \tag{6.1}$$

Equilibrium equation

$$\frac{\mathrm{d}}{\mathrm{d}r}(r\sigma_r) + \sigma_\vartheta = 0, \tag{6.2}$$

Creep flow rule for plane strain creep incompressibility

$$\begin{aligned}
\mathrm{d}\varepsilon_r^{\mathrm{c}} &= \frac{\mathrm{d}\varepsilon_{\mathrm{eq}}^{\mathrm{c}}}{\sigma_{\mathrm{eq}}}\left[\sigma_r - \frac{1}{2}(\sigma_\vartheta + \sigma_z)\right], \\
\mathrm{d}\varepsilon_\vartheta^{\mathrm{c}} &= \frac{\mathrm{d}\varepsilon_{\mathrm{eq}}^{\mathrm{c}}}{\sigma_{\mathrm{eq}}}\left[\sigma_\vartheta - \frac{1}{2}(\sigma_r + \sigma_z)\right], \qquad \mathrm{d}\varepsilon_z^{\mathrm{c}} = -\left(\mathrm{d}\varepsilon_r^{\mathrm{c}} + \mathrm{d}\varepsilon_\vartheta^{\mathrm{c}}\right),
\end{aligned} \tag{6.3}$$

Multiaxial time hardening hypothesis associated with the Kachanov–Galileo law

$$\mathrm{d}\varepsilon_{\mathrm{eq}}^{\mathrm{c}} = \left(\frac{\sigma_{\mathrm{eq}}}{1-D}\right)^m \dot{\mathsf{f}}(t)\mathrm{d}t, \qquad \mathrm{d}D = C\left\langle\frac{\sigma_\vartheta}{1-D}\right\rangle^r \mathrm{d}t, \tag{6.4}$$

where the equivalent stress σ_{eq} and the cumulative creep strain $\mathrm{d}\varepsilon_{\mathrm{eq}}^{\mathrm{c}}$ are defined as follows:

$$\begin{aligned}
\sigma_{\mathrm{eq}} &\stackrel{\mathrm{def}}{=} \frac{1}{\sqrt{2}}\left[(\sigma_r - \sigma_\vartheta)^2 + (\sigma_\vartheta - \sigma_z)^2 + (\sigma_z - \sigma_r)^2\right]^{1/2}, \\
\mathrm{d}\varepsilon_{\mathrm{eq}}^{\mathrm{c}} &\stackrel{\mathrm{def}}{=} \frac{2}{\sqrt{3}}\left[(\mathrm{d}\varepsilon_r^{\mathrm{c}})^2 + (\mathrm{d}\varepsilon_\vartheta^{\mathrm{c}})^2 + \mathrm{d}\varepsilon_r^{\mathrm{c}}\mathrm{d}\varepsilon_\vartheta^{\mathrm{c}}\right]^{1/2}.
\end{aligned} \tag{6.5}$$

Elimination of stresses from (6.1)–(6.5) and subsequent introduction of the dimensionless variables yields after some algebra the governing equations for the transient plane-strain creep problem in the dimensionless displacement (rates) formulation:

$$\frac{\partial}{\partial R}\left[\frac{1}{R}\frac{\partial}{\partial R}(RU)\right]=0 \qquad (\bar{t}=0),$$

$$\frac{\partial}{\partial R}\left[\frac{1}{R}\frac{\partial}{\partial R}(R\dot{U})\right]=\dot{F}+\frac{\partial\dot{G}}{\partial R} \qquad (\bar{t}>0), \tag{6.6}$$

$$\dot{F}=\frac{1-2\nu}{1-\nu}\left(\dot{E}^{c}_{r}-\dot{E}^{c}_{\vartheta}\right), \qquad \dot{G}=\frac{1-2\nu}{1-\nu}\dot{E}^{c}_{r},$$

where

$$\begin{aligned} &R=r/a, \quad U=u/a\varepsilon_0, \quad \bar{t}=E\sigma_0^{m-1}\mathsf{f}(t), \quad \varepsilon_0=\sigma_0/E,\\ &E_r=\varepsilon_r/\varepsilon_0, \quad E_\vartheta=\varepsilon_\vartheta/\varepsilon_0 \quad S_r=\sigma_r/\sigma_0, \quad S_\vartheta=\sigma_\vartheta/\sigma_0. \end{aligned} \tag{6.7}$$

Hence, finally, the general solution of the problem may be written as follows:

$$U=\frac{A_0}{2}R+\frac{B_0}{R} \qquad (\bar{t}=0),$$

$$\dot{U}=\frac{A_i}{2}R+\frac{B_i}{R}+\frac{\dot{I}_1}{R} \qquad (\bar{t}>0). \tag{6.8}$$

The dimensionless stress rates are

$$\dot{S}_r=\frac{1}{(1-2\nu)(1+\nu)}\left[-\frac{A_i}{2}-(1-2\nu)\frac{B_i}{R^2}+\dot{I}_2+\dot{I}_3\right],$$

$$\dot{S}_\vartheta=\frac{1}{(1-2\nu)(1+\nu)}\left[-\frac{A_i}{2}+(1-2\nu)\frac{B_i}{R^2}+\dot{I}_2+\dot{I}_3\right.$$

$$\left.-\nu\dot{G}-(1-2\nu)\dot{E}^{c}_{\vartheta}\right], \tag{6.9}$$

$$\dot{S}_z=\frac{1}{(1-2\nu)(1+\nu)}\left[-\nu A_i+2\nu\dot{I}_2+\nu\dot{G}-(1-2\nu)\dot{E}^{c}_{z}\right],$$

the dimensionless strain rates are

$$\dot{E}_r=-\frac{A_i}{2}-\frac{B_i}{R^2}+\dot{I}_2+\frac{\dot{I}_3}{1-2\nu},$$

$$\dot{E}_\vartheta=-\frac{A_i}{2}+\frac{B_i}{R^2}+\dot{I}_2-\frac{\dot{I}_3}{1-2\nu}, \tag{6.10}$$

$$\dot{E}_z=\dot{S}_z-\nu\left(\dot{S}_r+\dot{S}_\vartheta\right)+\dot{E}^{c}_{z}=\text{const},$$

and the auxiliary integrals $\dot{I}_1$, $\dot{I}_2$, $\dot{I}_2$ in (6.9) and (6.10) are defined as:

$$\dot{I}_1 = \frac{R^2}{2}\int_1^R \frac{\dot{F}}{\xi}\mathrm{d}\xi + \frac{1}{2}\int_1^R (2\dot{G} - \dot{F})\xi\mathrm{d}\xi,$$
$$\dot{I}_2 = \frac{1}{2}\int_1^R \frac{\dot{F}}{\xi}\mathrm{d}\xi, \qquad \dot{I}_3 = -\frac{1-2\nu}{2R^2}\int_1^R (2\dot{G} - \dot{F})\xi\mathrm{d}\xi. \tag{6.11}$$

The constants A_0, B_0 and A_i, B_i should be determined from the appropriate boundary conditions at $\bar{t} = 0$ (elastic state) and $\bar{t}_i > 0$ (creep state), respectively. Dimensionless creep strain rate components $\dot{E}_r^c$, $\dot{E}_\vartheta^c$, and $\dot{E}_z^c$ are to be obtained on each time step when the normalized creep constitutive and damage evolution equations (6.3) and (6.4) are simultaneously solved.

6.2.2 Results for pressurized tube

A detailed analysis of coupled creep-brittle damage to a long thick-walled tube with inner to outer radii ratio a/b, subjected to a uniform internal pressure p and under a plane strain state, was presented by Boyle and Spence (1983). In this section the numerical results obtained by Skrzypek (1993) for $b/a = 2$, $p/\sigma_0 = 0.2$, $m = 5$, $r = 3.5$ are briefly reported.

The redistribution of stresses σ_r and σ_ϑ with a normalized time t/t_I during the stage of latent failure from the initial elastic distribution (solid line) to the stress distribution at the initiation of fracture $t/t_I = 1$ (dot-dash line) is shown in Fig. 6.1a–d. First macrocracks appear here at the outer surface ($a/b = 1$), preceded by the damage accumulation which concentrates mainly in the outer part of the cross-section (Fig. 6.1e). A dot-dash line presents the steady state stress distribution for comparison. At rupture the hoop stress at the outermost fibers of the cross-section reaches zero. Then, the inwards propagation of the failure front begins, yielding finally the complete failure of the tube at $t = t_{II}$ (cf. Boyle and Spence, 1983).

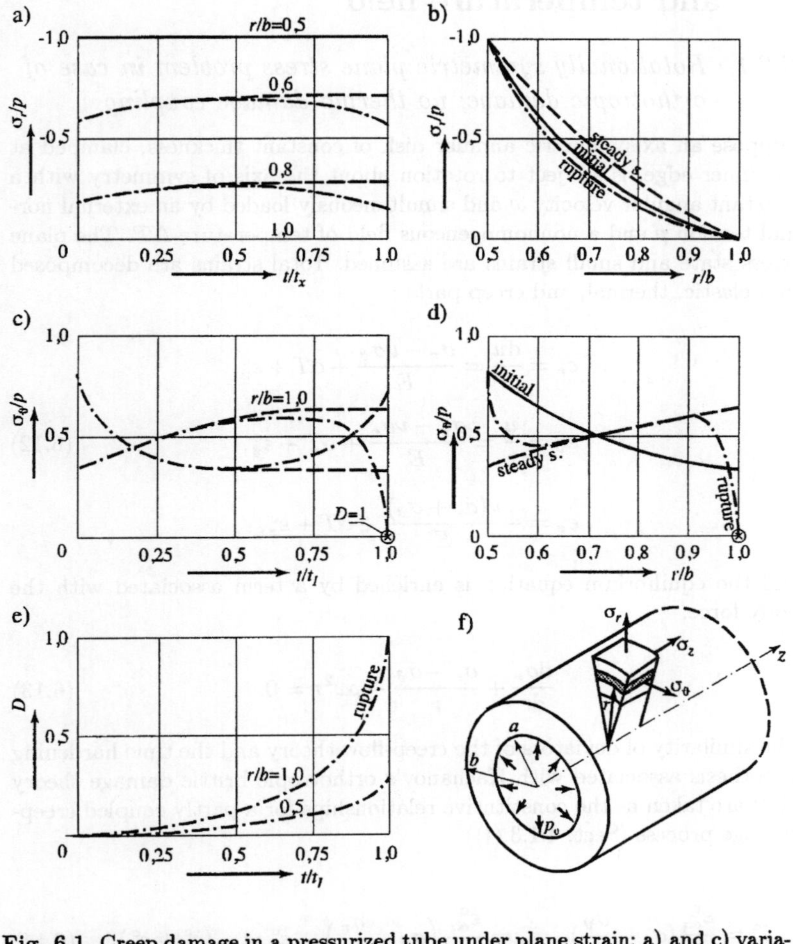

Fig. 6.1. Creep damage in a pressurized tube under plane strain: a) and c) variation of radial and circumferential stresses during failure, b) and d) redistribution of stresses, f) tube visualization; $b/a = 2.0$, $p/\sigma_0 = 0.2$, $m = 5$, $r = 3.5$ (after Skrzypek, 1993)

6.3 Example: Transient creep and creep failure of an annular disk of constant thickness subject to rotation, tension, and temperature field

6.3.1 Rotationally symmetric plane stress problem in case of orthotropic damage; no thermo-damage coupling

Suppose an axisymmetric annular disk of constant thickness, clamped at the inner edge, is subject to rotation about the axis of symmetry with a constant angular velocity ω and simultaneously loaded by an external normal tension p and a nonhomogeneous field of temperature ΔT. The plane stress state and small strains are assumed. Total strains are decomposed into elastic, thermal, and creep parts :

$$
\begin{aligned}
\varepsilon_r &= \frac{\mathrm{d}u}{\mathrm{d}r} = \frac{\sigma_r - \nu\sigma_\vartheta}{E} + \alpha T + \varepsilon_r^{\mathrm{c}}, \\
\varepsilon_\vartheta &= \frac{u}{r} = \frac{\sigma_\vartheta - \nu\sigma_r}{E} + \alpha T + \varepsilon_\vartheta^{\mathrm{c}}, \\
\varepsilon_z &= -\frac{\nu(\sigma_r + \sigma_\vartheta)}{E} + \alpha T + \varepsilon_z^{\mathrm{c}},
\end{aligned}
\tag{6.12}
$$

and the equilibrium equation is enriched by a term associated with the body force:

$$
\frac{\mathrm{d}\sigma_r}{\mathrm{d}r} + \frac{\sigma_r - \sigma_\vartheta}{r} + \rho\omega^2 r = 0. \tag{6.13}
$$

The similarity of deviators of the creep-flow theory and the time hardening hypothesis associated with Kachanov's orthotropic brittle damage theory (4.2) are taken as the constitutive relationships for a partly coupled creep-damage process (Sect. 4.1.3.(I))

$$
\dot\varepsilon_r^{\mathrm{c}} = \frac{\dot\varepsilon_{\mathrm{eq}}^{\mathrm{c}}}{\sigma_{\mathrm{eq}}}\left(\sigma_r - \frac{\sigma_\vartheta}{2}\right), \quad \dot\varepsilon_\vartheta^{\mathrm{c}} = \frac{\dot\varepsilon_{\mathrm{eq}}^{\mathrm{c}}}{\sigma_{\mathrm{eq}}}\left(\sigma_\vartheta - \frac{\sigma_r}{2}\right), \quad \dot\varepsilon_z^{\mathrm{c}} = -\left(\dot\varepsilon_r^{\mathrm{c}} + \dot\varepsilon_\vartheta^{\mathrm{c}}\right); \tag{6.14}
$$

$$
\dot\varepsilon_{\mathrm{eq}}^{\mathrm{c}} = (\widetilde\sigma_{\mathrm{eq}})^m\,\dot{\mathsf{f}}(t), \qquad \dot D_\nu = C\left\langle\frac{\sigma_\nu}{1 - D_\nu}\right\rangle^r. \tag{6.15}
$$

For the plane stress state, the Mises-type equivalent stress, the effective equivalent stress, and the equivalent creep strain rate are defined by the following formulas

$$\sigma_{\mathrm{eq}} \stackrel{\mathrm{def}}{=} \left[\sigma_r^2 + \sigma_\vartheta^2 - \sigma_r\sigma_\vartheta\right]^{1/2},$$

$$\widetilde{\sigma}_{\mathrm{eq}} \stackrel{\mathrm{def}}{=} \left[\left(\frac{\sigma_r}{1-D_r}\right)^2 + \left(\frac{\sigma_\vartheta}{1-D_\vartheta}\right)^2 - \left(\frac{\sigma_r}{1-D_r}\right)\left(\frac{\sigma_\vartheta}{1-D_\vartheta}\right)\right]^{1/2},$$

$$\dot{\varepsilon}^{\mathrm{c}}_{\mathrm{eq}} \stackrel{\mathrm{def}}{=} \frac{2}{\sqrt{3}}\left[\left(\dot{\varepsilon}^{\mathrm{c}}_r\right)^2 + \left(\dot{\varepsilon}^{\mathrm{c}}_\vartheta\right)^2 + \left(\dot{\varepsilon}^{\mathrm{c}}_r\right)\left(\dot{\varepsilon}^{\mathrm{c}}_\vartheta\right)\right]^{1/2}. \tag{6.16}$$

After elimination of stresses from (6.12)–(6.15) we arrive at the governing equations for initial (elastic) and transient (creep) problem:

$$\frac{\mathrm{d}^2u}{\mathrm{d}r^2} + \frac{1}{r}\frac{\mathrm{d}u}{\mathrm{d}r} - \frac{u}{r^2} = -kr + (1+\nu)\alpha\frac{\mathrm{d}T}{\mathrm{d}r} \qquad (t=0),$$

$$\frac{\mathrm{d}^2\dot{u}}{\mathrm{d}r^2} + \frac{1}{r}\frac{\mathrm{d}\dot{u}}{\mathrm{d}r} - \frac{\dot{u}}{r^2} = \frac{\dot{f}}{r} + \frac{\mathrm{d}\dot{g}}{\mathrm{d}r} \qquad (t>0), \tag{6.17}$$

$$\dot{f} = (1-\nu)\left(\dot{\varepsilon}^{\mathrm{c}}_r - \dot{\varepsilon}^{\mathrm{c}}_\vartheta\right), \qquad \dot{g} = \dot{\varepsilon}^{\mathrm{c}}_r + \nu\dot{\varepsilon}^{\mathrm{c}}_\vartheta, \quad k = \frac{1-\nu^2}{E}\rho\omega^2.$$

When Eqs. (6.17) are solved with the elastic solution ($t=0$) assumed as the initial condition for the transient creep we obtain (Skrzypek, 1993):

I. Elastic problem ($t=0$)

$$u^{\mathrm{e}} = \frac{A_0}{2}r + \frac{B_0}{r} - \frac{1-\nu^2}{8E}\rho\omega^2r^3 + (1+\nu)\frac{\alpha}{r}\int_a^r T\xi\mathrm{d}\xi,$$

$$\sigma^{\mathrm{e}}_r = \frac{E}{1-\nu^2}\left[\frac{1+\nu}{2}A_0 - (1-\nu)\frac{B_0}{r^2} - \frac{3+\nu}{8}\rho\omega^2r^2\right] - \frac{\alpha E}{r^2}\int_a^r T\xi\mathrm{d}\xi,$$

$$\sigma^{\mathrm{e}}_\vartheta = \frac{E}{1-\nu^2}\left[\frac{1+\nu}{2}A_0 + (1-\nu)\frac{B_0}{r^2} - \frac{1+3\nu}{8}\rho\omega^2r^2\right] + \frac{\alpha E}{r^2}\int_a^r T\xi\mathrm{d}\xi$$
$$-\alpha ET, \tag{6.18}$$

where, for the linear field of temperature $T(r) = T_a + \Delta T\dfrac{r-a}{b-a}$, the integrals (6.18) may be expressed in the form

$$\int_a^r T\xi\mathrm{d}\xi = \frac{\Delta T(r^3-a^3)}{3(b-a)} + \left(T_a - \frac{\Delta Ta}{b-a}\right)\frac{r^2-a^2}{2} \tag{6.19}$$

and the symbol T_a denotes the temperature at the inner radius of the disk (see Fig. 6.2).

II. Creep damage problem ($t > 0$)

$$\dot{u} = \frac{A_i}{2} r + \frac{B_i}{r} + \left(\frac{\dot{I}_1}{1+\nu} - \frac{\dot{I}_2}{1-\nu} \right) r,$$

$$\dot{\sigma}_r = \frac{E}{1-\nu^2} \left[\frac{1+\nu}{2} A_i - (1-\nu) \frac{B_i}{r^2} + \dot{I}_1 + \dot{I}_2 \right], \tag{6.20}$$

$$\dot{\sigma}_\vartheta = \frac{E}{1-\nu^2} \left[\frac{1+\nu}{2} A_i + (1-\nu) \frac{B_i}{r^2} + \dot{I}_1 - \dot{I}_2 \right] - E \dot{\varepsilon}_\vartheta^{\mathrm{c}},$$

where the integrals $\dot{I}_1$, $\dot{I}_2$ in (6.20) are defined as

$$\dot{I}_1 = \frac{1+\nu}{2} \int_a^r \frac{\dot{f}}{\xi} \mathrm{d}\xi, \quad \dot{I}_2 = -\frac{1-\nu}{2r^2} \int_a^r (2\dot{g} - \dot{f}) \xi \mathrm{d}\xi, \tag{6.21}$$

whereas $\dot{f}$ and $\dot{g}$ are expressed in terms of the creep strain rates (6.17)

$$\dot{\varepsilon}_r^{\mathrm{c}} = \frac{(\widetilde{\sigma}_{\mathrm{eq}})^m}{\sigma_{\mathrm{eq}}} \left(\sigma_r - \frac{\sigma_\vartheta}{2} \right) \dot{\mathsf{f}}(t), \qquad \dot{\varepsilon}_\vartheta^{\mathrm{c}} = \frac{(\widetilde{\sigma}_{\mathrm{eq}})^m}{\sigma_{\mathrm{eq}}} \left(\sigma_\vartheta - \frac{\sigma_r}{2} \right) \dot{\mathsf{f}}(t). \tag{6.22}$$

The constants A_0, B_0 and A_i, B_i are evaluated using the appropriate boundary conditions.

6.3.2 Solution of the creep problem with a propagation of failure front accounted for ($t_I \leq t \leq t_{II}$)

In the case under consideration the orthotropic failure criterion is checked independently for both principal damage components D_r and D_ϑ. If both $D_r < D_{\mathrm{crit}}^{(r)}$ and $D_\vartheta < D_{\mathrm{crit}}^{(\vartheta)}$ the damage growth period occurs when $t < t_I$. If one of D_r, D_ϑ reaches at the material point its critical value at $t = t_I$, say $D_\vartheta = D_{\mathrm{crit}}^{(\vartheta)}$, the corresponding macrocrack is initiated at this point such that the stress normal to the crack (unidirectional macrocrack orientation) is released, $\sigma_\vartheta = 0$. In this way, a partly failed zone may be formed in the structure when the failure front moves as time increases above t_I. At complete failure of the structure, the partly failed zone (with respect to one damage component) may extend all over the volume of the disk to make the structure unable to carry loadings at time t_{II}.

Admitting for the growth of the failure front with respect to the circumferential componentof the damage tensor that $D_\vartheta = 1$ and $D_r < 1$, the circumferential stress in the partly failed zone of the disk must drop to zero, $\sigma_\vartheta = 0$, and hence (6.16) and (6.22) take the new form:

$$\sigma_{\rm eq} = |\sigma_r|, \qquad \widetilde{\sigma}_{\rm eq} = \frac{|\sigma_r|}{1 - D_r}, \qquad (6.23)$$
$$\dot{\varepsilon}_r^{\rm c} = \frac{|\sigma_r|^{m-1}}{(1 - D_\vartheta)^m} \sigma_r \dot{\mathfrak{f}}(t), \quad \dot{\varepsilon}_\vartheta^{\rm c} = \dot{\varepsilon}_z^{\rm c} = -\tfrac{1}{2}\dot{\varepsilon}_r^{\rm c}.$$

Finally, the solution of the creep problem in the partly failed zone reduces to

$$\dot{u}^\dagger = C_j \ln r + D_j + \dot{I}_3, \quad \dot{\sigma}_r^\dagger = E\frac{C_j}{r}, \quad \dot{\sigma}_r^\dagger = 0, \quad \dot{I}_3 = \int_0^c \dot{\varepsilon}_r^{\rm c} {\rm d}\xi, \qquad (6.24)$$

where c denotes the radius of the failure front.

6.3.3 *Results for annular disk*

Boundary conditions for the initial (pre-critical) phase are as follows:

$$\begin{aligned} \sigma_r(a) &= 0 \quad \sigma_r(b) = p \qquad (t = 0), \\ \dot{\sigma}_r(a) &= 0 \quad \dot{\sigma}_r(b) = 0 \quad (0 < t_i < t_I) \end{aligned} \qquad (6.25)$$

enriched also by the continuity conditions for a post-critical phase

$$\dot{\sigma}_r^\dagger(c) = \dot{\sigma}_r(c) \quad \dot{u}^\dagger(c) = \dot{u}(c) \quad (t_I \le t_j \le t_{II}), \qquad (6.26)$$

enabling evaluation of the constants A_0, B_0, A_i, B_i, and C_j, D_j, respectively.

The numerical example is presented for a disk made of ASTM 321 stainless steel, the material data for which are:
$E = 1.77 \times 10^5$ MPa, $\nu = 0.3$, $\rho = 7.9 \times 10^3$ kg/m^3, $a = 0.02$ m, $b = 5a$, $h_0 = 0.002$ m, $\sigma_0 = 118$ MPa, $\alpha = 1.85 \times 10^{-5}$ K^{-1}, $p/\sigma_0 = 0.1$, $\omega = 100$ s^{-1}, $T_a = 773$ K, $\Delta T = 10$ K; whereas the temperature dependent material constants for creep rupture are shown in Table 6.2.

Table 6.2. Temperature dependent creep damage data for ASTM 321 stainless steel

Absolute temperature	C[Pa^{-r}s^{-1}]	r	m
773 K	2.13 $\times 10^{-42}$	3.90	5.60
783 K	1.35 $\times 10^{-41}$	3.82	5.49

During the stage of latent failure ($t < t_I$) the rapid accumulation of damage is observed mainly in the neighborhood of the inner edge where the initiation of failure occurs at t_I. Simultaneously the hoop stress σ_ϑ drops to zero at $r = a$ and, then, the failure profile begins to move outwards as

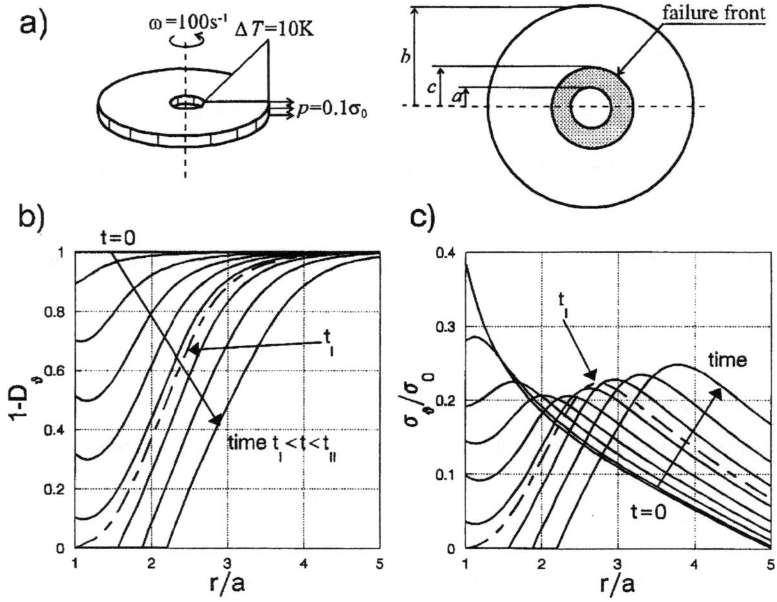

Fig. 6.2. Transient creep and creep failure of an annular disk: a) geometry and loadings, b) damage growth $0 < t < t_I$ and failure propagation $t_I < t < t_{II}$, c) circumferential stress redistribution at failure (after Skrzypek, 1993)

time increases above t_I. At complete failure, the failure profile meets the outer edge of the disk and, eventually, the structure becomes unserviceable.

The situation becomes slightly different when a disk clamped at the inner edge is considered. In this case, mixed boundary conditions must be used

$$\begin{aligned} &u(a) = 0 \quad \sigma_r(b) = 0 \qquad (t = 0)\,, \\ &\dot{u}(a) = 0 \quad \dot{\sigma}_r(b) = 0 \quad (0 < t_i < t_I) \end{aligned} \tag{6.27}$$

and the continuity conditions at the partly failed–nonfailed interface (6.26) holds. When the inverse temperature gradient is applied to the disk $\Delta T < 0$ (cf. Skrzypek, 1993), the damage process initiates at the outer edge with respect to the circumferential component D_ϑ and, as time increases, the circumferential damage accumulation runs faster than the radial one D_r. If the first macrocrack is initiated at $t = t_I$ in the outer fibers of the cross section $D_\vartheta\,(b) = 1$, the hoop stress in these fibers must drop to zero, $\sigma_\vartheta\,(b) = 0$. Then the failed zone begins to spread inwards and, eventually, it may occupy the whole disk at t_{II}, unless the prior accumulation of the radial damage at the inner edge of the disk causes disk separation (decohesion) from the shaft $D_r\,(a) = 1$ after the radial stress in this fiber has been released completely, $\sigma_r\,(a) = 0$.

7

Axisymmetric heat transfer problems in damaged cylinders and disks

7.1 Basic mechanical state equations of rotationally symmetric deformation under unsteady temperature field

Let us consider an axisymmetric problem in the displacement formulation which may describe plane stress (a disk of constant thickness) as well as the plane strain state (a cylinder). Applying the geometrically linear theory of small displacements and decomposing the total strains into elastic, creep, and thermal parts,

$$\boldsymbol{\varepsilon} = \boldsymbol{\varepsilon}^{\mathrm{e}} + \boldsymbol{\varepsilon}^{\mathrm{c}} + \boldsymbol{\varepsilon}^{\mathrm{th}}, \tag{7.1}$$

the problem may be expressed by the system of displacement (rate) equations as follows:

$$\begin{aligned} &\frac{\mathrm{d}^2 u}{\mathrm{d}r^2} + \frac{1}{r}\frac{\mathrm{d}u}{\mathrm{d}r} - \frac{u}{r^2} = h(1+\nu)\alpha\frac{\mathrm{d}T}{\mathrm{d}r} && (t=0), \\ &\frac{\mathrm{d}^2 \dot{u}}{\mathrm{d}r^2} + \frac{1}{r}\frac{\mathrm{d}\dot{u}}{\mathrm{d}r} - \frac{\dot{u}}{r^2} = \frac{\dot{f}}{r} + \frac{\mathrm{d}\dot{g}}{\mathrm{d}r} + h(1+\nu)\alpha\frac{\mathrm{d}\dot{T}}{\mathrm{d}r} && (t>0). \end{aligned} \tag{7.2}$$

The solution for displacements, stresses, and their rates takes the elementary form:

I. Elastic problem $(t=0)$

$$\begin{aligned} u &= c_1\frac{r}{2} + \frac{c_2}{2} + h(1+\nu)I_0, \\ \sigma_r &= \frac{E}{1+\nu}\left(k\frac{c_1}{2} - \frac{c_2}{r^2}\right) - h\frac{E}{r}I_0, \\ \sigma_\vartheta &= \frac{E}{1+\nu}\left(k\frac{c_1}{2} + \frac{c_2}{r^2}\right) + h\frac{E}{r}\left(I_0 - \alpha r T\right). \end{aligned} \tag{7.3}$$

II. Creep problem ($t > 0$)

$$
\begin{aligned}
\dot{u} &= c_3\frac{r}{2} + \frac{c_4}{2} + \dot{I}_1 + \dot{I}_2 + h(1+\nu)\dot{I}_0, \\
\dot{\sigma}_r &= \frac{E}{1+\nu}\left(k\frac{c_3}{2} - \frac{c_4}{r^2} + \frac{k\dot{I}_1 - \dot{I}_2}{r}\right) - h\frac{E}{r}\dot{I}_0, \\
\dot{\sigma}_\vartheta &= \frac{E}{1+\nu}\left(k\frac{c_3}{2} + \frac{c_4}{r^2} + \frac{k\dot{I}_1 + \dot{I}_2}{r} - \dot{g} + \dot{\varepsilon}_r^{\mathrm{c}} - \dot{\varepsilon}_\vartheta^{\mathrm{c}}\right) \\
&\quad + h\frac{E}{r}\dot{I}_0 - hE\alpha\dot{T}.
\end{aligned}
\tag{7.4}
$$

In the case of plane strain state and creep incompressibility,

$$
\dot{\varepsilon}_z = \frac{\left[\dot{\sigma}_z - \nu\left(\dot{\sigma}_r + \dot{\sigma}_\vartheta\right)\right]}{E} + \alpha\dot{T} - \dot{\varepsilon}_r^{\mathrm{c}} - \dot{\varepsilon}_\vartheta^{\mathrm{c}} = 0,
\tag{7.5}
$$

the appropriate axial stress and its rate ought to be taken into account:

$$
\begin{aligned}
\sigma_z &= \frac{E}{1+\nu}\nu k c_1 - hE\alpha T, && (t=0), \\
\dot{\sigma}_z &= \frac{E}{1+\nu}\left(\nu k c_3 + 2\nu k\frac{\dot{I}_1}{r} + h\dot{\varepsilon}_r^{\mathrm{c}} + \dot{\varepsilon}_\vartheta^{\mathrm{c}}\right) - hE\alpha\dot{T} && (t>0).
\end{aligned}
\tag{7.6}
$$

The auxiliary symbols in (7.3)–(7.6) are defined as follows (cf. Table 7.1):

Table 7.1. Auxiliary functions for basic rotationally symmetric deformation under unsteady temperature field (after Ganczarski and Skrzypek, 1995)

Quantity	Plane stress	Plane strain
$\dot{f}$	$(1-\nu)\left(\dot{\varepsilon}_r^{\mathrm{c}} - \dot{\varepsilon}_\vartheta^{\mathrm{c}}\right)$	$\frac{1-2\nu}{1-\nu}\left(\dot{\varepsilon}_r^{\mathrm{c}} - \dot{\varepsilon}_\vartheta^{\mathrm{c}}\right)$
$\dot{g}$	$\dot{\varepsilon}_r^{\mathrm{c}} + \nu\dot{\varepsilon}_\vartheta^{\mathrm{c}}$	$\frac{1-2\nu}{1-\nu}\dot{\varepsilon}_r^{\mathrm{c}}$
h	1	$\frac{1}{1-\nu}$
k	$\frac{1+\nu}{1-\nu}$	$\frac{1}{1-2\nu}$

$$
\begin{aligned}
I_0 &= \frac{\alpha}{r}\int_0^r T\xi \mathrm{d}\xi, \quad \dot{I}_0 = \frac{\alpha}{r}\int_0^r \dot{T}\xi \mathrm{d}\xi, \\
\dot{I}_1 &= \frac{r}{2}\int_0^r \frac{\dot{f}}{\xi}\mathrm{d}\xi, \quad \dot{I}_2 = \frac{1}{2r}\int_0^r (2\dot{g} - \dot{f})\xi \mathrm{d}\xi.
\end{aligned}
\tag{7.7}
$$

In the case of an axisymmetric problem, when the effect of isotropic deterioration on elastic moduli is to be analyzed, another formulation, based on the stress function, is more convenient:

$$\left.\begin{aligned}&\frac{\mathrm{d}^2\phi}{\mathrm{d}r^2}+\frac{1}{r}\frac{\mathrm{d}\phi}{\mathrm{d}r}-\frac{\phi}{r^2}=-\frac{E}{1-\nu}\alpha\frac{\mathrm{d}T}{\mathrm{d}r} \qquad (t=0)\,,\\&\frac{\mathrm{d}^2\dot{\phi}}{\mathrm{d}r^2}+\frac{1}{r}\frac{\mathrm{d}\dot{\phi}}{\mathrm{d}r}-\frac{\dot{\phi}}{r^2}+\left[\widetilde{E}\frac{\mathrm{d}}{\mathrm{d}r}\left(\frac{1}{\widetilde{E}}\right)\right.\\&\times\left.\left(\frac{\mathrm{d}\phi}{\mathrm{d}r}-\frac{\nu}{1-\nu}\frac{\phi}{r}\right)\right]^{\cdot}=-\left(\frac{\widetilde{E}}{1-\nu}\alpha\frac{\mathrm{d}T}{\mathrm{d}r}\right)^{\cdot}\\&+\left\{\frac{\widetilde{E}}{1-\nu^2}\left[\nu\frac{\mathrm{d}\varepsilon_r^{\mathrm{c}}}{\mathrm{d}r}-(1-\nu)\frac{\mathrm{d}\varepsilon_\vartheta^{\mathrm{c}}}{\mathrm{d}r}+\frac{\varepsilon_r^{\mathrm{c}}-\varepsilon_\vartheta^{\mathrm{c}}}{r}\right]\right\}^{\cdot}\end{aligned}\right\}(t>0)\,, \tag{7.8}$$

where, on the basis of the principle of strain equivalence, the effective Young's modulus $\widetilde{E}$ is expressed in terms of the continuity parameter ψ by the formula $\widetilde{E}=E\psi$. Note that in case of energy equivalence both Young's modulus and Poisson's ratio change with damage, and the microcrack growth influences both stress and strain, which is more realistic when compared to strain equivalence where drop of local stiffness results in local stress decrease only.

The stress components and their rates are now defined as follows:

$$\begin{aligned}&\sigma_r=\frac{\phi}{r},\quad \sigma_\vartheta=\frac{\mathrm{d}\phi}{\mathrm{d}r},\quad \sigma_z=\nu\left(\sigma_r+\sigma_\vartheta\right)-E\alpha T \qquad (t=0)\,,\\&\dot{\sigma}_r=\frac{\dot{\phi}}{r},\quad \dot{\sigma}_\vartheta=\frac{\mathrm{d}\dot{\phi}}{\mathrm{d}r},\quad \dot{\sigma}_z=\nu\left(\dot{\sigma}_r+\dot{\sigma}_\vartheta\right)-E\alpha\left(\psi T\right)^{\cdot}\\&\qquad +E\left[\psi\left(\varepsilon_r^{\mathrm{c}}+\varepsilon_\vartheta^{\mathrm{c}}\right)\right]^{\cdot} \qquad (t>0)\,.\end{aligned} \tag{7.9}$$

7.2 Constitutive equations

Two concepts of coupling between constitutive equations of creep and damage are formulated: fully coupled when damage orthotropy effect on creep flow is accounted for, or partly coupled when damage orthotropy effect on creep flow is disregarded.

7.2.1 Fully or partly creep-damage coupled approaches

I. Damage orthotropy effect on creep flow accounted for – fully coupled creepdamage approach

In general, when loadings are nonproportional, the orthotropic damage mechanism causes the creep process to be orthotropic as well (damage induced creep orthotropy). Hence, the fully coupled creep-damage approach, where the effective stress components are used and the time hardening hypothesis governs the creep strain-rate intensity (cf. Ganczarski and Skrzypek, 1994a), yields the following equations (cf. Sect. 4.1.3):

$$\dot{\varepsilon}^{\mathrm{c}}_{kl} = \frac{3}{2}\frac{\dot{\varepsilon}^{\mathrm{c}}_{\mathrm{eq}}}{\widetilde{\sigma}_{\mathrm{eq}}}\widetilde{s}_{kl}, \quad \dot{\varepsilon}^{\mathrm{c}}_{\mathrm{eq}} = (\widetilde{\sigma}_{\mathrm{eq}})^{m(T)}\dot{\mathrm{f}}(t), \quad \widetilde{s}_{kl} = \widetilde{\sigma}_{kl} - \frac{1}{3}\widetilde{\sigma}_{ii}\delta_{kl}, \tag{7.10}$$

where the effective stress $\widetilde{\sigma}_{kl}$ results from the appropriate equivalence principle and the equivalent stress σ_{eq}, effective equivalent stress $\widetilde{\sigma}_{\mathrm{eq}}$, and equivalent creep strain rate $\dot{\varepsilon}^{\mathrm{c}}_{\mathrm{eq}}$ are (cf. Ganczarski and Skrzypek, 1993):

$$\sigma_{\mathrm{eq}} = \sqrt{\frac{3}{2}s_{kl}s_{kl}}, \qquad \widetilde{\sigma}_{\mathrm{eq}} = \sqrt{\frac{3}{2}\widetilde{s}_{kl}\widetilde{s}_{kl}}, \qquad \dot{\varepsilon}^{\mathrm{c}}_{\mathrm{eq}} = \sqrt{\frac{2}{3}\dot{\varepsilon}^{\mathrm{c}}_{kl}\dot{\varepsilon}^{\mathrm{c}}_{kl}}. \tag{7.11}$$

The orthotropic damage-growth rule is applied to describe damage accumulation, Kachanov (1986):

$$\dot{D}_{ii} = C_k(T)\,\langle\widetilde{\sigma}_{ii}\rangle^{r_k(T)}, \qquad \widetilde{\sigma}_{ii} = \frac{\sigma_{ii}}{1 - D_{ii}}. \tag{7.12}$$

Symbols $\langle\rangle$ denote MacAuley brackets.

II. Damage orthotropy effect on creep flow disregarded – partly coupled creep-damage approach

In a simplified case the isotropic flow rule, instead of the orthotropic one, and the orthotropic damage growth rule are applied:

$$\dot{\varepsilon}^{\mathrm{c}}_{kl} = \frac{3}{2}\frac{\dot{\varepsilon}^{\mathrm{c}}_{\mathrm{eq}}}{\sigma_{\mathrm{eq}}}s_{kl}, \quad \dot{\varepsilon}^{\mathrm{c}}_{\mathrm{eq}} = (\widetilde{\sigma}_{\mathrm{eq}})^{m(T)}\dot{\mathrm{f}}(t), \quad s_{kl} = \sigma_{kl} - \frac{1}{3}\sigma_{ii}\delta_{kl}, \tag{7.13}$$

although, in general sense, such a formulation is inconsistent and may result in certain discrepancies when compared to the exact one.

7.2.2 Axisymmetric plane stress fully or partly coupled creep-damage problems

In a particular case of the axisymmetric plane stress, when terms associated with the z direction are neglected and the incompressibility of creep is assumed, we find:

$$
\left\{
\begin{aligned}
\dot{\varepsilon}_r^{\mathrm{c}} &= (\widetilde{\sigma}_{\mathrm{eq}})^{m(T)-1}\left[\frac{\sigma_r}{1-D_r} - \frac{\sigma_\vartheta}{2(1-D_\vartheta)}\right]\dot{\mathsf{f}}(t),\\
\dot{\varepsilon}_\vartheta^{\mathrm{c}} &= (\widetilde{\sigma}_{\mathrm{eq}})^{m(T)-1}\left[\frac{\sigma_\vartheta}{1-D_\vartheta} - \frac{\sigma_r}{2(1-D_r)}\right]\dot{\mathsf{f}}(t), \qquad \dot{\varepsilon}_z^{\mathrm{c}} = -\left(\dot{\varepsilon}_r^{\mathrm{c}} + \dot{\varepsilon}_\vartheta^{\mathrm{c}}\right),
\end{aligned}
\right.
\tag{7.14}
$$

or

$$
\left\{
\begin{aligned}
\dot{\varepsilon}_r^{\mathrm{c}} &= \frac{(\widetilde{\sigma}_{\mathrm{eq}})^{m(T)}}{\sigma_{\mathrm{eq}}}\left(\sigma_r - \frac{\sigma_\vartheta}{2}\right)\dot{\mathsf{f}}(t)\\
\dot{\varepsilon}_\vartheta^{\mathrm{c}} &= \frac{(\widetilde{\sigma}_{\mathrm{eq}})^{m(T)}}{\sigma_{\mathrm{eq}}}\left(\sigma_\vartheta - \frac{\sigma_r}{2}\right)\dot{\mathsf{f}}(t), \qquad \dot{\varepsilon}_z^{\mathrm{c}} = -\left(\dot{\varepsilon}_r^{\mathrm{c}} + \dot{\varepsilon}_\vartheta^{\mathrm{c}}\right),
\end{aligned}
\right.
\tag{7.15}
$$

in cases of the fully coupled (7.14) or partly coupled (7.15) approach, respectively, where

$$
\begin{aligned}
\sigma_{\mathrm{eq}} &= \sqrt{\sigma_r^2 + \sigma_\vartheta^2 - \sigma_r\sigma_\vartheta},\\
\widetilde{\sigma}_{\mathrm{eq}} &= \sqrt{\left(\frac{\sigma_r}{1-D_r}\right)^2 + \left(\frac{\sigma_\vartheta}{1-D_\vartheta}\right)^2 - \frac{\sigma_r\sigma_\vartheta}{(1-D_r)(1-D_\vartheta)}}.
\end{aligned}
\tag{7.16}
$$

In above formulations, the orthotropic or the isotropic creep law has been coupled with the orthotropic damage law. In general, the latter is described by different material functions $C_k(T)$, $r_k(T)$ and independently cumulating principal components of the continuity tensor ψ_k. In what follows, we consider the simplified case of material isotropy $C_r = C_\vartheta = C_z = C$ and $r_r = r_\vartheta = r_z = r$, but allow for the independent evolution of microcracks in each of principal directions ψ_r, ψ_ϑ, ψ_z. Another problem arises when the temperature dependence of creep rupture functions $m(T)$, $C(T)$, $r(T)$, which introduces material nonhomogeneity in the inelastic range, is considered (cf. Ganczarski and Skrzypek, 1995). The quantities $m(T)$, $r(T)$ must be linearly interpolated, whereas function $C(T)$, which strongly depends on a local temperature, must be logarithmically interpolated.

7.2.3 Axisymmetric plane strain coupled creep-isotropic damage problem

Relations (7.10)–(7.13) take a simple form when the creep incompressibility, the plane strain conditions, and the scalar formulation of the isotropic damage law $\mathbf{D} = \mathbf{1}D$, are assumed:

$$\dot{D} = C(T)\left\langle \frac{\sigma_I}{1-D} \right\rangle^{r(T)}, \qquad \widetilde{\sigma}_{\text{eq}} = \frac{\sigma_{\text{eq}}}{1-D},$$

$$\dot{\varepsilon}_r^{\text{c}} = \frac{\widetilde{\sigma}_{\text{eq}}^{m(T)-1}}{(1-D)^{m(T)}} \left(\sigma_r - \frac{\sigma_\vartheta + \sigma_z}{2} \right) \dot{\mathsf{f}}(t), \tag{7.17}$$

$$\dot{\varepsilon}_\vartheta^{\text{c}} = \frac{\widetilde{\sigma}_{\text{eq}}^{m(T)-1}}{(1-D)^{m(T)}} \left(\sigma_\vartheta - \frac{\sigma_r + \sigma_z}{2} \right) \dot{\mathsf{f}}(t),$$

where σ_I denotes the maximum principal stress, which refers to the Galileo hypothesis.

7.3 Thermo-mechanical rotationally symmetric boundary problems

7.3.1 Axisymmetric heat flow in cylinders or disks under thermo-creep-damage coupling conditions

Let us rewrite the heat transfer equations in damaged solids in the case of axisymmetric heat flow. To this aim the general equations (5.3), (5.7), (5.9) described in Sect. 5.1 using Cartesian coordinates must be transformed to cylindrical coordinates as follows:

Model A

$$\frac{1}{r}\frac{\mathrm{d}}{\mathrm{d}r}\left\{ r \left[\lambda_0 \left(1 - D(r,t)\right) \frac{\mathrm{d}T(r,t)}{\mathrm{d}r} \right] \right\} + \dot{q}_v = c_v \varrho \dot{T}, \tag{7.18}$$

Model B

$$\frac{1}{r}\frac{\mathrm{d}}{\mathrm{d}r}\left\{ r \left[\lambda_0 \left(1 - D(r,t)\right) \frac{\mathrm{d}T(r,t)}{\mathrm{d}r} - \sigma\epsilon_0 D(r,t)\, T^4(r,t) \right] \right\} \\ + \dot{q}_v = c_v \varrho \dot{T}, \tag{7.19}$$

Model C

$$\frac{1}{r}\frac{\mathrm{d}}{\mathrm{d}r}\left\{ r \left[\lambda_{\text{eq}}(r,t,T) \frac{\mathrm{d}T(r,t)}{\mathrm{d}r} \right] \right\} + \dot{q}_v = c_v \varrho \dot{T},$$

$$\lambda_{\text{eq}}(r,t,T) = \widetilde{\lambda}(r,t) + \sigma\epsilon_0 \left[4D + \frac{\mathrm{d}D/\mathrm{d}r}{\mathrm{d}T/\mathrm{d}r} T \right] T^3 \mathrm{d}r, \tag{7.20}$$

$$\widetilde{\lambda}(r,t) = \lambda_0 \left(1 - D(r,t)\right) \quad \text{or} \quad \widetilde{\lambda}(r,t) = \lambda_0 \left(1 - D(r,t)\right)^{1/2}.$$

Extension of (7.19) and (7.20) to rotationally symmetric disks of a variable thickness $h\,(r)$ yields:

Model B

$$\frac{1}{r}\frac{\mathrm{d}}{\mathrm{d}r}\left\{r\left[\lambda_0\psi\,(r,t)\,\frac{\mathrm{d}T\,(r,t)}{\mathrm{d}r}-\sigma\epsilon_0 D\,(r,t)\,T^4\,(r,t)\right]\right\}$$
$$+\frac{1}{h}\frac{\mathrm{d}h}{\mathrm{d}r}\left[\lambda_0\psi\,(r,t)\,\frac{\mathrm{d}T\,(r,t)}{\mathrm{d}r}-\sigma\epsilon_0 D\,(r,t)\,T^4\,(r,t)\right]+\dot{q}_v=c_v\varrho\dot{T}, \tag{7.21}$$

Model C

$$\frac{1}{r}\frac{\mathrm{d}}{\mathrm{d}r}\left(r\lambda_{\mathrm{eq}}\frac{\mathrm{d}T\,(r,t)}{\mathrm{d}r}\right)+\frac{1}{h}\frac{\mathrm{d}h}{\mathrm{d}r}\lambda_{\mathrm{eq}}\frac{\mathrm{d}T\,(r,t)}{\mathrm{d}r}+\dot{q}_v=c_v\varrho\dot{T},$$
$$\lambda_{\mathrm{eq}}\,(r,t,T)=\lambda_0\,(1-D\,(r,t))+\sigma\epsilon_0\left[4D+\frac{\mathrm{d}D/\mathrm{d}r}{\mathrm{d}T/\mathrm{d}r}T\right]T^3\mathrm{d}r, \tag{7.22}$$

7.3.2 Cylinder subject to a nonstationary radial temperature field under plane strain conditions

Let us consider a cylinder of inner and outer radii a and b, respectively, under the plane strain condition, subject to a nonstationary radial temperature gradient (Fig. 7.1).

Stresses and their rates satisfy (7.3) and (7.6) in case of the displacement formulation or (7.8) and (7.9) in case of the stress function formulation, respectively, as well as the homogeneous mechanical state boundary conditions:

$$\begin{array}{lll}\sigma_r(a)=0 & \sigma_r(b)=0 & (t=0)\,,\\ \dot{\sigma}_r(a)=0 & \dot{\sigma}_r(b)=0 & (t>0)\end{array} \tag{7.23}$$

or

$$\begin{array}{lll}\phi(a)=0 & \phi(b)=0 & (t=0)\,,\\ \dot{\phi}(a)=0 & \dot{\phi}(b)=0 & (t>0)\,.\end{array} \tag{7.24}$$

The temperature at both inner and outer edges of the cylinder is constant through the process; hence, the following boundary conditions for the temperature field have to be satisfied:

$$T(a)=T_a,\qquad T(b)=T_b. \tag{7.25}$$

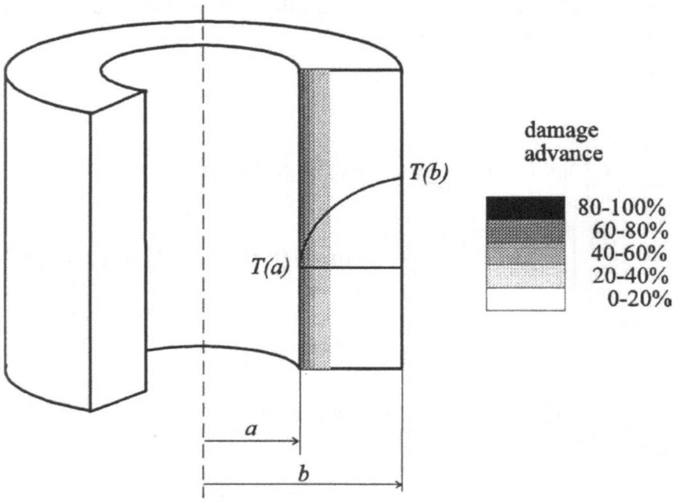

Fig. 7.1. Long cylindrical thick-walled tube subject to a nonstationary radial temperature gradient under plane strain conditions (after Ganczarski and Skrzypek, 1998b)

7.3.3 Thin circular disk subject to constant temperature at the edge and cooled through the faces under plane stress condition

Suppose a disk of constant thickness $h(r) = \text{const}$, which is thin enough to assume the plane state of stress, is considered. The disk is subject to constant temperature at the edge T_a and cooled through the faces by a fluid stream of temperature T_∞, as in a turbine rotor (cf. Fig. 7.2).

The mechanical state fulfills (7.3) and (7.4) and the homogeneous boundary conditions

$$\begin{array}{lll} u(0) = 0, & \sigma_r(R) = 0 & (t = 0), \\ \dot{u}(0) = 0, & \dot{\sigma}_r(R) = 0 & (t > 0), \end{array} \tag{7.26}$$

whereas the appropriate boundary conditions of the thermal state are

$$\begin{array}{lll} \dfrac{\mathrm{d}T(0)}{\mathrm{d}r} = 0, & T(R) = T_a & (t = 0), \\[2ex] \dfrac{\mathrm{d}\dot{T}(0)}{\mathrm{d}r} = 0, & \dot{T}(R) = 0 & (t > 0). \end{array} \tag{7.27}$$

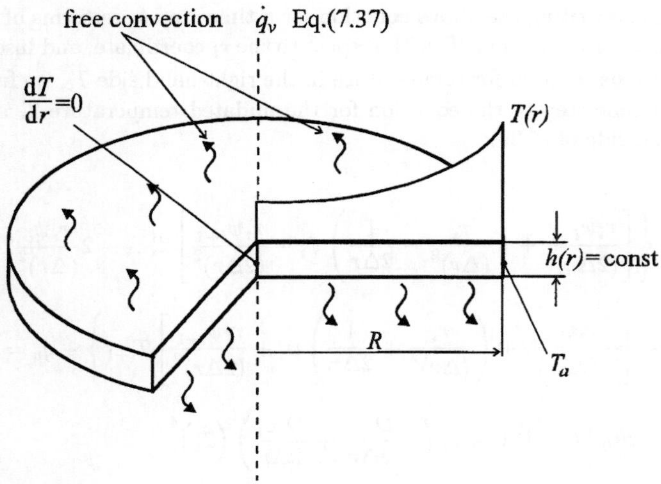

Fig. 7.2. Circular disk subject to constant temperature at the outer edge and cooled through the faces under the plane stress condition (after Ganczarski and Skrzypek, 1995)

7.3.4 Numerical procedure for the initial-boundary problem

To solve the coupled initial-boundary problem, we discretize time by inserting N time intervals Δt_k, where $t_0 = 0$, $\Delta t_k = t_k - t_{k-1}$ and $t_N = t_{\mathrm{R}}$ (rupture). Hence, the initial-boundary problem is reduced to a sequence of quasistatic boundary-value problems, the solution of which determines unknown functions at a given time t_k, e.g., $T(\mathbf{x}, t_k) = T^k(\mathbf{x})$, $\mathbf{D}(\mathbf{x}, t_k) = \mathbf{D}^k(\mathbf{x})$, etc. At each time step the Runge–Kutta II method is applied to yield updated functions T^{k+1}, $\mathbf{D}^{k+1}$, etc. To account for primary and tertiary creep regimes, a dynamically controlled time step Δt_k is required, the length of which is defined by the bounded maximum damage increment:

$$\Delta D^{\mathrm{lower}} \leq \max_{(i,j;\mathbf{x})} \left\{ \left[\dot{D}_{ij}^{k}(\mathbf{x}) - \dot{D}_{ij}^{k-1}(\mathbf{x}) \right] \Delta t_k \right\} \leq \Delta D^{\mathrm{upper}}. \tag{7.28}$$

Additionally, when a generally nonlinear heat transfer problem with respect to radial coordinate r is solved for quasistatic temperature changes ($\dot{T} = 0$), we have (Model B):

$$\frac{\mathrm{d}}{\mathrm{d}r} \left[r\lambda_0 \psi(r,t) \frac{\mathrm{d}T(r,t)}{\mathrm{d}r} \right] + \dot{q}_v = \frac{\mathrm{d}}{\mathrm{d}r} \left[r\sigma\epsilon_0 D(r,t)\, T^4(r,t) \right]. \tag{7.29}$$

The radiation-type term plays role of an additional nonhomogeneity if $D > 0$. Discretizing also radial coordinate r_i, by inserting equal mesh $\Delta r =$

$r_i - r_{i-1}$, rewriting the above equation for a time step t_k in terms of finite differences of ψ_i, D_i, and T_i with respect to the r_i coordinate, and inserting the previous solution for temperature in the right-hand side $\widetilde{T}_i$, we furnish at each time step t_k the equation for the updated temperature T_i at the left-hand side of (7.29):

$$\lambda_0 \left\{ \left[\frac{r_i \psi_{i-1}}{(2\Delta r)^2} + \left(\frac{r_i}{(\Delta r)^2} - \frac{1}{2\Delta r} \right) \psi_i - \frac{r_i \psi_{i+1}}{(2\Delta r)^2} \right] T_{i-1} - 2 \frac{r_i \psi_i}{(\Delta r)^2} T_i \right.$$

$$\left. + \left[- \frac{r_i \psi_{i-1}}{(2\Delta r)^2} + \left(\frac{r_i}{(\Delta r)^2} + \frac{1}{2\Delta r} \right) \psi_i + \frac{r_i \psi_{i+1}}{(2\Delta r)^2} \right] T_{i+1} \right\} + \dot{q}_v =$$

$$= \sigma \epsilon_0 \left[D_i \left(\widetilde{T}_i \right)^4 + r_i \left(- \frac{D_{i-1}}{2\Delta r} + \frac{D_{i+1}}{2\Delta r} \right) \left(\widetilde{T}_i \right)^4 \right.$$

$$\left. + 4 r_i D_i \left(\widetilde{T}_i \right)^3 \left(- \frac{\widetilde{T}_{i-1}}{2\Delta r} + \frac{\widetilde{T}_{i+1}}{2\Delta r} \right) \right] . \tag{7.30}$$

When (7.30) is solved, the new temperature distribution T_i is provided, considered next as a right-hand side nonhomogeneity for a subsequent temperature subiteration. The procedure is repeated until the calculated function T_i differs from $\widetilde{T}_i$ with a given accuracy. Equation (7.30) is solved by the FDM with the radial damage (continuity) component $D = D_r$, $\psi = \psi_r$ until the dominant damage reaches the critical level, $\max(D_r, D_\vartheta, D_z) = D_{\text{crit}}$.

7.3.5 Material data

Numerical examples deal with cylinders and disks made of the following materials:

i) Carbon steel (rolled, 0.40 Mn, 0.25 Si, 0.12 C, normalized, annealed at 850°C) the material data of which are (cf. Holman, 1990):

$E = 150$ GPa, $\sigma_{0.2} = 120$ MPa, $\nu = 0.3$, $\alpha = 1.4 \times 10^{-5}$ K^{-1}, $\lambda_0 = 43$ Wm^{-1}K^{-1}, $\beta = 14$ Wm^{-2}K^{-1}, $\sigma = 5.669 \times 10^{-8}$ Wm^{-1}K^{-4}, $\epsilon_0 = 0.60$, $a/b = 0.5$, $R = 1.0$ m, $T_\infty = 525$°C;

temperature dependent parameters are listed in Table 7.2.

ii) ASTM 321 stainless steel (rolled, 18 Cr, 0.45 Si, 0.4 M, 0.1 C, Ti/Nb stabilized, austenitic, annealed at 1070°C, air cooled) of the following data:

$E = 150$ GPa, $\sigma_{0.2} = 120$ MPa, $\alpha = 1.85 \times 10^{-5}$ K^{-1}, $\lambda_0 = 23$ Wm^{-1}K^{-1}, $\epsilon_0 = 0.50$;

temperature dependent parameters are listed in Table 7.3.

Table 7.2. Temperature affected material data for carbon steel (after Odqvist, 1966)

T (°C)	m	r	$\sigma^5_{C_B}$ (MPa)	C (Pa^{-r} s^{-1})
500	3.3	3.5	80	1.34×10^{-37}
550	2.5	2.3	40	2.75×10^{-27}
600	–	1.0	27	5.14×10^{-17}

Table 7.3. Temperature affected material data for ASTM 321 stainless steel (after Odqvist, 1966)

T (°C)	m	r	$\sigma^5_{C_B}$ (MPa)	C (Pa^{-r} s^{-1})
600	4.5	3.1	100	1.07×10^{-34}
650	4.0	2.8	60	1.21×10^{-31}
700	3.5	2.5	38	8.91×10^{-29}

7.4 Example: Thermo-damage coupling in a cylinder

7.4.1 *Thermo-damage coupling in a cylinder disregarded (stationary temperature field)*

Let us consider as a sample solution the case of a cylinder under stationary temperature gradient ΔT when the effect of damage accumulation on heat transfer is disregarded.

The classical Fourier heat transfer equation takes the form:

$$\frac{1}{r}\frac{\mathrm{d}}{\mathrm{d}r}\left(r\lambda_0\frac{\mathrm{d}T}{\mathrm{d}r}\right) = 0, \tag{7.31}$$

the solution of which is

$$T(r,t) = T(r,0) = \frac{\Delta T}{\ln(a/b)}\ln\frac{r}{a} + T_a. \tag{7.32}$$

Damage localization is observed along the circular line near the inner edge of the cylinder (Fig. 7.3a). The temperature field is stationary because of the absence of thermo-damage coupling (Fig. 7.3b), whereas the hoop stress relaxes with time to failure but not fast enough to overtake the damage accumulation in a cylinder (Fig. 7.3c). Consequently, the finite time of failure initiation $t^{(s)}_{I_{\mathrm{carbon}}}$ is reached.

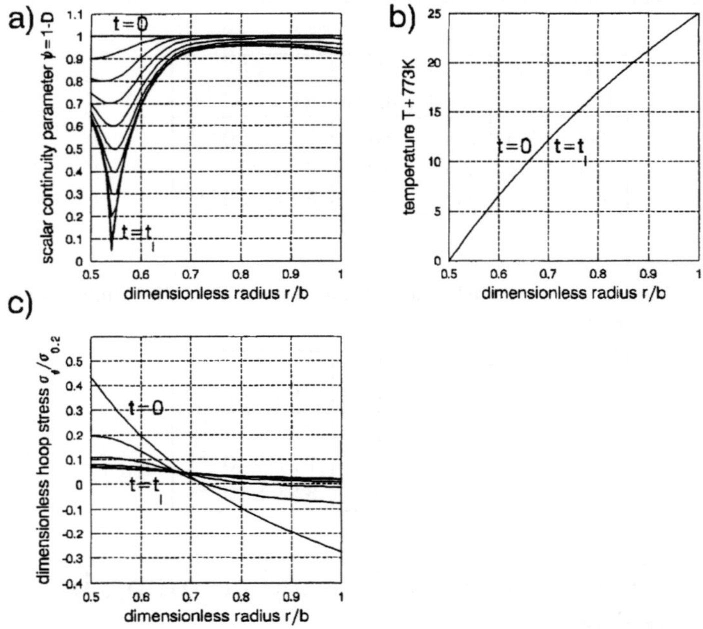

Fig. 7.3. A tube subject to creep under stationary temperature field (effect of thermo-damage coupling disregarded): a) scalar continuity parameter evolution, b) stationary temperature field, c) hoop stress redistribution (after Skrzypek and Ganczarski, 1998b)

7.4.2 Model A: Pure heat conductivity case

Consider a cylinder subject to a transient temperature field associated with constant temperature at both edges $\Delta T = T_b - T_a, T_a < T_b$, and the simplified equation of heat transfer (7.18) where the radiation through the damaged part of a cross section and the inner heat source are disregarded ($\epsilon_0 = 0$, $\dot{q}_v = 0$), whereas the temperature field changes in a quasistatic way ($\dot{T} = 0$):

$$\frac{1}{r}\frac{\mathrm{d}}{\mathrm{d}r}\left(r\lambda_0\Psi\frac{\mathrm{d}T}{\mathrm{d}r}\right) = 0. \tag{7.33}$$

Like the previous case, the damage accumulation also concentrates along the circular line near the inner edge (Fig. 7.4a). Consequently, as time increases, the conductivity across the damaged surface asymptotically approaches zero and, as a result, temperature and hoop stress jumps are formed (Fig. 7.4b,c).

The accompanying stress relaxation is not fast enough to prevent the structure from collapse. The corresponding lifetime $t_{I_{\mathrm{carbon}}}^{(\epsilon_0=0)} = 85\% t_{I_{\mathrm{carbon}}}^{(s)}$ is finite, and it is approximately 15% shorter compared to the case when thermo-damage coupling is disregarded.

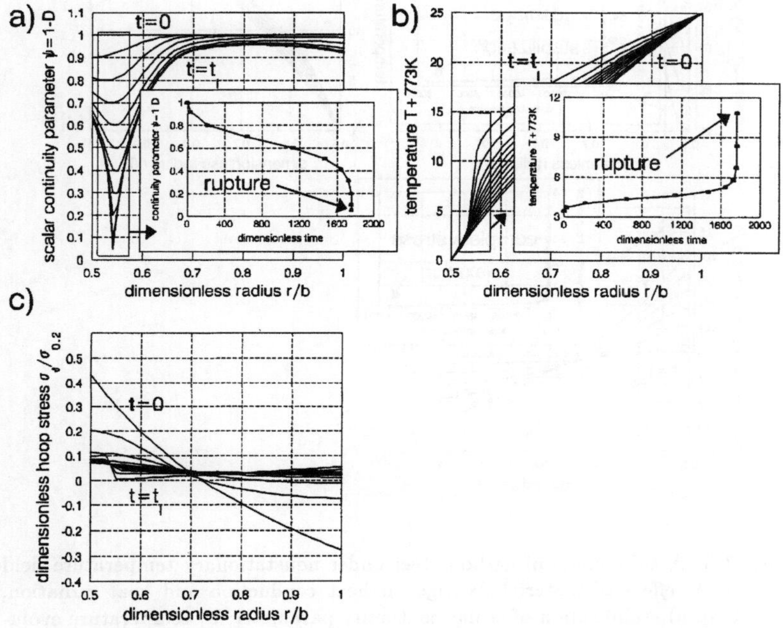

Fig. 7.4. A tube subject to creep under a nonstationary temperature field (Model A: effect of thermo-damage coupling incorporated, pure conductivity $\epsilon = 0$): a) scalar continuity parameter evolution, b) temperature field evolution resulting from damage accumulation, c) hoop stress redistribution (after Skrzypek and Ganczarski, 1998b)

7.4.3 Model B: Combined conductivity-radiation case

Taking the combined conductivity/radiation mechanism taken account

$$\frac{1}{r}\frac{\mathrm{d}}{\mathrm{d}r}\left[r\left(\lambda_0\psi\frac{\mathrm{d}T}{\mathrm{d}r} - \sigma\epsilon_0 T^4\right)\right] = 0 \tag{7.34}$$

may lead to two different mechanisms, depending on the material properties.

I. Complete stress relaxation mechanism

A cylinder made of the carbon steel, but with the combined conductivity-

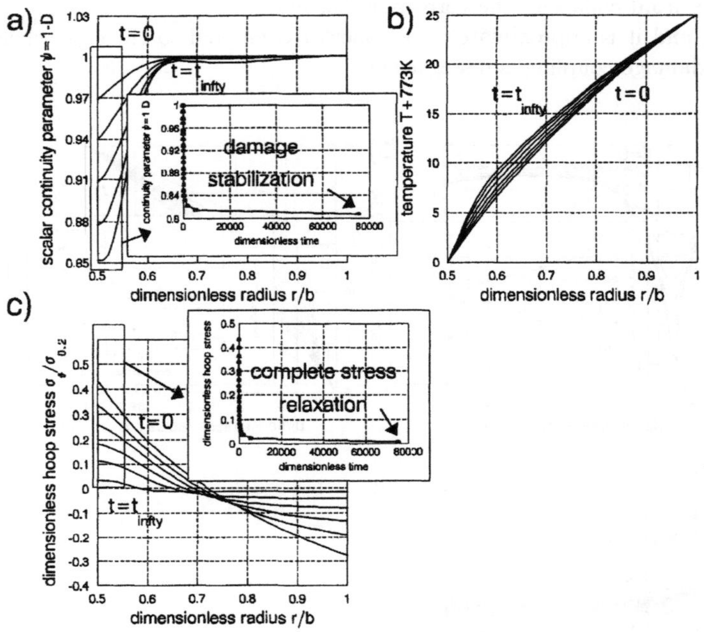

Fig. 7.5. A tube made of carbon steel under nonstationary temperature field (Model A: effect of material damage on heat conduction and heat radiation, $\epsilon_0 = 0.6$): a) stabilization of scalar continuity parameter, b) temperature evolution, c) complete hoop stress relaxation (Skrzypek and Ganczarski, 1998b)

radiation effect taken into account ($\epsilon_0 \neq 0$), exhibits a complete stress relaxation. The slower redistribution of temperature with time (Fig.7.5b) allows hoop stress to relax completely (Fig. 7.5c) and, in consequence, to prevent collapse (Fig. 7.5a). Hence, an infinite lifetime is predicted: $t_{I_{\mathrm{carbon}}}^{(\epsilon_0\neq 0)} \to \infty$.

II. Temperature saturation mechanism

When a cylinder made of stainless steel is concerned, the saturation of temperature precedes rupture. A phenomenon of temperature change due to the damage level increase may be observed despite the vanishing temperature gradient. Hence, an appropriate cutting-off procedure, to avoid thermodynamically inadmissible temperature fields (Fig. 7.6b, d), must be introduced. Formation of the temperature jump is visible in the inner zone that results in a change of sign of the hoop stress (Fig. 7.6c) and, eventually, the lower-band estimation of the lifetime $t_{I_{\mathrm{stainless}}}^{(\epsilon_0\neq 0)} = 38\% t_{I_{\mathrm{stainless}}}^{(s)}$.

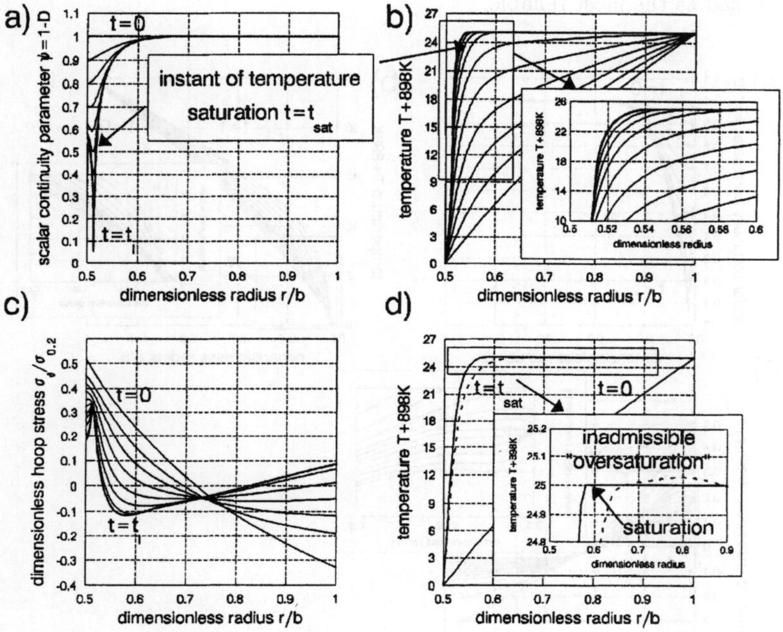

Fig. 7.6. Evolution of continuity parameter, temperature, and hoop stress in the case of combined conductivity-radation (Model B: saturation of temperature, stainless steel) (after Skrzypek and Ganczarski, 1998b)

7.4.4 Model C: Equivalent conductivity concept

The concept of equivalent conductivity-radiation exhibits essential differences depending on whether the derivative $\mathrm{d}D/\mathrm{d}T$ is disregarded or taken into account. When the exact formula (7.20) is applied for a stationary cylindrical heat flux with no inner heat sources and quasistatic temperature field changes, the following equation holds:

$$\frac{1}{r}\frac{\mathrm{d}}{\mathrm{d}r}\left[r\lambda^{\mathrm{eq}}(r,t,T)\frac{\mathrm{d}T}{\mathrm{d}r}\right]=0, \tag{7.35}$$

where:

$$\lambda^{\mathrm{eq}}(r,t,T)=\widetilde{\lambda}(r,t)+\sigma\epsilon_0\left[4D+\frac{\mathrm{d}D/\mathrm{d}r}{\mathrm{d}T/\mathrm{d}r}T\right]T^3\mathrm{d}r. \tag{7.36}$$

A characteristic hoop stress discontinuity is formed at the point of most advanced damage (Fig. 7.7c), and the lifetime is reached $t_{I_{\mathbf{stainless}}}^{(\lambda^{\mathrm{eq}})}=78\%$ $t_{I_{\mathbf{stainless}}}^{(s)}$. Concluding, the equivalent conductivity concept (7.35) is recommended as the most reliable.

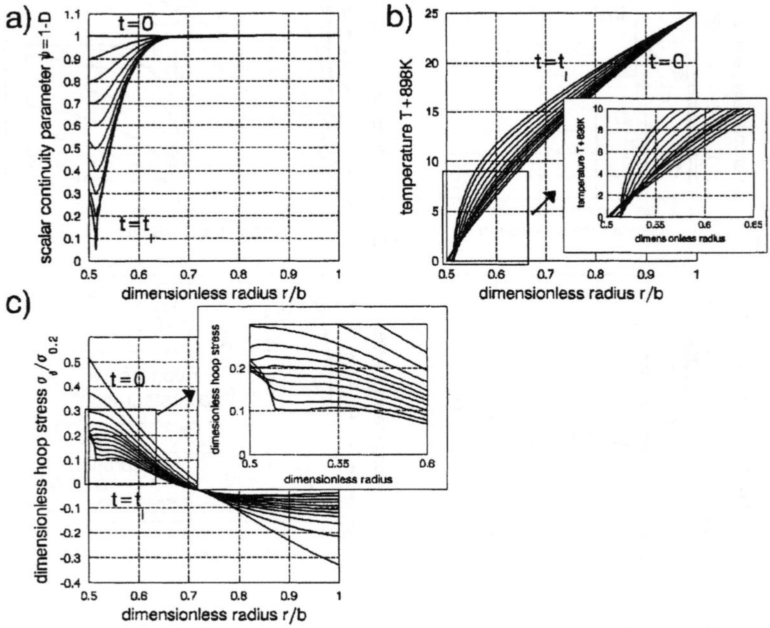

Fig. 7.7. Evolution of continuity parameter, temperature, and hoop stress in case of equivalent conductivity concept (Model C, stainless steel) (after Skrzypek and Ganczarski, 1998b)

7.5 Example: Complete stress relaxation in a disk

In the case of a disk of constant thickness h cooled through its faces by a fluid stream of temperature T_∞ (cf. Fig. 7.2), the equation of heat transfer (7.19) (Model B) requires an explicit formula for the inner heat source intensity:

$$\dot{q}_v = -2\frac{\beta}{h}\left(T - T_\infty\right). \tag{7.37}$$

Here, the heat transfer rate is related to the convection described by Newton's law of cooling. The quantity β is called the convection-transfer coefficient. Hence, assuming quasistatic temperature changes ($\dot{T} = 0$), the heat transfer equation takes the form:

$$\frac{1}{r}\frac{\mathrm{d}}{\mathrm{d}r}r\left[\left(\lambda_0\psi_r\frac{\mathrm{d}T}{\mathrm{d}r}\right)-\sigma\epsilon_0 D_r T^4\right]-2\frac{\beta}{h}\left(T-T_\infty\right)=0. \tag{7.38}$$

The most advanced damage accumulation appears at the center, where the two components of the continuity tensor are equal to each other (Fig. 7.8a,b). Dominant radial stress relaxes more quickly than the corresponding component of the continuity tensor as the latter approaches zero; therefore, the lifetime is infinite (Fig. 7.8c,d). In other words, no thermal failure occurs.

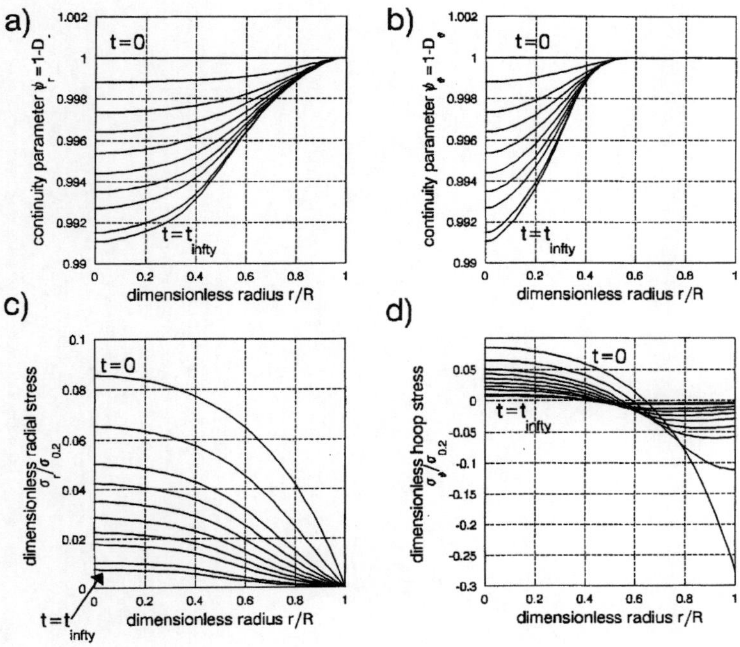

Fig. 7.8. A circular disk subject to constant temperature at the outer edge and cooled through faces: a) and b) saturation of radial and circumferential continuity parameters, respectively, c) and d) complete radial and hoop stress relaxation (no thermal failure) (cf. Ganczarski and Skrzypek, 1995)

8

Creep-damage and failure of axisymmetric disks with shear effect included

8.1 General formulation for basic mechanical state equations of plane stress–rotationally symmetric creep-damage process

8.1.1 Assumptions

Let us consider an annular disk rigidly fixed at the inner edge $r = a$. The disk is loaded by a system of general loadings which cause not only radial and circumferential stresses but also introduce in-plane shear effects (a particular nature of these loadings will be discussed in details in Sect. 8.2).

The following assumptions are used to account for the effect of rotation of principal directions of damage and stress tensors on the creep-damage process in disks:

i. Geometrically linear theory is applied to describe rotationally symmetric deformation of the disk; total strains (small) are decomposed into the elastic and creep portions:

$$\varepsilon_{r/\vartheta} = \varepsilon^{\mathrm{e}}_{r/\vartheta} + \varepsilon^{\mathrm{c}}_{r/\vartheta}, \qquad \gamma_{r\vartheta} = \gamma^{\mathrm{e}}_{r\vartheta} + \gamma^{\mathrm{c}}_{r\vartheta}. \tag{8.1}$$

ii. Elastic part is governed by Hooke's law (isotropic). No additional effect of the material deterioration on elastic properties is taken into account.

iii. Creep part is governed by either the isotropic or the modified orthotropic flow theory and by the time-hardening hypothesis (cf. Sect. 4.1.3).

iv. The Murakami–Ohno damage tensor $\mathbf{D}$ and its objective time-derivative $\overset{\nabla}{\mathbf{D}}$ are used (cf. Sect. 4.1.4).

v. Brittle damage is governed by the orthotropic void growth rule applied to current principal directions of stresses (rotation of principal axes of damage and stress tensors on creep-damage process in disks is accounted for) (cf. Sect. 4.1.2).

vi. Creep-damage coupling is formulated in alternative ways (Table 4.2):

– partly coupled approach (isotropic creep flow and orthotropic damage)

$$\dot{\varepsilon}^{c}_{IJ} = \frac{3}{2}\frac{\dot{\varepsilon}^{c}_{eq}}{\sigma_{eq}} s_{IJ}, \quad \dot{\varepsilon}^{c}_{eq} = (\widetilde{\sigma}_{eq})^{m}\dot{\mathsf{f}}(t), \quad \dot{D}_{IJ} = C_{IJ}\left\langle \frac{\sigma_{IJ}}{1 - D_{IJ}} \right\rangle^{r_{IJ}} \tag{8.2}$$

– fully coupled approach (modified orthotropic creep law for initially isotropic material and orthotropic damage).

$$\dot{\varepsilon}^{c}_{IJ} = \frac{3}{2}\frac{\dot{\varepsilon}^{c}_{eq}}{\widetilde{\sigma}_{eq}} \widetilde{s}_{IJ}, \quad \dot{\varepsilon}^{c}_{eq} = (\widetilde{\sigma}_{eq})^{m}\dot{\mathsf{f}}(t), \quad \dot{D}_{IJ} = C_{IJ}\left\langle \frac{\sigma_{IJ}}{1 - D_{IJ}} \right\rangle^{r_{IJ}} \tag{8.3}$$

vii. Plane stress–rotationally symmetric problems are accounted for; constant thickness of the annular disk rigidly fixed at the inner edge $r = a$ is assumed.

The classical orthotropic brittle damage law (Kachanov, 1958) is applicable for principal directions of the stress tensor. When shear stresses are accounted for, the principal directions of the stress tensor rotate with time and, hence, a tensorial formulation of the damage is required (Chow and Lu, 1992). In general, current principal directions of the stress tensor α_i and of the damage tensor β_i do not coincide, however, when the principal axes of stress rotate due to the shear effect, the principal axes of damage follow them. The symmetric second rank damage tensor $\mathbf{D}$ (Murakami and Ohno, 1981) is applied, and the objective derivative $\overset{\nabla}{\mathbf{D}}$ of the damage tensor is adopted, to account for the effect of rotation of principal directions on the damage accumulation process. Then, a current transformation to the global coordinate system (sampling coordinate space) is performed. The graphical interpretation of all auxiliary coordinate systems associated with the definition of the objective damage rate tensor $\overset{\nabla}{\mathbf{D}}$ in case of the plane stress rotationally symmetric deformation is shown in Fig. 8.1.

8.1.2 Reduced displacement mechanical state equations

The problem is formulated in displacements (cf. Penny and Marriott, 1995, also Ganczarski and Skrzypek, 1991):

$$\left.\begin{aligned} \mathcal{F}(u_r) &= -\frac{1-\nu^2}{E}\rho\omega_0^2 r \\ \mathcal{F}(u_\vartheta) &= \frac{\rho\varepsilon_0}{G} r \end{aligned}\right\} \quad (t=0)$$

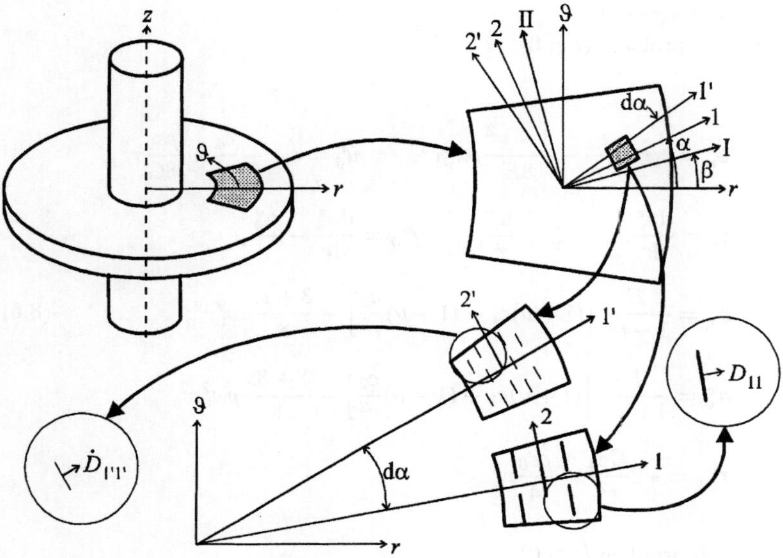

Fig. 8.1. Schematic creep damage accumulation of several orthotropic increments coincided with current principal stress axes (1, 2) and resulting rotation of current principal damage axes (I, II) in case of a disk (after Skrzypek and Ganczarski, 1998a)

$$\left.\begin{aligned}\mathcal{F}(\dot{u}_r) &= \frac{\mathrm{d}(2\dot{g}-\dot{f})}{\mathrm{d}r} + \frac{\dot{f}}{r} - 2\frac{(1-\nu^2)}{E}\rho\omega(t)\varepsilon(t)r \\ \mathcal{F}(\dot{u}_\vartheta) &= \frac{\mathrm{d}\dot{\gamma}^{\mathrm{c}}_{r\vartheta}}{\mathrm{d}r} + 2\frac{\dot{\gamma}^{\mathrm{c}}_{r\vartheta}}{r} + \frac{\rho\dot{\varepsilon}(t)}{G}r\end{aligned}\right\} \quad (t>0), \tag{8.4}$$

where the differential operator $\mathcal{F}[...]$, auxiliary symbols $\dot{g}$, $\dot{f}$, and the relationships involving the angular acceleration ε and the angular velocity ω take the form

$$\begin{aligned}&\mathcal{F}[\ldots] = \frac{\mathrm{d}^2\ldots}{\mathrm{d}r^2} + \frac{1}{r}\frac{\mathrm{d}\ldots}{\mathrm{d}r} - \frac{\ldots}{r^2}, \\ &\dot{g} = \dot{\varepsilon}^{\mathrm{c}}_r + \nu\dot{\varepsilon}^{\mathrm{c}}_\vartheta, \qquad \dot{f} = (1-\nu)\left(\dot{\varepsilon}^{\mathrm{c}}_r - \dot{\varepsilon}^{\mathrm{c}}_\vartheta\right), \qquad \varepsilon(t) = \dot{\omega}(t).\end{aligned} \tag{8.5}$$

8.1.3 Solution of mechanical state equations for constant angular acceleration

Assuming constant value of the angular acceleration $\varepsilon(t) = \pm\varepsilon_0$ or, in other words, a linear function of the angular velocity, the system (8.4) can

be solved explicitly.
I. Elastic problem ($t = 0$)

$$u_r^{\mathrm{e}} = c_1 r + \frac{c_2}{r} - \frac{1-\nu^2}{8E}\rho\omega_0^2 r^3, \qquad u_\vartheta^{\mathrm{e}} = c_3 r + \frac{c_4}{r} + \frac{\rho\varepsilon_0}{8G} r^3,$$

$$\varepsilon_r^{\mathrm{e}} = \frac{\mathrm{d}u_r^{\mathrm{e}}}{\mathrm{d}r}, \qquad \varepsilon_\vartheta^{\mathrm{e}} = \frac{u_r^{\mathrm{e}}}{r}, \qquad \gamma_{r\vartheta}^{\mathrm{e}} = \frac{\mathrm{d}u_\vartheta^{\mathrm{e}}}{\mathrm{d}r} - \frac{u_\vartheta^{\mathrm{e}}}{r},$$

$$\sigma_r^{\mathrm{e}} = \frac{E}{1-\nu^2}\left[(1+\nu)c_1 - (1-\nu)\frac{c_2}{r^2}\right] - \frac{3+\nu}{8}\rho\omega_0^2 r^2, \tag{8.6}$$

$$\sigma_\vartheta^{\mathrm{e}} = \frac{E}{1-\nu^2}\left[(1+\nu)c_1 + (1-\nu)\frac{c_2}{r^2}\right] - \frac{1+3\nu}{8}\rho\omega_0^2 r^2,$$

$$\tau_{r\vartheta}^{\mathrm{e}} = -2\frac{Gc_4}{r^2} + \frac{\rho\varepsilon_0}{4} r^2.$$

II. Creep problem ($t > 0$)

$$\dot{u}_r = \bar{c}_1 r + \frac{\bar{c}_2}{r} + \frac{r}{2}\int_a^r \frac{\dot{f}}{\xi}\mathrm{d}\xi + \frac{1}{2r}\int_a^r \xi(2\dot{g} - \dot{f})\mathrm{d}\xi$$

$$-\frac{1-\nu^2}{4E}\rho\varepsilon(t)\omega(t) r^3,$$

$$\dot{u}_\vartheta = \bar{c}_3 r + \frac{\bar{c}_4}{r} + r\int_a^r \frac{\dot{\gamma}_{r\vartheta}^{\mathrm{c}}}{\xi}\mathrm{d}\xi,$$

$$\dot{\varepsilon}_r = \frac{\mathrm{d}\dot{u}_r}{\mathrm{d}r} - \dot{\varepsilon}_r^{\mathrm{c}}, \qquad \dot{\varepsilon}_\vartheta = \frac{\dot{u}_r}{r} - \dot{\varepsilon}_\vartheta^{\mathrm{c}}, \qquad \dot{\gamma}_{r\vartheta} = \frac{\mathrm{d}\dot{u}_\vartheta}{\mathrm{d}r} - \frac{\dot{u}_\vartheta}{r} - \dot{\gamma}_{r\vartheta}^{\mathrm{c}}, \tag{8.7}$$

$$\dot{\sigma}_r = \frac{E}{1-\nu^2}\left[(1+\nu)\bar{c}_1 - (1-\nu)\frac{\bar{c}_2}{r^2} + \frac{1+\nu}{2}\int_a^r \frac{\dot{f}}{\xi}\mathrm{d}\xi\right.$$

$$\left.-\frac{1-\nu}{2r^2}\int_a^r \xi(2\dot{g} - \dot{f})\mathrm{d}\xi\right] - \frac{3+\nu}{4}\rho\varepsilon(t)\omega(t) r^2,$$

$$\dot{\sigma}_\vartheta = \frac{E}{1-\nu^2}\left[(1+\nu)\bar{c}_1 + (1-\nu)\frac{\bar{c}_2}{r^2} + \frac{1+\nu}{2}\int_a^r \frac{\dot{f}}{\xi}\mathrm{d}\xi\right.$$

$$+\frac{1-\nu}{2r^2}\int\limits_a^r \xi(2\dot{g}-\dot{f})\mathrm{d}\xi\Bigg] - E\dot{\varepsilon}^{\mathrm{c}}_{\vartheta} - \frac{1+3\nu}{4}\rho\varepsilon(t)\omega(t)r^2,$$

$$\dot{\tau}_{r\vartheta} = -2\frac{G\bar{c}_4}{r^2}.$$

8.1.4 Constitutive equations for coupled creep-damage problem in current principal stress directions

The creep strain rates derived on the base of isotropic or modified orthotropic flow theory associated with the time-hardening hypothesis, under additional assumptions of plane stress state and creep incompressibility, are expressed by the following formulae (cf. Ganczarski and Skrzypek, 1991):
Partly coupled

$$\dot{\varepsilon}^{\mathrm{c}}_{1/2} = \frac{(\widetilde{\sigma}_{\mathrm{eq}})^m}{\sigma_{\mathrm{eq}}}\left(\sigma_{1/2} - \frac{\sigma_{2/1}}{2}\right)\dot{\mathsf{f}}(t), \tag{8.8a}$$

Fully coupled

$$\dot{\varepsilon}^{\mathrm{c}}_{1/2} = (\widetilde{\sigma}_{\mathrm{eq}})^{m-1}\left(\frac{\sigma_{1/2}}{1-D_{1/2}} - \frac{1}{2}\frac{\sigma_{2/1}}{1-D_{2/1}}\right)\dot{\mathsf{f}}(t), \tag{8.8b}$$

$$\dot{\varepsilon}^{\mathrm{c}}_3 = -\dot{\varepsilon}^{\mathrm{c}}_1 - \dot{\varepsilon}^{\mathrm{c}}_2. \tag{8.8c}$$

The 2D objective derivative of the damage tensor takes the form (cf. Bathe, 1982):

$$\begin{bmatrix} \overset{\triangledown}{D}_{1'1'} & \overset{\triangledown}{D}_{1'2'} \\ \overset{\triangledown}{D}_{2'1'} & \overset{\triangledown}{D}_{2'2'} \end{bmatrix} = \begin{bmatrix} \dot{D}_{11} & 0 \\ 0 & \dot{D}_{22} \end{bmatrix} - \begin{bmatrix} D_{11} & D_{21} \\ D_{12} & D_{22} \end{bmatrix}\begin{bmatrix} 0 & \mathrm{d}\alpha \\ -\mathrm{d}\alpha & 0 \end{bmatrix} + \begin{bmatrix} 0 & -\mathrm{d}\alpha \\ \mathrm{d}\alpha & 0 \end{bmatrix}\begin{bmatrix} D_{11} & D_{12} \\ D_{21} & D_{22} \end{bmatrix}, \tag{8.9}$$

where non–objective damage rates are

$$\dot{D}_{11} = C_1\left\langle\frac{\sigma_{11}}{1-D_{11}}\right\rangle^{r_1}, \qquad \dot{D}_{22} = C_2\left\langle\frac{\sigma_{22}}{1-D_{22}}\right\rangle^{r_2}. \tag{8.10}$$

When the objective damage rate tensor $\overset{\triangledown}{D}_{IJ}$ (8.9) is transformed from current principal directions of the stress tensor (IJ) to the sampling coordinates (ij) $\overset{\triangledown}{D}_{ij}$, the new damage tensor $D_{ij}(t+\Delta t)$ is achieved:

$$\begin{aligned} &\overset{\nabla}{D}_{IJ} \overset{\text{transf.}}{\longrightarrow} \overset{\nabla}{D}_{ij}, \\ &D_{ij}(t+\Delta t) = D_{ij}(t) + \overset{\nabla}{D}_{ij}(t)\Delta t. \end{aligned} \tag{8.11}$$

Consequently, the creep strain rates (8.8) referring to the global coordinate system (ij) are obtained via the transformation of the creep strain rates written in current principal directions of the stress tensor $(1, 2)$:

$$\begin{cases} \dot{\varepsilon}_r^{\mathrm{c}} = \dfrac{\dot{\varepsilon}_1^{\mathrm{c}} + \dot{\varepsilon}_2^{\mathrm{c}}}{2} + \dfrac{\dot{\varepsilon}_1^{\mathrm{c}} - \dot{\varepsilon}_2^{\mathrm{c}}}{2}\cos 2\alpha, \\ \dot{\varepsilon}_\vartheta^{\mathrm{c}} = \dfrac{\dot{\varepsilon}_1^{\mathrm{c}} + \dot{\varepsilon}_2^{\mathrm{c}}}{2} - \dfrac{\dot{\varepsilon}_1^{\mathrm{c}} - \dot{\varepsilon}_2^{\mathrm{c}}}{2}\cos 2\alpha, \\ \dot{\gamma}_{r\vartheta}^{\mathrm{c}} = (\dot{\varepsilon}_1^{\mathrm{c}} - \dot{\varepsilon}_2^{\mathrm{c}})\sin 2\alpha. \end{cases} \tag{8.12}$$

The intensities of the stress, the effective stress, and the strain rates are defined by the following formulas (cf. Ganczarski and Skrzypek, 1991):

$$\begin{aligned} &\sigma_{\mathrm{eq}} = \sqrt{\sigma_1^2 + \sigma_2^2 - \sigma_1\sigma_2}, \\ &\widetilde{\sigma}_{\mathrm{eq}} = \sqrt{\left(\frac{\sigma_1}{1-D_1}\right)^2 + \left(\frac{\sigma_2}{1-D_2}\right)^2 - \frac{\sigma_1}{(1-D_1)}\frac{\sigma_2}{(1-D_2)}}, \\ &\dot{\varepsilon}_{\mathrm{eq}}^{\mathrm{c}} = \frac{2}{\sqrt{3}}\sqrt{\dot{\varepsilon}_1^{\mathrm{c}^2} + \dot{\varepsilon}_2^{\mathrm{c}^2} + \dot{\varepsilon}_1^{\mathrm{c}}\dot{\varepsilon}_2^{\mathrm{c}}}. \end{aligned} \tag{8.13}$$

8.2 Boundary problems for creep damage in annular disks in case of rotating principal axes

Example A: Disk under steady tension and steady torsion

Let us consider a disk under the steady tension and torsion (Fig. 8.2). Displacements, strains, stresses and their rates satisfy (8.6)–(8.7) and the following boundary conditions:

$$\left.\begin{aligned} u_r^{\mathrm{e}}(a) = 0, \quad u_\vartheta^{\mathrm{e}}(a) = 0, \\ \sigma_r^{\mathrm{e}}(b) = p, \quad \tau_{r\vartheta}^{\mathrm{e}}(b) = s \end{aligned}\right\} \quad (t=0), \qquad \left.\begin{aligned} \dot{u}_r(a) = 0, \quad \dot{u}_\vartheta(a) = 0 \\ \dot{\sigma}_r(b) = 0, \quad \dot{\tau}_{r\vartheta}(b) = 0 \end{aligned}\right\} \quad (t>0). \tag{8.14}$$

Example B: Disk under steady tension and multiple reverse torsion

In case of a multiple reverse torsion (Fig. 8.3) displacements, strains, stresses and their rates satisfy (8.6)–(8.7) and the following boundary conditions:

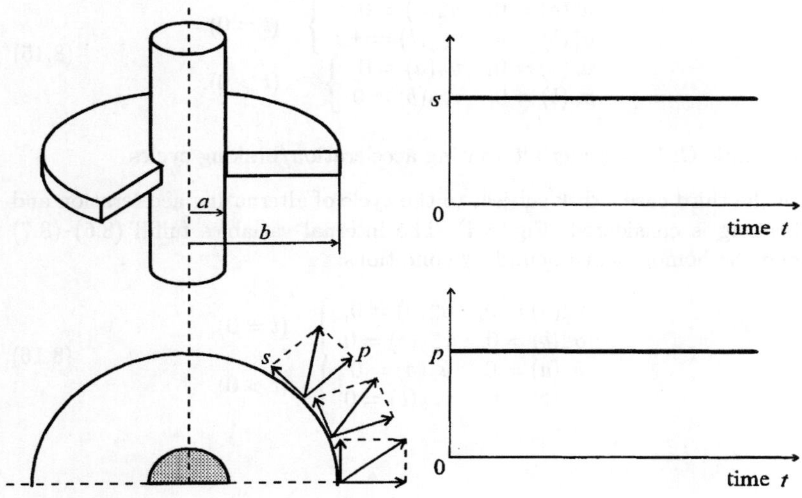

Fig. 8.2. Layout of a circular disk subject to creep damage under steady peripheral tension p and torsion s (cf. Skrzypek and Ganczarski, 1998a)

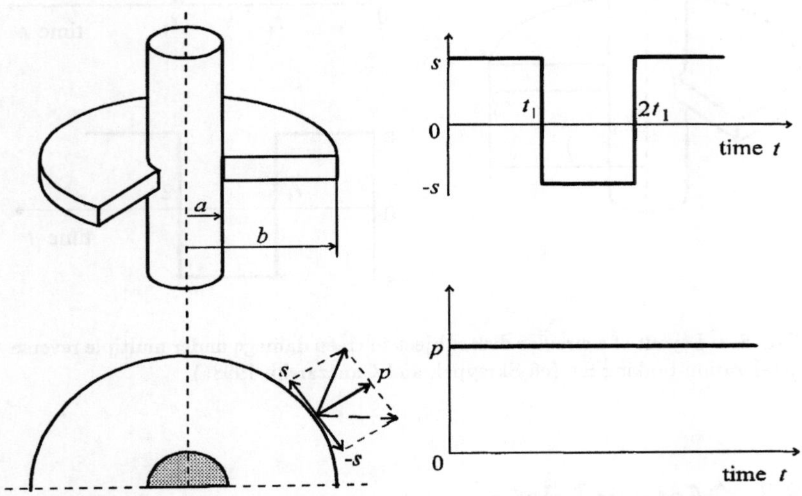

Fig. 8.3. Layout of a circular disk subject to creep damage under steady peripheral tension p and multiple reverse torsion $\pm s$ (cf. Skrzypek and Ganczarski, 1998a)

$$\left.\begin{array}{ll} u_r^e(a)=0, & u_\vartheta^e(a)=0, \\ \sigma_r^e(b)=p, & \tau_{r\vartheta}^e(b)=\pm s \end{array}\right\} \quad (t=0), \\ \left.\begin{array}{ll} \dot{u}_r(a)=0, & \dot{u}_\vartheta(a)=0 \\ \dot{\sigma}_r(b)=0, & \dot{\tau}_{r\vartheta}(b)=0 \end{array}\right\} \quad (t>0). \tag{8.15}$$

Example C: Disk under alternating acceleration/braking cycles

In the third case a disk subject to the cycle of alternating acceleration and braking is considered (Fig. 8.4). The internal variables fulfill (8.6)–(8.7) and the homogeneous boundary conditions:

$$\left.\begin{array}{ll} u_r^e(a)=0, & u_\vartheta^e(a)=0, \\ \sigma_r^e(b)=0, & \tau_{r\vartheta}^e(b)=0 \end{array}\right\} \quad (t=0), \\ \left.\begin{array}{ll} \dot{u}_r(a)=0, & \dot{u}_\vartheta(a)=0 \\ \dot{\sigma}_r(b)=0, & \dot{\tau}_{r\vartheta}(b)=0 \end{array}\right\} \quad (t>0). \tag{8.16}$$

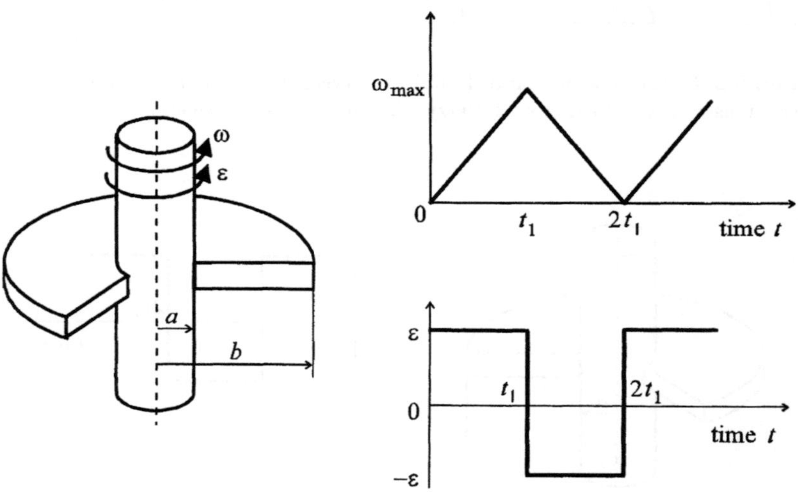

Fig. 8.4. Layout of a circular disk subject to creep damage under multiple reverse acceleration-braking $\pm\varepsilon$ (cf. Skrzypek and Ganczarski, 1998a)

8.3 Material data

First two boundary problems A and B formulated in Sect. 8.2 deal with disks made of stainless steel (rolled 18 Cr 8 Ni 0.45 Si 0.4 Mn 0.1 C Ti/Nb stabilized, austenite annealed at 1070°C, air cooled (ASTM 321)) with the following properties at temperature 500°C (cf. Odqvist, 1966): $E = 180$ GPa, $\sigma_{0.2} = 120$ MPa, $\nu = 0.3$, $m = 5.6$, $r = 3.9$, $\sigma^5_{C_B} = 210$ MPa, where $\sigma^5_{C_B}$ denotes the stress causing creep rupture in 10^5 hr. Magnitudes of load are: $p = 0.2 \times \sigma_{0.2}$, $s = p/20$.

In case of problem C in Sect. 8.2 the disk is made of carbon steel (rolled 0.40 Mn 0.25 Si 0.12 C normalized, annealed at 840°C) with the following properties at temperature 500°C: $E = 170$ GPa, $\sigma_{0.2} = 120$ MPa, $\nu = 0.3$, $m = 3.3$, $r = 3.5$, $\sigma^5_{C_B} = 80$ MPa, $\varrho = 7850$ kg/m^3. The disk is subject to angular acceleration/braking $\varepsilon = \pm 10$ s^{-2}, whereas the maximal angular velocity is $\omega_{\max} = 50$ s$^{-1} = 3000$ min^{-1}.

8.4 Numerical results

8.4.1 *Example A: Creep-damage accumulation and shear-type failure mechanism in disks under steady tension and steady torsion*

A representative distribution of the dominant current principal component of the continuity tensor ψ_I is presented in Fig. 8.5a. The first macrocrack appears around the inner edge ($r = a$). The damage zone is narrow and limited to the closest neighborhood of the fixed disk edge.

Consider the evolution of angles of principal directions of α-stress, and β-damage tensors at the point of the first macrocrack initiation. In the partly coupled case of (8.2), the isotropic flow rule introduces similarity of the stress deviator s_{ij} and the creep strain rate deviator $\dot{\varepsilon}^c_{ij}$. The angles of principal directions of the stress tensor $\boldsymbol{\sigma}$ (α) and the damage tensor $\mathbf{D}$ (β) slightly differ from one another during the primary creep phase (Fig. 8.5c). However, the principal direction of the stress tensor precedes the principal direction of the strain tensor when process enters the secondary and tertiary creep. On the primary creep, when damage is not advanced, components of the damage tensor $\mathbf{D}$ strongly depend on the corresponding components of the objective damage rate tensor $\overset{\triangledown}{\mathbf{D}}$. Therefore, the principal direction of the damage tensor (β) slightly precedes the principal direction of the stress tensor (α). However, during secondary and tertiary phases, when damage reaches a more advanced level, and an influence of the objective damage rate tensor on the damage tensor is not so strong, the opposite sign discrepancy between principal directions of both tensors, increasing with time, is observed. When the fully coupled approach is assumed (8.3), the orthotropic flow rule introduces similarity of the effective stress deviator $\tilde{s}_{ij}$ and the strain rate deviator $\dot{\varepsilon}^c_{ij}$. In this case, final magnitudes of all principal angles reach a lower level than in the previously discussed case, and differences between them are more noticeable (Fig. 8.5c, d). However, in the case of the orthotropic flow rule only a 0.3% increase of the lifetime is observed when compared to the lifetime obtained for the isotropic formulation.

The shear type failure mechanism is strictly associated with the hoop displacement discontinuity around the inner edge (Figs. 8.5b and 8.6).

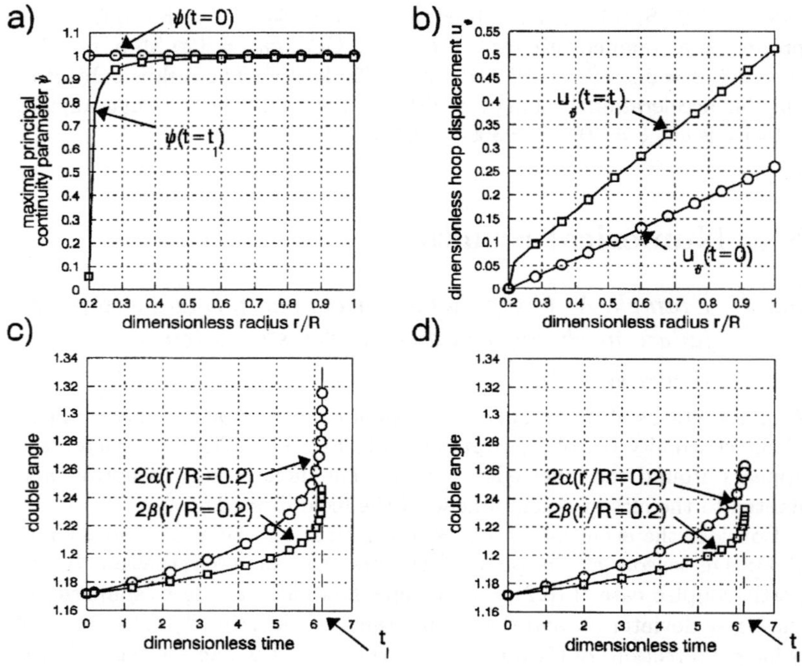

Fig. 8.5. Disk under steady tension and torsion: a) damage evolution with time to failure, b) formation of the hoop displacement discontinuity, c) and d) rotation of principal stress axes α, and principal damage axes β in case of scalar and tensorial creep-damage coupling, respectively (at inner disk edge $r/R = 0.2$) (after Skrzypek and Ganczarski, 1998a)

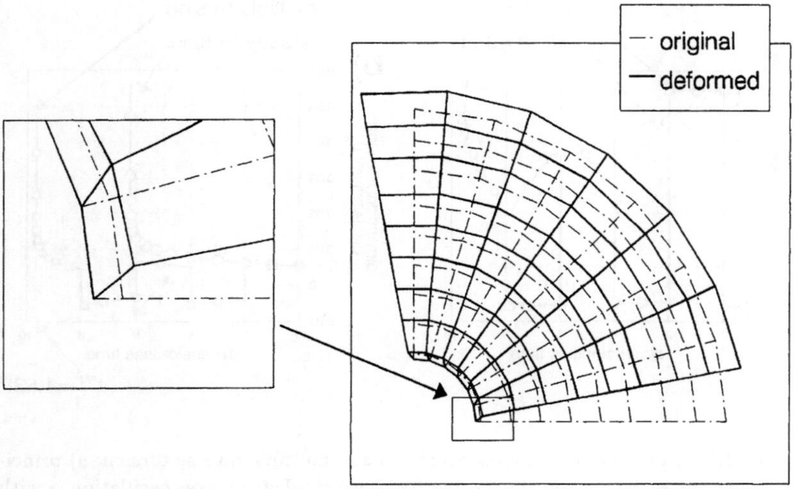

Fig. 8.6. Schematic illustration of the shear-type failure mechanism of a disk under steady tension and torsion (after Skrzypek and Ganczarski, 1998a)

8.4.2 Example B: Alternative creep-damage accumulation and the shear type failure mechanism in a disk under multiple reverse torsion

Alternating torsion causes reverse jumps of the principal axes of the stress tensor around the $\alpha = 0$ direction. Consequently, the principal axes of the damage tensor (β) also undergo rotations. However, the changes of β are not as sharp as those of α, and they oscillate nonsymmetrically around the direction $\beta = 0$ with an inclination to the direction corresponding to the first loading cycle. On the tertiary creep phase a slope of the β angle versus time rapidly increases preceding a shear type rupture mechanism in disk. Due to the alternating torsion, the damage accumulation process develops in reverse material fibres in an alternative manner which produces a characteristic response of damage freezing during each even loading cycle (Fig. 8.7b) and, eventually, a 53% increase in lifetime is observed when compared to the steady torsion case.

The shear-type failure mechanism in a disk also in this case corresponds to the hoop displacement discontinuity (Fig. 8.6). After a number of oscillations around the zero value, the hoop displacement u_ϑ rapidly increases in the direction corresponding to the first loading cycle.

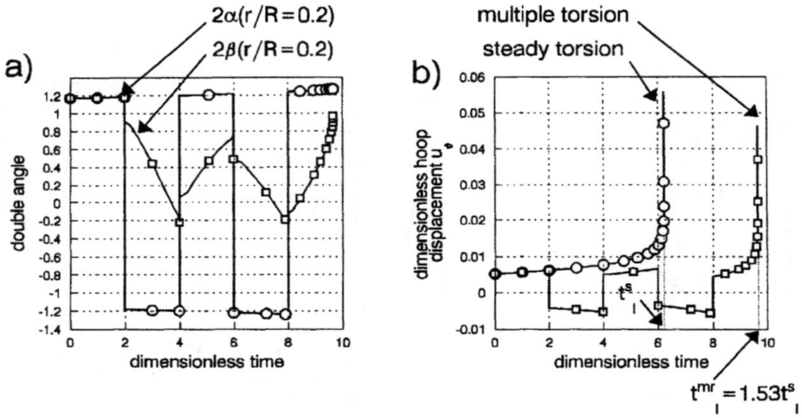

Fig. 8.7. A disk subject to steady tension and multiple reverse torsion: a) principal damage axes rotation β resulting from principal stress axes oscillation α with time to failure, b) formation of bilateral hoop displacement discontinuity with time to failure in case of multiple reverse torsion versus steady torsion

8.4.3 Example C: Accumulation of creep-damage and the decohesion-type failure mechanism in a disk under alternating acceleration-braking cycles

In the case of a disk subjected to alternating body forces due to acceleration/braking cycles, the loading path differs from the proportional one so essentially that the simple isotropic flow rule (partly coupled) is no longer sufficient to describe the creep-damage interaction. Therefore, in this case, the modified orthotropic flow rule (fully coupled) is applied. It turns out, however, that the radial component of the body forces is dominant when compared to the hoop component.

Therefore, the first macrocrack appears around the inner edge with the normal of the radial direction, whereas other components of the damage tensor are less advanced (Fig. 8.8a). The alternating nature of body forces causes that all principal axes of: the stress (α) and the damage (β) tensors oscillate around the zero value (Fig. 8.8b). During all phases of the creep process principal angles of the damage (β) tensor precedes the principal angle of the stress tensor (α). Amplitudes of oscillations of principal angles decrease from one cycle of acceleration/braking to the other. The only exception is the terminal phase of the tertiary creep when a significant increase of magnitudes of the principal angles of the stress (α) and the damage (β) tensors is observed.

The decohesion-type failure mechanism corresponds to formation of the

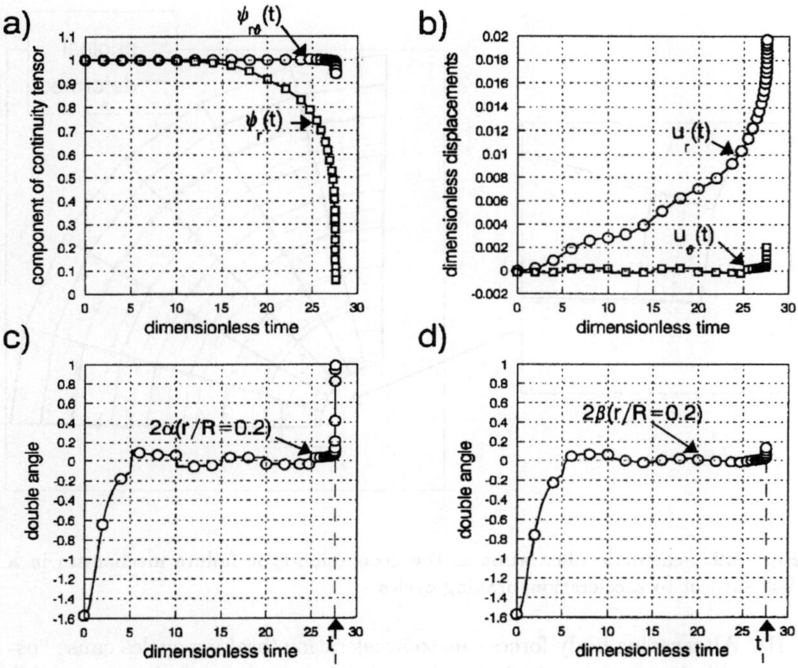

Fig. 8.8. A disk subject to acceleration/braking cycles: a) damage evolution, b) radial and hoop displacement evolution with time to failure, c) variation of principal direction of the stress tensor α, d) evolution of principal direction of the damage tensor β

radial displacement discontinuity at the braking phase of the loading cycle and refers to the dominant macrocrack of the radial normal (Fig. 8.9).

8.5 Conclusions

I. When the principal directions of the stress rotate, the principal directions of the damage tensor (β) follow the principal directions of the stress tensor (α). The more the loading path differs from the proportional path, the stronger the observed differences between the angles (α) and (β).

II. Multiple reverse torsion leads to a significant increase of the lifetime compared to the case of steady torsion. A shear-type failure mechanism, due to the dominant hoop component of the damage tensor, accompanied with the hoop displacement discontinuity, is observed in both cases.

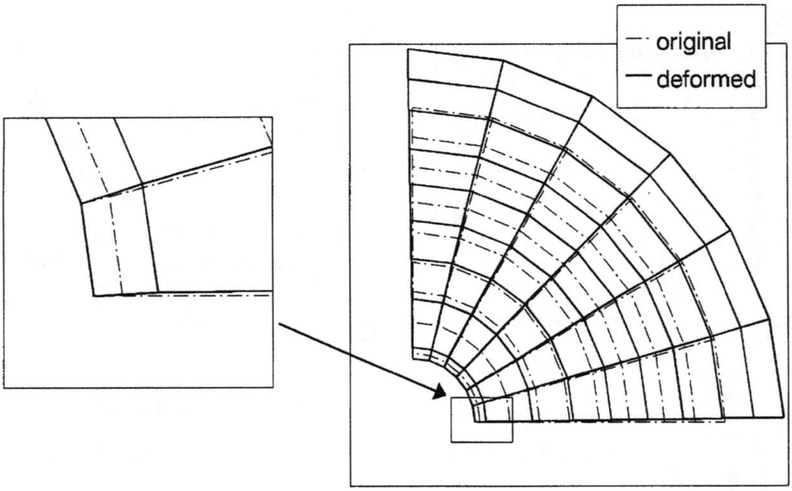

Fig. 8.9. Schematic illustration of the decohesion-type failure mechanism in a disk subject to acceleration/braking cycles

III. Alternating body forces due to acceleration/braking cycles cause "oscillations" of principal directions of the stress (α) and the damage (β) tensors around the zero value. Hence, after a finite number of cycles at the instant of rupture, both principal angles approach the zero value. Consequently, a decohesion-type failure mechanism, due to the dominant radial component of the damage tensor accompanied with the radial displacement discontinuity, is observed.

9

Creep damage and failure analysis of thin axisymmetric plates

9.1 Basic state equations for axisymmetric Love–Kirchhoff plates of variable thickness under arbitrary loadings

Let us consider an axisymmetric plate of variable thickness loaded by an external pressure, body forces and a temperature field (Fig. 9.1). When a cylindrical coordinate system is defined, the problem can be written in displacements. Let us assume that:

i. loadings are reduced to the middle surface,

ii. the theory of small displacements with geometry changes accounted for (second-order theory) is applied,

iii. the Love–Kirchhoff hypothesis of straight and normal segments is postulated,

$$\varepsilon_{r/\vartheta} = \lambda_{r/\vartheta} + \kappa_{r/\vartheta} z, \qquad \varepsilon_{r\vartheta} = \lambda_{r\vartheta} + \kappa_{r\vartheta} z, \tag{9.1}$$

iv. small strain decomposition holds,

$$\varepsilon_{r/\vartheta} = \varepsilon^{\mathrm{e}}_{r/\vartheta} + \varepsilon^{\mathrm{c}}_{r/\vartheta} + \alpha T, \qquad \varepsilon_{r\vartheta} = \varepsilon^{\mathrm{e}}_{r\vartheta} + \varepsilon^{\mathrm{c}}_{r\vartheta}, \tag{9.2}$$

v. plane stress state ($\sigma_z = 0$) holds,

vi. thickness and temperature depend on the radial coordinate r only.

The state variables for the general coupled elastic-creep problem must fulfil the following system of equations:
Membrane and bending equilibrium

$$\frac{\partial(n_r r)}{\partial r} + \frac{\partial n_{r\vartheta}}{\partial \vartheta} - n_\vartheta + q_r r = 0,$$

$$\frac{\partial(n_{r\vartheta} r)}{\partial r} + \frac{\partial n_\vartheta}{\partial \vartheta} + n_{r\vartheta} + q_\vartheta r = 0,$$

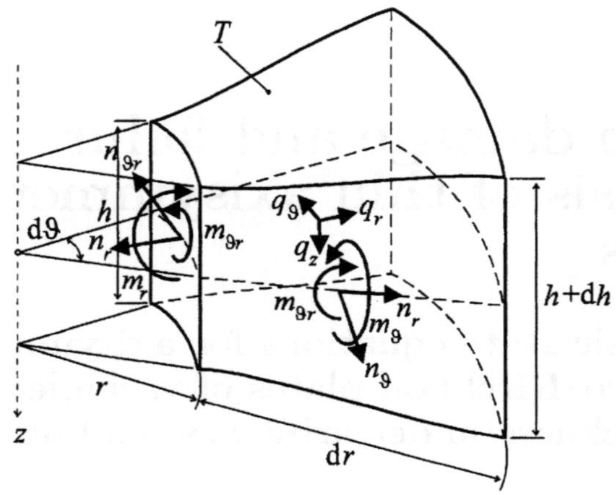

Fig. 9.1. Element of an axisymmetric plate of variable thickness subjected to external pressure, body forces, and temperature field

$$\begin{aligned}&\frac{1}{r}\frac{\partial^2(m_r r)}{\partial r^2} - \frac{1}{r}\frac{\partial m_\vartheta}{\partial r} + \frac{1}{r^2}\frac{\partial^2 m_\vartheta}{\partial \vartheta^2} + \frac{2}{r^2}\frac{\partial^2(m_{r\vartheta} r)}{\partial r \partial \vartheta} - \frac{1}{r}\frac{\partial}{\partial r}(n_r r \varphi_r)\\&-\frac{1}{r}\frac{\partial}{\partial \vartheta}(n_\vartheta \varphi_\vartheta) - \frac{1}{r}\frac{\partial}{\partial r}(n_{r\vartheta} r \varphi_\vartheta) - \frac{1}{r}\frac{\partial}{\partial \vartheta}(n_{r\vartheta}\varphi_r) + q_z = 0;\end{aligned} \tag{9.3}$$

Cauchy geometric relations

$$\begin{aligned}\varphi_r &= -\frac{\partial w}{\partial r}, \qquad \varphi_\vartheta = -\frac{1}{r}\frac{\partial w}{\partial \vartheta},\\ \kappa_r &= \frac{\partial \varphi_r}{\partial r}, \qquad \kappa_\vartheta = \frac{\varphi_r}{r} + \frac{1}{r}\frac{\partial \varphi_\vartheta}{\partial \vartheta},\\ \kappa_{r\vartheta} &= \frac{1}{2}\left[\frac{1}{r}\left(\frac{\partial \varphi_r}{\partial \vartheta} - \varphi_\vartheta\right) + \frac{\partial \varphi_\vartheta}{\partial r}\right],\\ \lambda_r &= \frac{\partial u}{\partial r}, \qquad \lambda_\vartheta = \frac{u}{r} + \frac{1}{r}\frac{\partial v}{\partial \vartheta},\\ \lambda_{r\vartheta} &= \frac{1}{2}\left[\frac{1}{r}\frac{\partial u}{\partial \vartheta} - \frac{v}{r} + \frac{\partial v}{\partial r}\right];\end{aligned} \tag{9.4}$$

Constitutive equations (cf. Penny and Marriott, 1995)

$$\sigma_{r/\vartheta} = \frac{E}{1-\nu^2}\left[\left(\varepsilon_{r/\vartheta} + \nu\varepsilon_{\vartheta/r}\right) - \left(\varepsilon^{\mathrm{c}}_{r/\vartheta} + \nu\varepsilon^{\mathrm{c}}_{\vartheta/r}\right) - (1+\nu)\alpha T\right],$$

$$\sigma_{r\vartheta} = \frac{E}{1+\nu}\left(\varepsilon_{r\vartheta} - \varepsilon^{\mathrm{c}}_{r\vartheta}\right),$$

$$n_{r/\vartheta} = \mathcal{B}(\lambda_{r/\vartheta} + \nu\lambda_{\vartheta/r}) - n^{\mathrm{c}}_{r/\vartheta} - \mathcal{B}(1+\nu)\alpha T,$$

$$n_{r\vartheta} = \mathcal{B}(1-\nu)\lambda_{r\vartheta} - n^{\mathrm{c}}_{r\vartheta},$$

$$m_{r/\vartheta} = \mathcal{D}(\kappa_{r/\vartheta} + \nu\kappa_{\vartheta/r}) - m^{\mathrm{c}}_{r/\vartheta}, \qquad m_{r\vartheta} = \mathcal{D}(1-\nu)\kappa_{r\vartheta} - m^{\mathrm{c}}_{r\vartheta}, \tag{9.5}$$

where the following definitions of inelastic generalized stresses hold:

$$n^{\mathrm{c}}_{r/\vartheta} = \frac{E}{1-\nu^2}\int_{-h/2}^{h/2}\left(\varepsilon^{\mathrm{c}}_{r/\vartheta} + \nu\varepsilon^{\mathrm{c}}_{\vartheta/r}\right)\mathrm{d}z,$$

$$m^{\mathrm{c}}_{r/\vartheta} = \frac{E}{1-\nu^2}\int_{-h/2}^{h/2}\left(\varepsilon^{\mathrm{c}}_{r/\vartheta} + \nu\varepsilon^{\mathrm{c}}_{\vartheta/r}\right)z\mathrm{d}z, \tag{9.6}$$

$$n^{\mathrm{c}}_{r\vartheta} = \frac{E}{1-\nu^2}\int_{-h/2}^{h/2}\varepsilon^{\mathrm{c}}_{r\vartheta}\mathrm{d}z,$$

$$m^{\mathrm{c}}_{r\vartheta} = \frac{E}{1-\nu^2}\int_{-h/2}^{h/2}\varepsilon^{\mathrm{c}}_{r\vartheta}z\mathrm{d}z$$

and membrane and bending stiffnesses are

$$\mathcal{D}(r) = \frac{Eh^3(r)}{12(1-\nu^2)}, \qquad \mathcal{B}(r) = \frac{Eh(r)}{1-\nu^2}. \tag{9.7}$$

9.2 Reduced membrane-bending equations for plates under axisymmetric loadings

9.2.1 Unilaterally coupled Kármán system extended to visco-elastic plates of variable thickness

Elimination of strains (9.1) in terms of displacements (9.4) from constitutive equations (9.6) and from equilibrium equations (9.3) leads to the basic system of partial differential equations, which may be a reduced to system of ordinary differential equations when the following assumptions hold:

i. Loadings are assumed to be axisymmetric:

$$n_{r\vartheta} = n^{\mathrm{c}}_{r\vartheta} = m^{\mathrm{c}}_{r\vartheta} = q_\vartheta = v = 0, \qquad q_r = -\frac{\mathrm{d}U}{\mathrm{d}r}. \tag{9.8}$$

ii. The Fourier expansions of displacements are used:

$$w(r,\vartheta) = f(r)\cos k\vartheta. \tag{9.9}$$

iii. The Airy function defines generalized stresses:

$$n_r = \frac{1}{r}\frac{\mathrm{d}F}{\mathrm{d}r} + U, \qquad n_\vartheta = \frac{\mathrm{d}^2F}{\mathrm{d}r^2} + U. \tag{9.10}$$

Hence, the reduced governing displacement-type equations take the form

$$\begin{aligned} &\mathcal{B}\left(\frac{\mathrm{d}^2u}{\mathrm{d}r^2} + \frac{1}{r}\frac{\mathrm{d}u}{\mathrm{d}r} - \frac{u}{r^2}\right) + \frac{\mathrm{d}\mathcal{B}}{\mathrm{d}r}\left(\frac{\mathrm{d}u}{\mathrm{d}r} + \frac{\nu}{r}u\right) \\ &= -q_r + \frac{\mathrm{d}n^{\mathrm{c}}_r}{\mathrm{d}r} + \frac{n^{\mathrm{c}}_r - n^{\mathrm{c}}_\vartheta}{r} + (1+\nu)\alpha\left(\frac{\mathrm{d}\mathcal{B}}{\mathrm{d}r}T + \mathcal{B}\frac{\mathrm{d}T}{\mathrm{d}r}\right), \end{aligned} \tag{9.11}$$

$$\begin{aligned} &\mathcal{D}\nabla^4_r f + \frac{\mathrm{d}\mathcal{D}}{\mathrm{d}r}\left(2\frac{\mathrm{d}^3f}{\mathrm{d}r^3} + \frac{2+\nu}{r}\frac{\mathrm{d}^2f}{\mathrm{d}r^2} - \frac{(1+2k^2)}{r^2}\frac{\mathrm{d}f}{\mathrm{d}r} + 3\frac{k^2}{r^3}f\right) \\ &+\frac{\mathrm{d}^2\mathcal{D}}{\mathrm{d}r^2}\left(\frac{\mathrm{d}^2f}{\mathrm{d}r^2} + \frac{\nu}{r}\frac{\mathrm{d}f}{\mathrm{d}r} - \nu\frac{k^2}{r^2}f\right) - n_r\frac{\mathrm{d}^2f}{\mathrm{d}r^2} - \frac{n_\vartheta}{r}\left(\frac{\mathrm{d}f}{\mathrm{d}r} - \frac{k^2}{r}f\right) \\ &= q_z - q_r\frac{\mathrm{d}f}{\mathrm{d}r} - \frac{\mathrm{d}^2m^{\mathrm{c}}_r}{\mathrm{d}r^2} - \frac{1}{r}\frac{\mathrm{d}\left(2m^{\mathrm{c}}_r - m^{\mathrm{c}}_\vartheta\right)}{\mathrm{d}r}. \end{aligned} \tag{9.12}$$

It turns out, however, that another mixed approach may be more convenient in some particular cases. Let us rewrite the equation of the membrane state by using the Airy function, applying the compatibility condition, where elongations of the middle surface are described by inverted equations (9.1) and generalized stresses by the Airy function, to obtain (cf. Timoshenko, 1951):

$$\begin{aligned} &\frac{1}{\mathcal{B}}\nabla^4_r F + \frac{\mathrm{d}}{\mathrm{d}r}\left(\frac{1}{\mathcal{B}}\right)\left(2\frac{\mathrm{d}^3F}{\mathrm{d}r^3} + \frac{2-\nu}{r}\frac{\mathrm{d}^2F}{\mathrm{d}r^2} - \frac{1}{r^2}\frac{\mathrm{d}F}{\mathrm{d}r}\right) \\ &+\frac{\mathrm{d}^2}{\mathrm{d}r^2}\left(\frac{1}{\mathcal{B}}\right)\left(\frac{\mathrm{d}^2F}{\mathrm{d}r^2} - \frac{\nu}{r}\frac{\mathrm{d}F}{\mathrm{d}r}\right) + (1-\nu^2)\alpha\nabla^2_r T + (1-\nu)\nabla^2_r\left(\frac{U}{\mathcal{B}}\right) \\ &= -\nabla^2_r\left(\frac{n^{\mathrm{c}}_\vartheta - \nu n^{\mathrm{c}}_r}{\mathcal{B}}\right) - \frac{1+\nu}{r}\frac{\mathrm{d}}{\mathrm{d}r}\left(\frac{n^{\mathrm{c}}_\vartheta - n^{\mathrm{c}}_r}{\mathcal{B}}\right), \end{aligned} \tag{9.13}$$

$$
\begin{aligned}
&\mathcal{D}\nabla_r^4 f + \frac{\mathrm{d}\mathcal{D}}{\mathrm{d}r}\left(2\frac{\mathrm{d}^3 f}{\mathrm{d}r^3} + \frac{2+\nu}{r}\frac{\mathrm{d}^2 f}{\mathrm{d}r^2} - \frac{(1+2k^2)}{r^2}\frac{\mathrm{d}f}{\mathrm{d}r} + 3\frac{k^2}{r^3}f\right)\\
&\quad + \frac{\mathrm{d}^2\mathcal{D}}{\mathrm{d}r^2}\left(\frac{\mathrm{d}^2 f}{\mathrm{d}r^2} + \frac{\nu}{r}\frac{\mathrm{d}f}{\mathrm{d}r} - \nu\frac{k^2}{r^2}f\right) - \left(\frac{1}{r}\frac{\mathrm{d}F}{\mathrm{d}r} + U\right)\frac{\mathrm{d}^2 f}{\mathrm{d}r^2}\\
&\quad - \left(\frac{\mathrm{d}^2 F}{\mathrm{d}r^2} + U\right)\left(\frac{1}{r}\frac{\mathrm{d}f}{\mathrm{d}r} - \frac{k^2}{r^2}f\right)\\
&\quad = q_z - q_r\frac{\mathrm{d}f}{\mathrm{d}r} - \frac{\mathrm{d}^2 m_r^{\mathrm{c}}}{\mathrm{d}r^2} - \frac{1}{r}\frac{\mathrm{d}\left(2m_r^{\mathrm{c}} - m_\vartheta^{\mathrm{c}}\right)}{\mathrm{d}r}.
\end{aligned}
\tag{9.14}
$$

Although the equation of the membrane state takes a more complicated form than in the previous approach, this formulation is frequently quoted because of formal similarities of both operators.

Each of derived systems of equations (9.11)–(9.14) is a unilaterally coupled Kármán system extended to the case of the visco-elastic plate of variable thickness. Note that in a classical Kármán formulation the fully coupled equations of membrane and bending states hold (cf. von Kármán, 1910, also Fung, 1969) and, then, additional nonlinear terms associated with the Gaussian curvature appear in the equation of bending state (third order-theory). In the case under consideration the unilaterally coupled systems (9.11)–(9.14) allow consideration of the equation of the membrane state independently from the equation of the bending state. After the membrane state equation is solved, the bending state equation can be solved when the generalized membrane forces taken from the previous one are introduced.

9.2.2 Finite difference method approach

One of the methods to solve boundary differential problems consists in replacing the differential operators, entering both differential equations and boundary conditions, by the appropriate finite differences. This approach leads to a finite system of algebraic equations instead of the differential ones.

Although the finite difference operators expressed in the Cartesian system of coordinates are well known, their transformation to the cylindrical coordinates system requires the plate Laplace operators to be used,

$$
\begin{aligned}
&\nabla_r^2 .. = \frac{\mathrm{d}^2 ..}{\mathrm{d}r^2} + \frac{1}{r}\frac{\mathrm{d}..}{\mathrm{d}r} - \frac{k^2}{r^2}..\\
&\nabla_r^4 .. = \frac{\mathrm{d}^4 ..}{\mathrm{d}r^4} + \frac{2}{r}\frac{\mathrm{d}^3 ..}{\mathrm{d}r^3} - \frac{1+2k^2}{r^2}\frac{\mathrm{d}^2 ..}{\mathrm{d}r^2} + \frac{1+2k^2}{r^3}\frac{\mathrm{d}..}{\mathrm{d}r} + \frac{k^2(k^2-4)}{r^4}..,
\end{aligned}
\tag{9.15}
$$

where $k = 0, 1, \ldots, N$ denotes a number of half-waves, which decides whether the deformation is symmetric or nonsymmetric where $k = 0$ denotes a fundamental symmetric mode.

Laplace's differential operators ∇_r^2 and ∇_r^4 independent of angular coordinate ϑ may be replaced by finite differences (cf. Benda, 1964, also Kączkowski, 1980):

$$\hat{\nabla}_r^2 f \cong \left[\frac{1}{(\Delta r)^2} - \frac{1}{2r\Delta r}\right] f_{i-1} - \left[\frac{2}{(\Delta r)^2} + \frac{k^2}{r^2}\right] f_i + \left[\frac{1}{(\Delta r)^2} + \frac{1}{2r\Delta r}\right] f_{i+1}$$

$$\begin{aligned}\hat{\nabla}_r^4 f \cong & \left[\frac{1}{(\Delta r)^4} - \frac{1}{2r(\Delta r)^3} - \frac{1}{2(r-\Delta r)(\Delta r)^3} + \frac{1}{4r(r-\Delta r)(\Delta r)^2}\right] f_{i-2} \\ & + \left[-\frac{4}{(\Delta r)^4} + \frac{2}{r(\Delta r)^3} - k^2\left(\frac{1}{(r-\Delta r)^2} + \frac{1}{r^2}\right)\left(\frac{1}{(\Delta r)^2} - \frac{1}{2r\Delta r}\right)\right] f_{i-1} \\ & + \left[\frac{6}{(\Delta r)^4} - \frac{1}{2(r+\Delta r)(\Delta r)^3} + \frac{1}{2(r-\Delta r)(\Delta r)^3}\right. \\ & \left. - \frac{1}{4r(r+\Delta r)(\Delta r)^2} - \frac{1}{4r(r-\Delta r)(\Delta r)^3} + \frac{k^2}{r^2}\left(\frac{4}{(\Delta r)^2} + \frac{k^2}{r^2}\right)\right] f_i \\ & + \left[-\frac{4}{(\Delta r)^4} - \frac{2}{r(\Delta r)^3} - k^2\left(\frac{1}{(r+\Delta r)^2} + \frac{1}{r^2}\right)\left(\frac{1}{(\Delta r)^2} + \frac{1}{2r\Delta r}\right)\right] f_{i+1} \\ & + \left[\frac{1}{(\Delta r)^4} + \frac{1}{2r(\Delta r)^3} + \frac{1}{2(r+\Delta r)(\Delta r)^3} + \frac{1}{4r(r+\Delta r)(\Delta r)^2}\right] f_{i+2}.\end{aligned} \tag{9.16}$$

The above finite difference operators (9.16), like the differential operators (9.15) from which they are obtained, exhibit singularities at the central point, $r = 0$ and neighboring $r = \Delta r$ which may be omitted by the following formulas (Kączkowski, 1980):

$$\begin{aligned}\hat{\nabla}_r^2 f(r=0) &\cong -\frac{4}{(\Delta r)^2} f_1 - \frac{4}{(\Delta r)^2} f_2, \\ \hat{\nabla}_r^4 f(r=0) &\cong \frac{18}{(\Delta r)^4} f_1 - \frac{24}{(\Delta r)^4} f_2 + \frac{6}{(\Delta r)^4} f_3, \\ \hat{\nabla}_r^2 f(r=\Delta r) &\cong \frac{f_1}{2(\Delta r)^2} - 2\frac{f_2}{(\Delta r)^2} + \frac{3f_3}{2(\Delta r)^2}, \\ \hat{\nabla}_r^4 f(r=\Delta r) &\cong -\frac{3}{(\Delta r)^4} f_1 + \left(\frac{41}{8} + 2\right)\frac{f_2}{(\Delta r)^4} - \frac{6}{(\Delta r)^4} f_3 + \frac{15}{8}\frac{f_4}{(\Delta r)^4}.\end{aligned} \tag{9.17}$$

9.3 Reduced membrane-bending equations for prestressed sandwich axisymmetric plates of variable thickness

Let us consider a sandwich plate composed of three layers: two working layers of thickness g_s and a core of depth $h_s - g_s$ (Fig. 9.2). In this approximation the uniform cross-section of the plate may be treated as a double-point substitutive section and, hence, the process of integration of stresses through the thickness is reduced to simply summing them up. The introduced substitutive section is statically determined (cf. Życzkowski, 1981) and requires redefinition of membrane and bending stiffnesses (9.7) (cf. Armand, 1972)

$$\mathcal{B}_s(r) = 2\frac{Eg_s(r)}{(1-\nu^2)}, \qquad \mathcal{D}_s(r) = \frac{Eh_s^2(r)\, g_s(r)}{2(1-\nu^2)}. \tag{9.18}$$

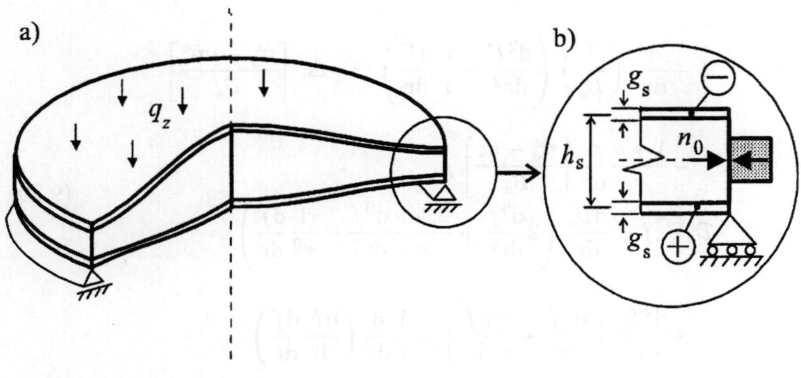

Fig. 9.2. Substitutive sandwich section for plate element: a) sandwich section of variable thickness, b) a simply supported plate prestressed in-plane by the elastic ring (cf. Ganczarski and Skrzypek, 1993)

9.3.1 Kármán equations extended to viscoelasticity

Let us take into account the most general problem of a sandwich plate of variable thickness where thicknesses of the core $h_s - g_s$ and the working layers g_s may change. In what follows we confine ourselves to the mixed-type formulation (9.13) and (9.14) which is more convenient for further analysis. After formally differentiating (9.13) and (9.14) with respect to time and assuming of stationary behavior of body forces and temperature fields, we arrive at the following system of equations ($k = 0$):

I. Elastic problem ($t = 0$)

$$\begin{aligned}
&\frac{1}{\mathcal{B}_s}\widehat{\nabla}_r^4 F + \frac{\mathrm{d}}{\mathrm{d}r}\left(\frac{1}{\mathcal{B}_s}\right)\left(2\frac{\mathrm{d}^3 F}{\mathrm{d}r^3} + \frac{2-\nu}{r}\frac{\mathrm{d}^2 F}{\mathrm{d}r^2} - \frac{1}{r^2}\frac{\mathrm{d}F}{\mathrm{d}r}\right)\\
&+\frac{\mathrm{d}^2}{\mathrm{d}r^2}\left(\frac{1}{\mathcal{B}_s}\right)\left(\frac{\mathrm{d}^2 F}{\mathrm{d}r^2} - \frac{\nu}{r}\frac{\mathrm{d}F}{\mathrm{d}r}\right) = 0,\\
&\mathcal{D}_s\widehat{\nabla}_r^4 f - \frac{\mathrm{d}\mathcal{D}_s}{\mathrm{d}r}\left(2\frac{\mathrm{d}^3 f}{\mathrm{d}r^3} + \frac{2+\nu}{r}\frac{\mathrm{d}^2 f}{\mathrm{d}r^2} - \frac{1}{r^2}\frac{\mathrm{d}f}{\mathrm{d}r}\right)\\
&+\frac{\mathrm{d}^2\mathcal{D}_s}{\mathrm{d}r^2}\left(\frac{\mathrm{d}^2 f}{\mathrm{d}r^2} + \frac{\nu}{r}\frac{\mathrm{d}f}{\mathrm{d}r}\right) - \frac{1}{r}\frac{\mathrm{d}}{\mathrm{d}r}\left(\frac{\mathrm{d}F}{\mathrm{d}r}\frac{\mathrm{d}f}{\mathrm{d}r}\right) = q.
\end{aligned} \tag{9.19}$$

II. Creep problem ($t > 0$)

$$\begin{aligned}
&\frac{1}{\mathcal{B}_s}\widehat{\nabla}_r^4 \dot{F} + \frac{\mathrm{d}}{\mathrm{d}r}\left(\frac{1}{\mathcal{B}_s}\right)\left(2\frac{\mathrm{d}^3 \dot{F}}{\mathrm{d}r^3} + \frac{2-\nu}{r}\frac{\mathrm{d}^2 \dot{F}}{\mathrm{d}r^2} - \frac{1}{r^2}\frac{\mathrm{d}\dot{F}}{\mathrm{d}r}\right)\\
&+\frac{\mathrm{d}^2}{\mathrm{d}r^2}\left(\frac{1}{\mathcal{B}_s}\right)\left(\frac{\mathrm{d}^2 \dot{F}}{\mathrm{d}r^2} - \frac{\nu}{r}\frac{\mathrm{d}\dot{F}}{\mathrm{d}r}\right) = -\Delta_r\left[\frac{\dot{n}_\vartheta^c - \nu\dot{n}_r^c}{\mathcal{B}_s}\right]\\
&-\frac{1+\nu}{r}\frac{\mathrm{d}}{\mathrm{d}r}\left[\frac{\dot{n}_\vartheta^c - \dot{n}_r^c}{\mathcal{B}_s}\right],\\
&\mathcal{D}_s\widehat{\nabla}_r^4 \dot{f} + \frac{\mathrm{d}\mathcal{D}_s}{\mathrm{d}r}\left(2\frac{\mathrm{d}^3 \dot{f}}{\mathrm{d}r^3} + \frac{2+\nu}{r}\frac{\mathrm{d}^2 \dot{f}}{\mathrm{d}r^2} - \frac{1}{r^2}\frac{\mathrm{d}\dot{f}}{\mathrm{d}r}\right)\\
&+\frac{\mathrm{d}^2\mathcal{D}_s}{\mathrm{d}r^2}\left(\frac{\mathrm{d}^2 \dot{f}}{\mathrm{d}r^2} + \frac{\nu}{r}\frac{\mathrm{d}\dot{f}}{\mathrm{d}r}\right) - \frac{1}{r}\frac{\mathrm{d}}{\mathrm{d}r}\left(\frac{\mathrm{d}F}{\mathrm{d}r}\frac{\mathrm{d}f}{\mathrm{d}r}\right)^{\cdot}\\
&= -\frac{\mathrm{d}^2\dot{m}_r^c}{\mathrm{d}r^2} - \frac{1}{r}\frac{\mathrm{d}(2\dot{m}_r^c - \dot{m}_\vartheta^c)}{\mathrm{d}r},
\end{aligned} \tag{9.20}$$

where generalized membrane stresses are expressed by the following differential operators of the Airy function:

$$n_r = \frac{1}{r}\frac{\mathrm{d}F}{\mathrm{d}r}, \qquad n_\vartheta = \frac{\mathrm{d}^2 F}{\mathrm{d}r^2}, \tag{9.21}$$

whereas definitions of inelastic generalized stresses (9.6) take a simplified form:

$$\begin{aligned}
n_{r/\vartheta}^c &= \frac{\mathcal{B}_s}{2}\left[\varepsilon_{r/\vartheta}^{c^+} + \varepsilon_{r/\vartheta}^{c^-} + \nu\left(\varepsilon_{\vartheta/r}^{c^+} + \varepsilon_{\vartheta/r}^{c^-}\right)\right],\\
m_{r/\vartheta}^c &= \frac{\mathcal{D}_s}{2}\left[\varepsilon_{r/\vartheta}^{c^+} - \varepsilon_{r/\vartheta}^{c^-} + \nu\left(\varepsilon_{\vartheta/r}^{c^+} - \varepsilon_{\vartheta/r}^{c^-}\right)\right].
\end{aligned} \tag{9.22}$$

Consequently, the following extension of the Love–Kirchhoff hypothesis (9.1) holds:

$$\varepsilon^{\pm}_{r/\vartheta} = \lambda_{r/\vartheta} \pm \kappa_{r/\vartheta} \frac{h_{\mathrm{s}}}{2}. \tag{9.23}$$

9.3.2 *Constitutive equations for coupled creep-damage problems*

Constitutive equations (9.5) when applied to the sandwich section take the form

$$\sigma^{\pm}_{r/\vartheta} = \frac{E}{1-\nu^2}\left[\left(\lambda_{r/\vartheta} + \nu\lambda_{\vartheta/r}\right) \pm \left(\kappa_{r/\vartheta} + \nu\kappa_{\vartheta/r}\right)\frac{h_{\mathrm{s}}}{2} - \left(\varepsilon^{\mathrm{c}\pm}_{r/\vartheta} + \nu\varepsilon^{\mathrm{c}\pm}_{\vartheta/r}\right)\right], \tag{9.24}$$

where inelastic strains $\varepsilon^{\mathrm{c}+}_{r/\vartheta}$ have to be specified by the use of creep-damage constitutive state equations.

Assuming the similarity of deviators based on the flow theory and the time hardening hypothesis associated with the Kachanov orthotropic brittle damage law (cf. Kachanov, 1986, also Ganczarski and Skrzypek, 1992), the following system of partly coupled constitutive equations for creep and damage (isotropic flow rule and orthotropic damage growth rule) is formulated:

$$\begin{aligned} \dot{\varepsilon}^{\mathrm{c}\pm}_{kl} &= \frac{3}{2}\frac{\dot{\varepsilon}^{\mathrm{c}\pm}_{\mathrm{eq}}}{\sigma^{\pm}_{\mathrm{eq}}} s^{\pm}_{kl}, \qquad \dot{\varepsilon}^{\mathrm{c}\pm}_{\mathrm{eq}} = \left(\widetilde{\sigma}^{\pm}_{\mathrm{eq}}\right)^{m(T)} \dot{\mathsf{f}}(t), \\ \dot{D}^{\pm}_{\nu} &= C_{\nu}(T)\left\langle \frac{\sigma^{\pm}_{\nu}}{1-D^{\pm}_{\nu}} \right\rangle^{r_{\nu}(T)}. \end{aligned} \tag{9.25}$$

Additionally, assuming plane stress and creep incompressibility, the intensities of the stress, the effective stress, and the creep strain rates are defined by the following formulas:

$$\begin{aligned} \sigma^{\pm}_{\mathrm{eq}} &= \sqrt{\frac{3}{2} s^{\pm}_{kl} s^{\pm}_{kl}}, \\ \widetilde{\sigma}^{\pm}_{\mathrm{eq}} &= \sqrt{\left(\frac{\sigma^{\pm}_{r}}{1-D^{\pm}_{r}}\right)^2 + \left(\frac{\sigma^{\pm}_{\vartheta}}{1-D^{\pm}_{\vartheta}}\right)^2 - \frac{\sigma^{\pm}_{r}\sigma^{\pm}_{\vartheta}}{\left(1-D^{\pm}_{r}\right)\left(1-D^{\pm}_{\vartheta}\right)}}, \\ \dot{\varepsilon}^{\mathrm{c}\pm}_{r/\vartheta} &= \frac{\left(\widetilde{\sigma}^{\pm}_{\mathrm{eq}}\right)^m}{\sigma^{\pm}_{\mathrm{eq}}}\left(\sigma^{\pm}_{r/\vartheta} - \frac{1}{2}\sigma^{\pm}_{\vartheta/r}\right)\dot{\mathsf{f}}(t), \qquad \dot{\varepsilon}^{\mathrm{c}\pm}_{z} = -\left(\dot{\varepsilon}^{\mathrm{c}\pm}_{r} + \dot{\varepsilon}^{\mathrm{c}\pm}_{\vartheta}\right). \end{aligned} \tag{9.26}$$

However, in the case of strongly nonproportional loadings, it is more reasonable to couple the effective stress deviator and the creep strain deviator in the orthotropic flow rule which leads to the fully coupled approach:

$$\dot{\varepsilon}^{c\pm}_{kl} = \frac{3}{2}\frac{\dot{\varepsilon}^{c\pm}_{eq}}{\widetilde{\sigma}^{\pm}_{eq}}\widetilde{s}^{\pm}_{kl}. \tag{9.27}$$

In the case of plane stress and creep incompressibility, by setting $C_{r/\vartheta}(T) = C$, $r_{r/\vartheta} = r$, the following representation of (9.25) through (9.26) holds:

$$\begin{gathered}
\dot{\varepsilon}^{c\pm}_{r/\vartheta} = \left(\widetilde{\sigma}^{\pm}_{r/\vartheta}\right)^{m-1}\left[\frac{\sigma^{\pm}_{r/\vartheta}}{1-D^{\pm}_{r/\vartheta}} - \frac{1}{2}\frac{\sigma^{\pm}_{\vartheta/r}}{1-D^{\pm}_{\vartheta/r}}\right]\dot{\mathsf{f}}(t), \\
\dot{\varepsilon}^{c\pm}_{z} = -\left(\dot{\varepsilon}^{c\pm}_{r} + \dot{\varepsilon}^{c\pm}_{\vartheta}\right), \qquad \dot{D}^{\pm}_{r/\vartheta} = C\left\langle\frac{\sigma^{\pm}_{r/\vartheta}}{1-D^{\pm}_{r/\vartheta}}\right\rangle^{r},
\end{gathered} \tag{9.28}$$

where the effective deviatoric stress components are given by

$$\widetilde{s}^{\pm}_{r/\vartheta} = \frac{2}{3}\left(\frac{\sigma^{\pm}_{r/\vartheta}}{1-D^{\pm}_{r/\vartheta}} - \frac{1}{2}\frac{\sigma^{\pm}_{\vartheta/r}}{1-D^{\pm}_{\vartheta/r}}\right). \tag{9.29}$$

9.3.3 Membrane state equation in a particular case of constant thickness of the working layers

Let us return to the displacement formulation of the membrane state equations (9.11). In case when the membrane stiffness $\mathcal{B}_s$ for sandwich section (9.18) is independent of the distance between the working layers $h_s(r)$, the temperature field is considered as stationary, and there are no radial body force components,

$$\frac{d\mathcal{B}_s}{dr} = 0, \qquad \dot{T} = 0, \qquad q_r = 0, \tag{9.30}$$

the membrane state equations reduce to the Euler-type equations

$$\begin{aligned}
&\frac{d^2u}{dr^2} + \frac{1}{r}\frac{du}{dr} - \frac{u}{r^2} = (1+\nu)\alpha\frac{dT}{dr} && (t=0), \\
&\frac{d^2\dot{u}}{dr^2} + \frac{1}{r}\frac{d\dot{u}}{dr} - \frac{\dot{u}}{r^2} = \frac{1}{\mathcal{B}_s}\left(\frac{d\dot{u}^c_r}{dr} + \frac{\dot{n}^c_r - \dot{n}^c_\vartheta}{r}\right) && (t>0),
\end{aligned} \tag{9.31}$$

the analytical solution of which is furnished as follows:

$$
\left.
\begin{aligned}
&u = C_1\frac{r}{2} + \frac{C_2}{r} + (1+\nu)\frac{\alpha}{r}\int_0^r T\xi \mathrm{d}\xi, \\
&n_r = \mathcal{B}_s\left[\frac{1+\nu}{2}C_1 - (1-\nu)\frac{C_2}{r^2} - (1-\nu^2)\frac{\alpha}{r^2}\int_0^r T\xi \mathrm{d}\xi\right], \\
&n_\vartheta = \mathcal{B}_s\left[\frac{1+\nu}{2}C_1 + (1-\nu)\frac{C_2}{r^2} + (1-\nu^2)\frac{\alpha}{r^2}\int_0^r T\xi \mathrm{d}\xi \right.\\
&\qquad \left. -(1-\nu^2)\alpha T\right];
\end{aligned}
\right\} (t=0)
$$

$$
\left.
\begin{aligned}
&\dot{u} = C_3\frac{r}{2} + \frac{C_4}{r} + \frac{r}{\mathcal{B}_s}\left(\frac{\dot{I}_1}{1-\nu} + \frac{\dot{I}_2}{1+\nu}\right), \\
&\dot{n}_r = \mathcal{B}_s\left[\frac{1+\nu}{2}C_3 - (1-\nu)\frac{C_4}{r^2}\right] - \dot{I}_1 + \dot{I}_2, \\
&\dot{n}_\vartheta = \mathcal{B}_s\left[\frac{1+\nu}{2}C_3 + (1-\nu)\frac{C_4}{r^2}\right] + \dot{I}_1 + \dot{I}_2 - \dot{n}_\vartheta^c + \nu\dot{n}_r^c,
\end{aligned}
\right\} (t>0)
\tag{9.32}
$$

where the auxiliary integrals $\dot{I}_1$ and $\dot{I}_2$ are

$$
\dot{I}_1 = \frac{1-\nu}{2r^2}\int_0^r \left(\dot{n}_r^c + \dot{n}_\vartheta^c\right)\xi \mathrm{d}\xi, \qquad \dot{I}_2 = \frac{1+\nu}{2}\int_0^r \frac{\dot{n}_r^c - \dot{n}_\vartheta^c}{\xi}\mathrm{d}\xi. \tag{9.33}
$$

9.3.4 Bending state equations in a case of the rigidification principle (no coupling between the membrane and bending states)

Systems of equations (9.19) and (9.20) become uncoupled when the rigidification principle is applied. Hence, products of membrane forces and bending displacements vanish and the systems (9.19) and (9.20) reduce to two independent systems of equations. The equations of bending state take a classical form when basic mode of deformation $k = 0$ is assumed and the thickness is constant $h_s = \text{const}$, $g_s = \text{const}$

$$
\begin{aligned}
&\mathcal{D}_s\nabla_r^4 f = q_z && (t=0), \\
&\mathcal{D}_s\nabla_r^4 \dot{f} = -\frac{\mathrm{d}^2\dot{m}_r^c}{\mathrm{d}r^2} - \frac{1}{r}\frac{\mathrm{d}(2\dot{m}_r^c - \dot{m}_\vartheta^c)}{\mathrm{d}r} && (t>0).
\end{aligned}
\tag{9.34}
$$

When definitions of the angle of slope and the shear force are introduced in (9.34) as follows:

$$\varphi = -\frac{\mathrm{d}f}{\mathrm{d}r}, \qquad Q = \frac{1}{r}\int_0^r q_z \xi \mathrm{d}\xi \tag{9.35}$$

the Euler-type equations (Penny and Marriott, 1995, Ganczarski, 1992) are obtained

$$\begin{aligned} &\frac{\mathrm{d}^2\varphi}{\mathrm{d}r^2} + \frac{1}{r}\frac{\mathrm{d}\varphi}{\mathrm{d}r} - \frac{\varphi}{r^2} = -\frac{Q}{\mathcal{D}_\mathrm{s}} && (t=0),\\ &\frac{\mathrm{d}^2\dot{\varphi}}{\mathrm{d}r^2} + \frac{1}{r}\frac{\mathrm{d}\dot{\varphi}}{\mathrm{d}r} - \frac{\dot{\varphi}}{r^2} = \frac{1}{\mathcal{D}_\mathrm{s}}\left(\frac{\mathrm{d}\dot{m}_r^\mathrm{c}}{\mathrm{d}r} + \frac{\dot{m}_r^\mathrm{c} - \dot{m}_\vartheta^\mathrm{c}}{r}\right) && (t>0), \end{aligned} \tag{9.36}$$

the solution of which takes the following form (Ganczarski and Skrzypek, 1993):

$$\left.\begin{aligned} \varphi &= \frac{1}{\mathcal{D}_\mathrm{s}}\left(\frac{C_1}{2}r + \frac{C_2}{r} - \frac{r}{2}\int_0^r Q\mathrm{d}\xi + \frac{1}{2r}\int_0^r Q\xi^2\mathrm{d}\xi\right),\\ m_r &= \frac{1+\nu}{2}C_1 - (1-\nu)\frac{C_2}{r^2} - \frac{1+\nu}{2}\int_0^r Q\mathrm{d}\xi\\ &\quad -\frac{1-\nu}{2r^2}\int_0^r Q\xi^2\mathrm{d}\xi,\\ m_\vartheta &= \frac{1+\nu}{2}C_1 + (1-\nu)\frac{C_2}{r^2} - \frac{1+\nu}{2}\int_0^r Q\mathrm{d}\xi\\ &\quad +\frac{1-\nu}{2r^2}\int_0^r Q\xi^2\mathrm{d}\xi; \end{aligned}\right\} \quad (t=0)$$

$$\left.\begin{aligned} \dot{\varphi} &= \frac{1}{\mathcal{D}_\mathrm{s}}\left[\frac{C_3}{2}r + \frac{C_4}{r} + \left(\frac{\dot{I}_3}{1-\nu} + \frac{\dot{I}_4}{1+\nu}\right)r\right],\\ \dot{m}_r &= \frac{1+\nu}{2}C_3 - (1-\nu)\frac{C_4}{r^2} + \dot{I}_3 + \dot{I}_4,\\ \dot{m}_\vartheta &= \frac{1+\nu}{2}C_3 + (1-\nu)\frac{C_4}{r^2} - \dot{I}_3 + \dot{I}_4\\ &\quad -\dot{m}_\vartheta^\mathrm{c} + \nu\dot{m}_r^\mathrm{c}, \end{aligned}\right\} \quad (t>0) \tag{9.37}$$

where the auxiliary integrals are defined as follows:

$$\dot{I}_3 = \frac{1-\nu}{2r^2} \int_0^r (\dot{m}_r^c + \dot{m}_\vartheta^c)\,\xi \mathrm{d}\xi, \qquad \dot{I}_4 = \frac{1+\nu}{2} \int_0^r \frac{\dot{m}_r^c - \dot{m}_\vartheta^c}{\xi} \mathrm{d}\xi.$$

9.4 Thermally prestressed sandwich plates of constant thickness

9.4.1 Basic unilaterally coupled membrane-bending Kármán equations visco-thermo-elasticity

In a case when a plate is simultaneously loaded by out-of-plane forces (due to bending) and in-plane forces (due to membrane prestressing), the equations for the membrane and bending states are unilaterally coupled in such a sense that in the bending state equation an additional term affected by the membrane force appears. Hence, (9.31) and (9.34) are no longer uncoupled and, therefore, they must be solved simultaneously. In order to formulate coupled membrane-bending equations, it is more convenient here to use the displacement formulation for both the membrane and the bending state.

Let us consider an axisymmetric sandwich plate of constant thickness under the following assumptions:

i. two-point substitutive sandwich section obeys the Love–Kirchhoff hypothesis,

ii. the plate is loaded by a uniform pressure q and a stationary temperature field $T(r)$,

iii. the creep-damage properties are described by the flow theory, the time hardening hypothesis and the orthotropic damage growth rule when the partly coupled formulation is used (Table 4.2),

iv. the initial prestressing is imposed by the elastic ring or the cylindrical shell,

v. the displacement formulation of the unilaterally coupled membrane -bending equations is applied,

vi. a constant plate thickness is assumed, $\mathrm{d}\mathcal{B}_s/\mathrm{d}r = 0$, $\mathrm{d}\mathcal{D}_s/\mathrm{d}r = 0$ (Fig. 9.3),

vii. the fundamental symmetric deformation mode is assumed, $w(r,\vartheta) = w(r)$.

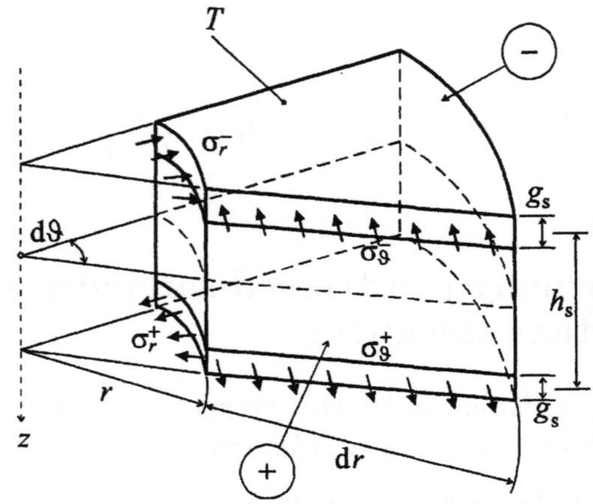

Fig. 9.3. Substitutive sandwich section for a plate element

When the geometrically linear theory of small displacements is applied with the geometry changes introduced, the problem can be formulated as unilaterally coupled, where terms associated with the Gaussian curvature are disregarded and the von Kármán coupled equation system extended to visco-thermo-elasticity can be used:

$$\left.\begin{array}{l}\dfrac{\mathrm{d}^2u}{\mathrm{d}r^2}+\dfrac{1}{r}\dfrac{\mathrm{d}u}{\mathrm{d}r}-\dfrac{u}{r^2}=(1+\nu)\alpha\dfrac{\mathrm{d}T}{\mathrm{d}r}\\ \mathcal{D}_\mathrm{s}\nabla_r^4w-\dfrac{1}{r}\dfrac{\mathrm{d}(n_r r\dfrac{\mathrm{d}w}{\mathrm{d}r})}{\mathrm{d}r}=q\end{array}\right\}\quad(t=0),$$

$$\left.\begin{array}{l}\mathcal{B}_\mathrm{s}\left(\dfrac{\mathrm{d}^2\dot{u}}{\mathrm{d}r^2}+\dfrac{1}{r}\dfrac{\mathrm{d}\dot{u}}{\mathrm{d}r}-\dfrac{\dot{u}}{r^2}\right)=\dfrac{\mathrm{d}\dot{n}_r^\mathrm{c}}{\mathrm{d}r}+\dfrac{\dot{n}_r^\mathrm{c}-\dot{n}_\vartheta^\mathrm{c}}{r}\\ \mathcal{D}_\mathrm{s}\nabla_r^4\dot{w}-\dfrac{1}{r}\left[\dfrac{\mathrm{d}\left(n_r r\dfrac{\mathrm{d}w}{\mathrm{d}r}\right)}{\mathrm{d}r}\right]^{\cdot}=-\dfrac{\mathrm{d}^2\dot{m}_r^\mathrm{c}}{\mathrm{d}r^2}\\ \qquad-\dfrac{1}{r}\dfrac{\mathrm{d}(2\dot{m}_r^\mathrm{c}-\dot{m}_\vartheta^\mathrm{c})}{\mathrm{d}r}\end{array}\right\}\quad(t>0).\qquad(9.38)$$

The elementary solutions of the equations of the membrane state (9.38) are given by (9.32):

$$\left.\begin{aligned} u &= C_1\frac{r}{2} + \frac{C_2}{r} + (1+\nu)\alpha T\frac{r}{2} \\ n_{r/\vartheta} &= \mathcal{B}_\text{s}\left[\frac{(1+\nu)}{2}C_1 \pm (1-\nu)\frac{C_2}{r^2} \mp (1-\nu^2)\alpha\frac{T}{2}\right] \end{aligned}\right\} (t=0), \quad (9.39)$$

$$\left.\begin{aligned} \dot{u} &= C_3\frac{r}{2} + \frac{C_4}{r} + \frac{r}{\mathcal{B}_s}\left[\frac{\dot{I}_1}{1-\nu} + \frac{\dot{I}_2}{1+\nu}\right] \\ \dot{n}_r &= \mathcal{B}_\text{s}\left[\frac{(1+\nu)}{2}C_3 - (1-\nu)\frac{C_4}{r^2}\right] - \dot{I}_1 + \dot{I}_2 \\ \dot{n}_\vartheta &= \mathcal{B}_\text{s}\left[\frac{(1+\nu)}{2}C_3 + (1-\nu)\frac{C_4}{r^2}\right] + \dot{I}_1 + \dot{I}_2 - \dot{n}^\text{c}_\vartheta + \nu\dot{n}^\text{c}_r \end{aligned}\right\} (t>0),$$

where

$$\dot{I}_1 = \frac{1-\nu}{2r^2}\int_0^r \left(\dot{n}^\text{c}_r + \dot{n}^\text{c}_\vartheta\right)\xi \,\text{d}\xi, \quad \dot{I}_2 = \frac{1+\nu}{2}\int_0^r \frac{\dot{n}^\text{c}_r - \dot{n}^\text{c}_\vartheta}{\xi}\text{d}\xi, \quad (9.40)$$

The equations of the bending state must be solved numerically (e.g., by the FDM) with the previously obtained solution of the membrane state introduced as the coupling. The stresses, their rates, and the generalized inelastic stresses are defined as follows:

$$\left.\begin{aligned} \sigma^\pm_{r/\vartheta} &= \pm\frac{m_{r/\vartheta}}{hg} + \frac{n_{r/\vartheta}}{2g}, \\ \dot{\sigma}^\pm_{r/\vartheta} &= \pm\frac{\dot{m}_{r/\vartheta} + \dot{m}^\text{c}_{r/\vartheta}}{hg} + \frac{\dot{n}_{r/\vartheta} + \dot{n}^\text{c}_{r/\vartheta}}{2g} - \frac{\mathcal{B}_\text{s}}{2g}\left(\dot{\varepsilon}^{\text{c}\pm}_{r/\vartheta} + \nu\dot{\varepsilon}^{\text{c}\pm}_{\vartheta/r}\right); \\ n^\text{c}_{r/\vartheta} &= \frac{\mathcal{B}_\text{s}}{2}\left[\varepsilon^{\text{c}+}_{r/\vartheta} + \varepsilon^{\text{c}-}_{r/\vartheta} + \nu\left(\varepsilon^{\text{c}+}_{\vartheta/r} + \varepsilon^{\text{c}-}_{\vartheta/r}\right)\right], \\ m^\text{c}_{r/\vartheta} &= \frac{\mathcal{D}_\text{s}}{h}\left[\varepsilon^{\text{c}+}_{r/\vartheta} - \varepsilon^{\text{c}-}_{r/\vartheta} + \nu\left(\varepsilon^{\text{c}+}_{\vartheta/r} - \varepsilon^{\text{c}-}_{\vartheta/r}\right)\right]. \end{aligned}\right. \quad (9.41)$$

The total strains are decomposed into elastic (superscript e), creep (superscript c), and thermal parts: $\varepsilon^\pm_{r/\vartheta} = \varepsilon^{\text{e}\pm}_{r/\vartheta} + \varepsilon^{\text{c}\pm}_{r/\vartheta} + \alpha t$, where superscripts + or − refer to the lower (exterior) or the upper (interior) sandwich layers, respectively. The similarity of deviators, based on the flow theory (partly coupled approach), and the time hardening hypothesis associated with Kachanov's orthotropic brittle damage theory, are taken as the constitutive relationships for creep (cf. Boyle and Spence, 1983, Kachanov,

1986). For the plane stress state, additionally assuming that the creep incompressibility, the strain rates, and the intensities of the stress and of the effective stress are defined by the following formulas (cf. Ganczarski and Skrzypek, 1991):

$$
\begin{aligned}
\dot{\varepsilon}^{c\pm}_{r/\vartheta} &= \frac{\left(\widetilde{\sigma}^{\pm}_{\mathrm{eq}}\right)^{m}}{\sigma^{\pm}_{\mathrm{eq}}}\left(\sigma^{\pm}_{r/\vartheta} - \frac{\sigma^{\pm}_{\vartheta/r}}{2}\right)\dot{\mathsf{f}}(t),\\
\dot{\varepsilon}^{c\pm}_{z} &= -\dot{\varepsilon}^{c\pm}_{r} - \dot{\varepsilon}^{c\pm}_{\vartheta},\\
\sigma^{\pm}_{\mathrm{eq}} &= \sqrt{\frac{3}{2}s^{\pm}_{kl}s^{\pm}_{kl}} = \sqrt{\sigma^{\pm 2}_{r} + \sigma^{\pm 2}_{\vartheta} - \sigma^{\pm}_{r}\sigma^{\pm}_{\vartheta}},\\
\widetilde{\sigma}^{\pm}_{\mathrm{eq}} &= \sqrt{\left(\frac{\sigma^{\pm}_{r}}{1-D^{\pm}_{r}}\right)^{2} + \left(\frac{\sigma^{\pm}_{\vartheta}}{1-D^{\pm}_{\vartheta}}\right)^{2} - \left(\frac{\sigma^{\pm}_{r}}{1-D^{\pm}_{r}}\right)\left(\frac{\sigma^{\pm}_{\vartheta}}{1-D^{\pm}_{\vartheta}}\right)}.
\end{aligned}
\tag{9.42}
$$

9.4.2 Example: Built-in plate fitted into the cylindrical shell

An elastically built-in plate fitted into a cylindrical shell, with the initial fit δ imposed, is considered (Fig. 9.4).

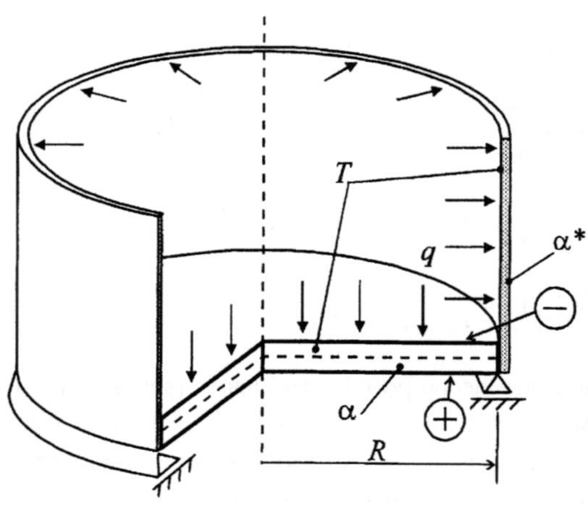

Fig. 9.4. A built-in plate prestressed by the elastic cylindrical shell (cf. Ganczarski and Skrzypek, 1993b)

$$\left.\begin{aligned} n_r(R) &= -N \\ m_r(R) &= -M \\ u(R) - \bar{w}(0) &= \delta \\ \frac{\mathrm{d}w(R)}{\mathrm{d}r} &= \bar{\varphi}(0) \end{aligned}\right\} (t=0) \qquad \left.\begin{aligned} \dot{n}_r(R)\mathrm{d}r &= -\mathrm{d}N \\ \dot{m}_r(R)\mathrm{d}r &= -\mathrm{d}M \\ \dot{u}(R)\mathrm{d}t - \mathrm{d}\bar{w}(0) &= 0 \\ \frac{\mathrm{d}\dot{w}(R)}{\mathrm{d}r}\mathrm{d}t &= \mathrm{d}\bar{\varphi}(0) \end{aligned}\right\} (t>0). \tag{9.43}$$

The cylindrical shell is described by the classical equation:

$$\frac{\mathrm{d}^4\bar{w}}{\mathrm{d}x^4} + 4\bar{k}^4\bar{w} = \frac{2-\bar{\nu}}{2}\frac{q}{\bar{D}} + \frac{\bar{E}\bar{h}}{R\bar{D}}\bar{\alpha}T \tag{9.44}$$

the solution of which (for the half-infinite structure) takes the form:

$$\begin{aligned} \bar{w}(x) &= \frac{M}{2\bar{D}\bar{k}^2}e^{-\bar{k}x}\left[\cos(\bar{k}x) - \sin(\bar{k}x)\right] \\ &\quad + \frac{N}{2\bar{D}\bar{k}^3}e^{-\bar{k}x}\cos(\bar{k}x) + \frac{2-\bar{\nu}}{2}\frac{R^2}{\bar{E}\bar{h}}q + R\bar{\alpha}T, \\ \bar{\varphi}(x) &= \frac{\mathrm{d}\bar{w}(x)}{\mathrm{d}x} = -\frac{M}{\bar{D}\bar{k}}e^{-\bar{k}x}\cos(\bar{k}x) - \frac{N}{2\bar{D}\bar{k}^2}e^{-\bar{k}x}[\cos(\bar{k}x) - \sin(\bar{k}x)], \\ \bar{k}^4 &= \frac{3(1-\bar{\nu}^2)}{R^2\bar{h}^2}, \qquad \bar{D} = \frac{\bar{E}\bar{h}^3}{12(1-\bar{\nu}^2)}. \end{aligned} \tag{9.45}$$

The double-point failure criterion is formulated as follows.

The initial fit δ that produces the peripheral prestressing radial force N and the corresponding bending moment M, such that the failure simultaneously occurs both in the central plate region at the exterior circumferential fibers $D_\vartheta^+(0) \to 1$ and along the periphery at the interior radial fibers $D_r^-(R) \to 1$, is sought for.

$$t = t_I: \qquad \sup[D^{\pm}_{r/\vartheta}(\delta)]_{r\cong 0} = \sup[D^{\pm}_{r/\vartheta}(\delta)]_{r=R} = 1. \tag{9.46}$$

The maximum value of the damage components $D_r^{\pm}$ and $D_\vartheta^{\pm}$ is found when both sandwich layers are examined (cf. Ganczarski, 1992). Note that in this case the prestressing force N and the moment M are the dependent quantities since they both depend on one prestressing parameter δ. Hence, when the transient creep process is solved, one of these quantities, say N, is to be determined by the additional iteration loop in order to satisfy the current plate-shell interaction.

The numerical example deals with a plate made of ASTM 321 stainless steel: $E = 177$ GPa, $\sigma_0 = 118$ MPa, $\nu = 0.3$, $\alpha = 1.8 \times 10^{-5}$ K^{-1}, $R = 0.5$ m, $h_s = 0.025$ m, $g_s = 0.005$ m, $q = 118$ kPa; the temperature dependent material constants for creep rupture at temperature 783 K are (cf. Odqvist, 1966): $C = 2.13 \times 10^{-42}$ Pa^{-r}/s, $r = 3.9$, $m = 5.6$, whereas material constants for the cylindrical shell made of ASTM 310 stainless steel are as follows: $\bar{E} = E$, $\bar{\nu} = \nu$, $\bar{\alpha} = 1.7 \times 10^{-5}$ K^{-1}, $\bar{h} = \sqrt[3]{\frac{6}{5}\frac{h_s}{2.5}}$.

The dimensionless lifetime of the plate fitted into the cylindrical shell versus the fit δ is shown in Fig. 9.5. The maximum lifetime appears in a characteristic "switch" point, at the intersection of the curves that correspond to two different failure mechanisms, failure due to macrocracks in the circumferential fibers in the central region in the interior layer and failure due to macrocracks in the radial fibers along the periphery in the exterior layer (Fig. 9.6). The maximum lifetime corresponds to the positive value of fit $\delta \geq 0$, which means that the shell must initially be "too loose".

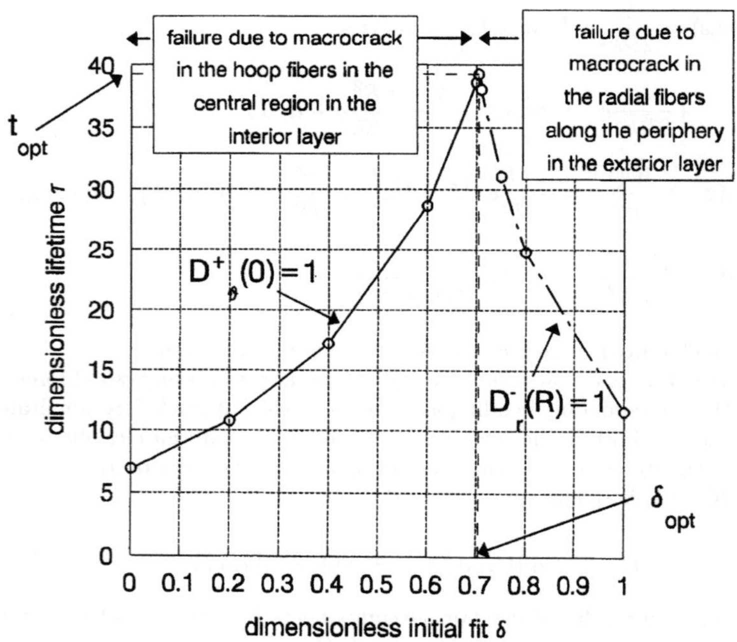

Fig. 9.5. Lifetime of the plate fitted into the cylindrical shell versus the initial fit δ

The corresponding distributions of the continuity components $\psi^{\pm}_{r/\vartheta}$ at the instant of failure for the optimal solutions are presented in Fig. 9.7. The time to rupture prediction for the discussed case is $t_{\rm opt} = 2.97t_I$, compared to the lifetime of a simply supported plate in a pure bending state without the initial prestressing $t_{\rm b} = t_I$.

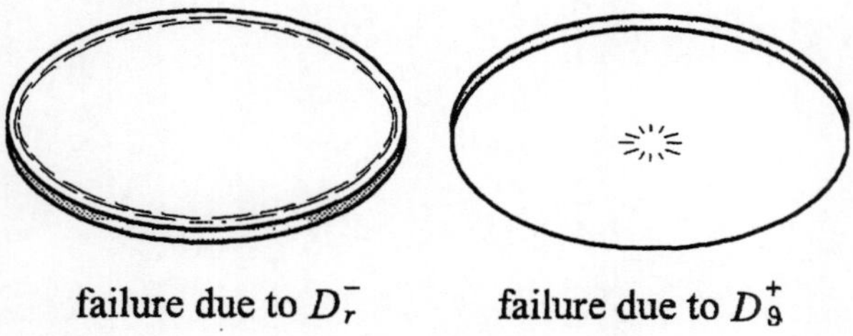

Fig. 9.6. Failure mechanisms for optimal prestressing $\delta_{\rm opt}$

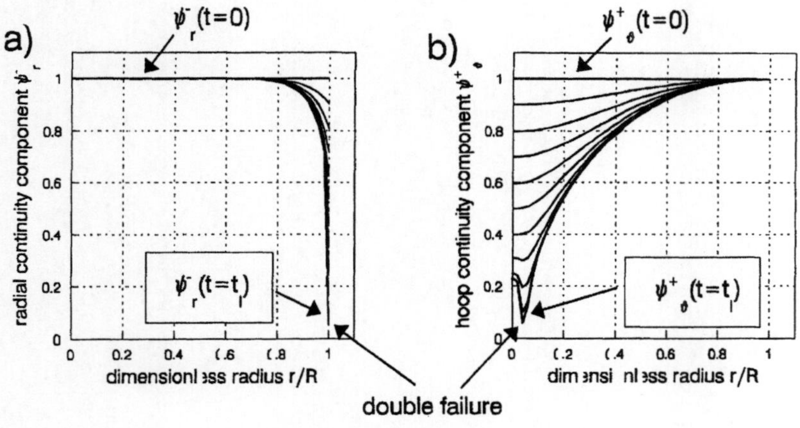

Fig. 9.7. Damage evolution with time to failure according to the orthotropic damage growth in an built-in plate of constant thickness prestressed by the elastic shell: a) the radial component (peripheral exterior), b) the hoop component (central interior) (cf. Ganczarski and Skrzypek, 1993)

10

Two-dimensional coupled anisotropic creep-brittle damage and elastic-brittle failure problems

10.1 Orthotropic coupled creep-brittle damage of Reissner's plates under in-plane and out-of-plane loadings

10.1.1 General equations

In the frame of the classical theory of thin plates (cf. Chap. 9), the effect of shear deformation due to the transverse stress is disregarded, which is equivalent to assuming of the shear modulus is equal to infinity. A more accurate and realitic theory is due to Reissner.

Let us consider an element of a plate of moderate thickness $h(x_1, x_2)$ subjected to an external transversal load $q\mathrm{d}x_1\mathrm{d}x_2$ and to a system of stress components (Fig. 10.1).

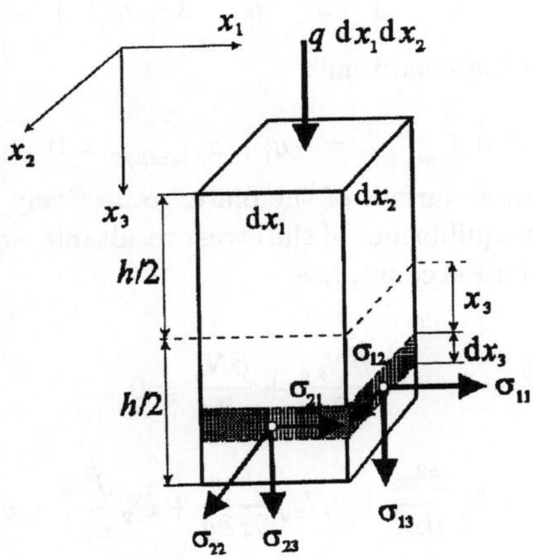

Fig. 10.1. Loadings imposed to an element of Reissner's plate

Assuming that small total strains are decomposed into elastic and creep components $\varepsilon_{ij} = \varepsilon^{\mathrm{e}}_{ij} + \varepsilon^{\mathrm{c}}_{ij}$, the displacements u_0, v_0, w_0 of any point across the thickness of the plate fulfill relations:

$$\frac{\partial u_0}{\partial x} = \frac{1}{E}[\sigma_x - \nu(\sigma_y + \sigma_z)] + \varepsilon_x^c,$$

$$\frac{\partial v_0}{\partial y} = \frac{1}{E}[\sigma_y - \nu(\sigma_x + \sigma_z)] + \varepsilon_y^c,$$

$$\frac{\partial u_0}{\partial y} + \frac{\partial v_0}{\partial x} = \frac{\tau_{xy}}{G} + \gamma_{xy}^c, \tag{10.1}$$

$$\frac{\partial u_0}{\partial z} + \frac{\partial w_0}{\partial x} = \frac{\tau_{xz}}{G} + \gamma_{xz}^c,$$

$$\frac{\partial v_0}{\partial z} + \frac{\partial w_0}{\partial y} = \frac{\tau_{yz}}{G} + \gamma_{yz}^c.$$

The equation $\partial w_0/\partial z = [\sigma_z - \nu(\sigma_x + \sigma_y)]/E + \varepsilon_z^c$ is not used, as it contradicts the assumed linear law of the stress components distribution $\sigma_x, \sigma_y, \tau_{xy}$. According to the classical Reissner–Mindlin moderate thickness plate theory, the straight and normal segment to the mid-plane before deformation remains straight but not necessarily normal to the mid-plane after deformation. On the other hand, inextensibility of the normal segment is usually assumed $\partial w_0/\partial z = 0$, although, in what follows, the stress component σ_z is also accounted for in the form (cf. Love, 1944):

$$\sigma_z = -\frac{3q}{4}\left[\frac{2}{3} - \frac{2z}{h} + \frac{1}{3}\left(\frac{2z}{h}\right)^3\right], \tag{10.2}$$

satisfying the following conditions:

$$\sigma_z\big|_{z=-h/2} = -q, \quad \sigma_z\big|_{z=h/2} = 0 \tag{10.3}$$

at the upper and lower surface of the plate, respectively.

The equations of equilibrium of the stress resultants, when the geometry changes are taken into account, are:

$$\frac{\partial N_x}{\partial x} + \frac{\partial N_{xy}}{\partial y} = 0, \qquad \frac{\partial N_{xy}}{\partial x} + \frac{\partial N_y}{\partial y} = 0,$$

$$\frac{\partial Q_x}{\partial x} + \frac{\partial Q_y}{\partial y} + N_x\frac{\partial^2 w}{\partial x^2} + 2N_{xy}\frac{\partial^2 w}{\partial x \partial y} + N_y\frac{\partial^2 w}{\partial y^2} + q = 0, \tag{10.4}$$

$$\frac{\partial M_x}{\partial x} - \frac{\partial M_{xy}}{\partial y} - Q_x = 0, \qquad -\frac{\partial M_{xy}}{\partial x} + \frac{\partial M_y}{\partial y} - Q_y = 0.$$

The average value w of the transverse displacement, taken over the thickness of the plate, as well as average values φ_x, φ_y of the rotation angles and in-plane displacements u, v result from equating the work of resultant couples on the average rotations and the work of resultant forces on average displacements with the work of the corresponding stresses on actual displacements u_0, v_0, w_0 in the same section:

$$w = \frac{3}{2h}\int_{-h/2}^{h/2} w_0\left[1-\left(\frac{2z}{h}\right)^2\right]dz,$$

$$\varphi_x = \frac{12}{h^2}\int_{-h/2}^{h/2}\frac{u_0 z}{h}dz, \quad \varphi_y = \frac{12}{h^2}\int_{-h/2}^{h/2}\frac{v_0 z}{h}dz, \tag{10.5}$$

$$u = \int_{-h/2}^{h/2}\frac{u_0}{h}dz, \quad v = \int_{-h/2}^{h/2}\frac{v_0}{h}dz.$$

Expressing average displacements by resultant forces and resultant couples in equilibrium equation (10.4), the following system of three equations is obtained (cf. Love, 1944):

$$\begin{aligned}
&\nabla^2 u + \frac{1+\nu}{2}\frac{\partial}{\partial y}\left(\frac{\partial v}{\partial x} - \frac{\partial u}{\partial y}\right) = \frac{1}{\mathcal{B}}\frac{\partial N_x^c}{\partial x},\\
&\nabla^2 v + \frac{1+\nu}{2}\frac{\partial}{\partial x}\left(\frac{\partial u}{\partial y} - \frac{\partial v}{\partial x}\right) = \frac{1}{\mathcal{B}}\frac{\partial N_y^c}{\partial y},\\
&\mathcal{D}\nabla^4 w + N_x\frac{\partial^2 w}{\partial x^2} + 2N_{xy}\frac{\partial^2 w}{\partial x\partial y} + N_y\frac{\partial^2 w}{\partial y^2} = q - \frac{h^2}{10}\frac{2-\nu}{1-\nu}\nabla^2 q\\
&+\mathcal{D}\nabla^2\left(\frac{\partial Q_x^c}{\partial x} + \frac{\partial Q_y^c}{\partial y}\right) - \frac{\partial^2 M_x^c}{\partial x^2} - 2\frac{\partial^2 M_{xy}^c}{\partial x\partial y} - \frac{\partial^2 M_y^c}{\partial y^2},
\end{aligned} \tag{10.6}$$

where the following definitions of the isotropic membrane and the bending stiffnesses:

$$\mathcal{B} = \frac{Eh}{1-\nu^2}, \qquad \mathcal{D} = \frac{Eh^3}{12(1-\nu^2)} \tag{10.7}$$

and the generalized inelastic forces are introduced:

$$N_{x/y}^c = \frac{E}{1-\nu^2}\int_{-h/2}^{h/2}\left(\varepsilon_{x/y}^c + \nu\varepsilon_{y/x}^c\right)dz,$$

$$N_{xy}^c = \frac{E}{2(1+\nu)}\int_{-h/2}^{h/2}\gamma_{xy}^c dz,$$

$$Q^{\mathrm{c}}_{x/y} = \frac{3}{2h}\int_{-h/2}^{h/2} \gamma^{\mathrm{c}}_{xz/yz}\left[1-\left(\frac{2z}{h}\right)^2\right] dz,$$

$$M^{\mathrm{c}}_{x/y} = \frac{E}{1-\nu^2}\int_{-h/2}^{h/2}\left(\varepsilon^{\mathrm{c}}_{x/y} + \nu\varepsilon^{\mathrm{c}}_{y/x}\right) z dz, \tag{10.8}$$

$$M^{\mathrm{c}}_{xy} = \frac{E}{2(1+\nu)}\int_{-h/2}^{h/2} \gamma^{\mathrm{c}}_{xy} z dz.$$

The derived system of equations (10.6) is the simplified unilaterally coupled Kármán system extended to the case of visco-elastic plate of moderate thickness. In the Kármán formulation, the fully coupled equations of membrane and bending states occur where additional nonlinear terms associated with Gaussian curvature appear in the equations of membrane state expressed in terms of the Airy stress function (the third-order theory).

10.1.2 Basic equations of axisymmetric plate

Assumption of an axisymmetric problem allows elimination of the displacement in the circumferential direction v. Expanding transverse displacement and bending moments in trigonometric series, the basic system of equations (10.6), transformed to cylindrical coordinates r, ϑ, z, takes the form when the engineering notation is used:
I. Elastic state ($t = 0$)

$$\begin{aligned}
&\nabla^2 u - \frac{u}{r^2} = 0,\\
&\mathcal{D}\nabla^4 w - N_r\frac{\mathrm{d}^2 w}{\mathrm{d}r^2} - N_\vartheta\frac{1}{r}\frac{\mathrm{d}w}{\mathrm{d}r} + N_\vartheta\frac{k^2}{r^2}w = q,
\end{aligned} \tag{10.9a}$$

II. Creep state ($t > 0$)

$$\begin{aligned}
&\nabla^2\dot{u} - \frac{\dot{u}}{r^2} = \frac{1}{\mathcal{B}}\left(\frac{\mathrm{d}\dot{N}^{\mathrm{c}}_r}{\mathrm{d}r} + \frac{\dot{N}^{\mathrm{c}}_r - \dot{N}^{\mathrm{c}}_\vartheta}{r}\right),\\
&\mathcal{D}\nabla^4\dot{w} - \left(N_r\frac{\mathrm{d}^2 w}{\mathrm{d}r^2}\right)^{\cdot} - \left(N_\vartheta\frac{1}{r}\frac{\mathrm{d}w}{\mathrm{d}r}\right)^{\cdot} + \left(N_\vartheta\frac{k^2}{r^2}w\right)^{\cdot}\\
&= \mathcal{D}\nabla^2\left(\frac{\mathrm{d}\dot{Q}^{\mathrm{c}}_{rz}}{\mathrm{d}r}\right) - \frac{\mathrm{d}^2\dot{M}^{\mathrm{c}}_r}{\mathrm{d}r^2} - \frac{1}{r}\frac{\mathrm{d}}{\mathrm{d}r}\left(2\dot{M}^{\mathrm{c}}_r - \dot{M}^{\mathrm{c}}_\vartheta\right) + \frac{k^2}{r^2}\dot{M}^{\mathrm{c}}_\vartheta,
\end{aligned} \tag{10.9b}$$

under additional assumptions that $M_{r\vartheta} = 0, N_{r\vartheta} = 0, N^{\mathrm{c}}_{r\vartheta} = 0$ and $q =$ const. In the further analysis only the fundamental mode $k = 0$ is considered.

Suppose that the vector of displacements $\{u, w\}$ is found; then, all internal variables are expressed by the following formulas:
I. Elastic solution ($t = 0$)

$$\varphi = -\frac{\mathrm{d}w}{\mathrm{d}r} - \frac{6}{5}\frac{1+\nu}{Eh}qr,$$

$$\kappa_r = \frac{\mathrm{d}^2 w}{\mathrm{d}r^2}, \qquad \kappa_\vartheta = \frac{1}{r}\frac{\mathrm{d}w}{\mathrm{d}r}, \qquad \lambda_r = \frac{\mathrm{d}u}{\mathrm{d}r}, \qquad \lambda_\vartheta = \frac{u}{r},$$

$$m_{r/\vartheta} = \mathcal{D}\left(\kappa_{r/\vartheta} + \nu\kappa_{\vartheta/r}\right), \qquad n_{r/\vartheta} = \mathcal{B}\left(\lambda_{r/\vartheta} + \nu\lambda_{\vartheta/r}\right),$$

$$\sigma_{r/\vartheta} = \frac{12 m_{r/\vartheta}}{h^3}z + \frac{n_{r/\vartheta}}{h} + \frac{\nu}{1-\nu}\sigma_z,$$

$$\sigma_z = -\frac{3q}{4}\left[\frac{2}{3} - 2\frac{z}{h} + \frac{1}{3}\left(\frac{2z}{h}\right)^3\right], \qquad \tau_{rz} = -\frac{3q}{4}\frac{r}{h}\left[1 - \left(\frac{2z}{h}\right)^2\right], \tag{10.10a}$$

II. Creep solution ($t > 0$)

$$\dot\varphi = -\frac{\mathrm{d}\dot w}{\mathrm{d}r} + \dot Q^{\mathrm{c}},$$

$$\dot m_{r/\vartheta} = \mathcal{D}\left(\dot\kappa_{r/\vartheta} + \nu\dot\kappa_{\vartheta/r} + \frac{\mathrm{d}\dot Q^{\mathrm{c}}}{\mathrm{d}r} + \nu\frac{\dot Q^{\mathrm{c}}}{r}\right) - \dot m^{\mathrm{c}}_{r/\vartheta},$$

$$\dot n_{r/\vartheta} = \mathcal{B}\left(\dot\lambda_{r/\vartheta} + \nu\dot\lambda_{\vartheta/r}\right) - \dot n^{\mathrm{c}}_r,$$

$$\dot\sigma_{r/\vartheta} = \frac{12(\dot m_{r/\vartheta} + \dot m^{\mathrm{c}}_{r/\vartheta})}{h^3}z + \frac{\dot n_{r/\vartheta} + \dot n^{\mathrm{c}}_{r/\vartheta}}{h} + \frac{E}{1-\nu^2}\left(\dot\varepsilon^{\mathrm{c}}_{r/\vartheta} + \dot\varepsilon^{\mathrm{c}}_{\vartheta/r}\right),$$

$$\dot\sigma_z = 0, \qquad \dot\tau_{rz} = 0. \tag{10.10b}$$

Note that in the particular case of an infinitely thin plate, expressions for the bending moments coincide with the classical thin-plate theory.

10.1.3 Constitutive equations

In the case when the transverse shear effects are taken into account, the principal directions $\alpha_i(I, J)$ of the stress tensor undergo plane rotation with time $\mathrm{d}\alpha_i(I, J)$ and, consequently, the principal directions of microcracks $\beta_i(I, J)$ follow them. All constitutive equations, the flow rule, the time hardening hypothesis, and the Kachanov-type orthotropic brittle rupture law are employed for current coordinate system referring to the principal directions $\alpha_i(I, J)$ of the stress tensor ($I = J$):

$$\dot{\varepsilon}^{c}_{IJ} = \frac{3}{2}\frac{\dot{\varepsilon}^{c}_{eq}}{\sigma_{eq}} s_{IJ}, \text{ partly coupled approach,} \tag{10.11a}$$

$$\dot{\varepsilon}^{c}_{IJ} = \frac{3}{2}\frac{\dot{\varepsilon}^{c}_{eq}}{\widetilde{\sigma}_{eq}} \widetilde{s}_{IJ}, \text{ fully coupled approach,} \tag{10.11b}$$

$$\dot{\varepsilon}^{c}_{eq} = \left(\widetilde{\sigma}_{eq}\right)^{m} \dot{f}(t), \qquad \dot{D}_{IJ} = C\left\langle \frac{\sigma_{IJ}}{1-D_{IJ}} \right\rangle^{r}, \tag{10.11c}$$

where the actual state of damage is represented by a second-rank symmetric tensor D_{IJ}. Depending on the partly or fully coupled creep-damage approach, the principal directions of the creep strain rates deviator coincide with the principal directions of either the stress deviator or the effective stress deviator, respectively and the following definitions hold:

$$\sigma_{eq} = \sqrt{\frac{3}{2} s_{IJ} s_{IJ}}, \quad \widetilde{\sigma}_{eq} = \sqrt{\frac{3}{2} \widetilde{s}_{IJ} \widetilde{s}_{IJ}}, \quad \dot{\varepsilon}^{c}_{eq} = \sqrt{\frac{2}{3} \dot{\varepsilon}^{c}_{IJ} \dot{\varepsilon}^{c}_{IJ}}. \tag{10.12}$$

Introducing the brittle damage law defines nonobjective measure of the damage rate tensor $\dot{\mathbf{D}}$ (the effect of rotation of the principal directions is disregarded). The objective measure based on the definition of the Zaremba–Jaumann derivative on the plane rotation (r, z) is defined as:

$$\overset{\nabla}{D}_{IJ} = \dot{D}_{IJ} - D^{T}_{IJ} \begin{bmatrix} 0 & 0 & -\mathrm{d}\alpha \\ 0 & 0 & 0 \\ \mathrm{d}\alpha & 0 & 0 \end{bmatrix} - \begin{bmatrix} 0 & 0 & \mathrm{d}\alpha \\ 0 & 0 & 0 \\ -\mathrm{d}\alpha & 0 & 0 \end{bmatrix} D_{IJ}. \tag{10.13}$$

When the objective damage rate tensor $\overset{\nabla}{D}_{IJ}$ (10.13) is transformed from current principal directions of the stress tensor $\alpha_i(IJ)$ to the sampling coordinates (ij) $\overset{\nabla}{D}_{ij}$, the new representation of the damage tensor $D_{ij}(t + \Delta t)$ is achieved:

$$D_{ij}(t+\Delta t) = D_{ij}(t) + \overset{\nabla}{D}_{ij}(t)\Delta t. \tag{10.14}$$

The graphical interpretation of all auxiliary coordinate systems associated with the definition of the objective damage rate tensor $\overset{\nabla}{D}_{ij}$ in case of the axisymmetric plane stress state is shown in Fig. 10.2.

10.1.4 Initial and boundary conditions

Two boundary problems are considered:

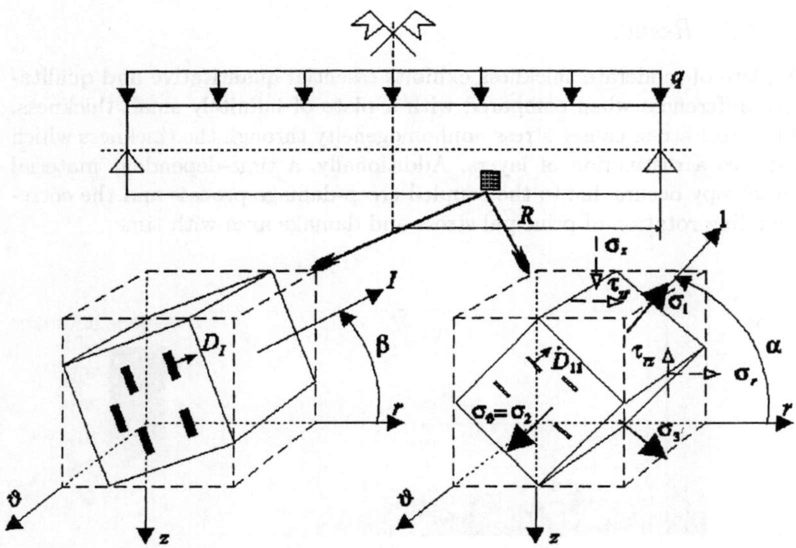

Fig. 10.2. Corotational coordiante systems coincided with locally principal directions of damage or stress tensors

Example A: Prestressed simply supported plate

$$\begin{array}{ll} \text{for } t=0 & \text{for } t>0 \\ u(0)=0 & \dot{u}(0)=0 \\ n_r(R)=-n_0 & \dot{n}_r(R)=0 \\ \varphi(0)=0 & \dot{\varphi}(0)=0 \\ m_r(R)=0 & \dot{m}_r(R)=0 \\ w(R)=0 & \dot{w}(R)=0. \end{array} \tag{10.15}$$

Example B: Prestressed clamped plate

$$\begin{array}{ll} \text{for } t=0 & \text{for } t>0 \\ u(0)=0 & \dot{u}(0)=0 \\ n_r(R)=-n_0 & \dot{n}_r(R)=0 \\ \varphi(0)=0 & \dot{\varphi}(0)=0 \\ \varphi(R)=0 & \dot{\varphi}(R)=0 \\ w(R)=0 & \dot{w}(R)=0. \end{array} \tag{10.16}$$

The plates are made of ASTM 321 stainless steel (rolled 18 Cr 8 Ni 0.45 Si 0.4 Mn 0.1 C Ti/Nb stabilized, austenite annealed at 1070°C, air cooled) with the following properties at temperature 500°C (cf. Odqvist, 1974): E = 180 GPa, $\sigma_{0.2}$ = 120 MPa, ν = 0.3, m = 5.6, r = 3.9, $\sigma^5_{C_B}$ = 210 MPa, where $\sigma^5_{C_B}$ denotes the stress causing creep rupture in 10^5 hr. Plate thickness to diameter ratio is $h/2R = 0.1$, and $q = 0.01 \times \sigma_{0.2}$.

10.1.5 Results

A plate of moderate thickness exhibits essential quantitative and qualitative differences when compared with a plate of infinitely small thickness. The shear stress causes stress nonhomogeneity through the thickness which requires a distinction of layers. Additionally, a time-dependent material anisotropy occurs due to the coupled creep-damage process and the corresponding rotation of principal stress and damage axes with time.

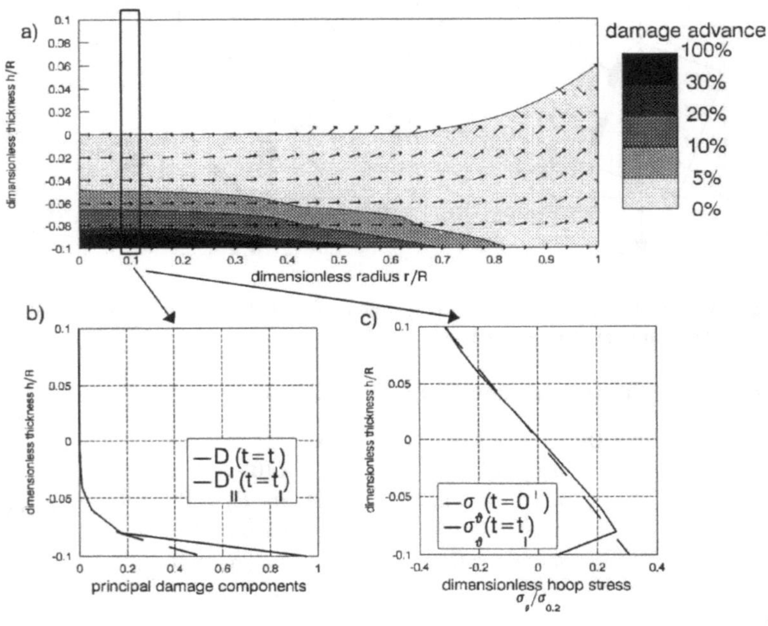

Fig. 10.3. Example A: Simply supported unprestressed plate: a) map of damage advance and rotation of principal directions of damage, b) distribution of principal components of damage, c) hoop stress relaxation at point of first microcrack

In case of the simply supported unprestressed plate ($n_0 = 0$) the tensile stresses are dominant at the center of the plate on the bottom external fibers (Fig. 10.3a,b), nevertheless, combined creep relaxation and damage processes cause the first macrocrack with respect to the hoop direction D_I to appear at a certain distance from the plate center (cf. Ganczarski and Skrzypek, 1993, 1994). The corresponding hoop stress component rapidly relaxes in the damaged zone (Fig. 10.3c). The rotation of principal directions of damage, which follow current principal directions of tensile stresses, is particularly clear in the inner zone around the neutral axis. At the instant of load imposition, $t = 0^+$, they exhibit a slope of 45° which gradually decreases with time to reach 0° at the instant of first macrocrack $t = t_I$ (Fig. 10.3a).

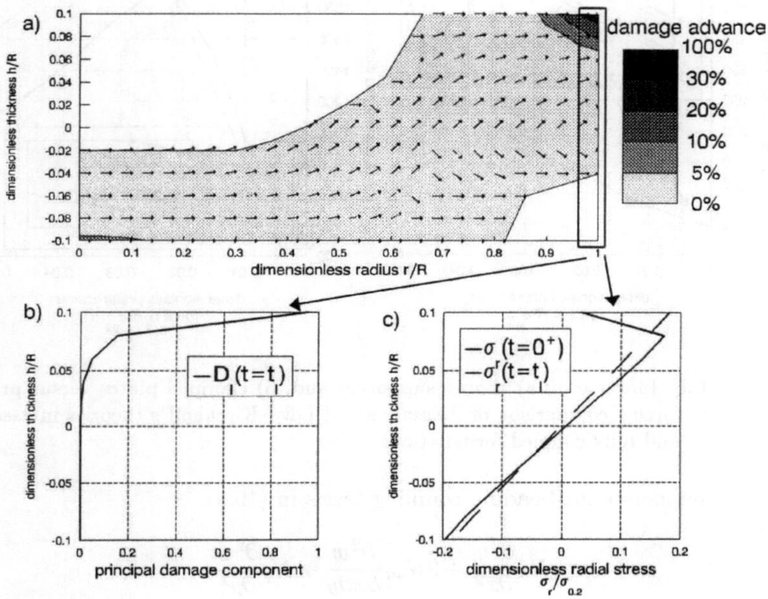

Fig. 10.4. Example B: Clamped unprestressed plate: a) map of damage advance and rotation of principal direction of damage, b) distribution of principal component of damage, c) radial stress relaxation at point of first microcrack

In contrast to the above described mode of support, in the case of the clamped unprestressed plate ($n_0 = 0$) there exist two zones of tensile stresses due to bending moments changing signs: one in the central bottom fibers of low advance of damage (5%) and the other, dominant, at the peripheral top fibers (Fig. 10.4a,b). Therefore, the radial stress relaxes there quickly (Fig. 10.4c). The field of principal directions of damage exhibits characteristic perturbation around the abscissa $r/R = 0.6$ according to the change of signs of bending moments, and the above mentioned effect of straightening with time of principal directions is observed (Fig. 10.4a).

An essential improvement of the plate lifetime is obtained when the prestressing force $n_0 \neq 0$ is imposed in the plate mid-surface (Fig. 10.5a,b). In both considered cases, an optimal control of prestressing, decreasing tensile stresses, turns out to be a powerful technique for lifetime improvement

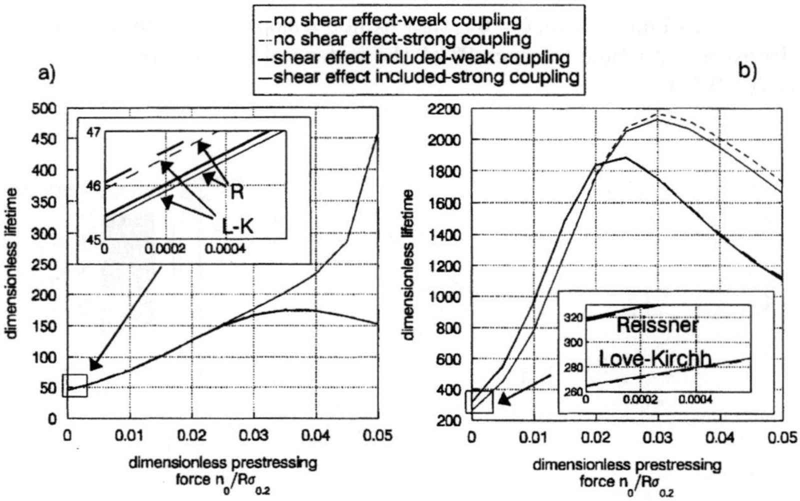

Fig. 10.5. Lifetime of: a) simply supported and, b) clamped plates versus prestressing force – comparison of Reissner's and Love–Kirchhoff's theories in cases of partly and fully coupled formulations

until the membrane-bending coupling terms in (10.6),

$$N_x \frac{\partial^2 w}{\partial x^2} + 2N_{xy} \frac{\partial^2 w}{\partial x \partial y} + N_y \frac{\partial^2 w}{\partial y^2},$$

begin to dominate. In the case of a simply supported plate, small and moderate magnitudes of prestressing ($n_0/R\sigma_{0.2} \leq 0.025$) do not result in the essential differences between Reissner's theory and the classical Love–Kirchhoff theory, whereas in the clamped plate case Reissner's theory yields up to 20% improvement of lifetime, decreasing with the prestressing growth.

Precise analysis of the lifetime of prestressed clamped plate allows one to observe quantitative differences in time to rupture between the isotropic (partly coupled) (10.11a) and the orthotropic (fully coupled) (10.11b) formulations of the flow rule. For advanced prestressing ($n_0/R\sigma_{0.2} \geq 0.03$), when paths of loading are strongly nonproportional, the relative improvement of lifetime for fully coupled formulations may reach 4.5%. Other cases do not confirm such clear differences, which are comparable with rounding errors.

10.2 2D CDM approach to coupled damage-fracture of plates under in-plane loadings

10.2.1 Geometry, loadings governing equations and boundary conditions of a structure

A simply supported, clamped 2D structure subjected to in-plane uniform load as it is shown in Fig. 10.6 is analyzed. The solution is considered in the domain $\mathfrak{D} = \{(x_1, x_2) \in \mathbb{R}^2 : x_1 \in \langle 0, 5 \rangle, x_2 \in \langle 0, 1 \rangle\}$, where x_i, $i = 1, 2$ denote dimensionless independent variables $x_1 = x/w$, $x_2 = y/w$, w is the structure width, and x, y are Cartesian coordinates.

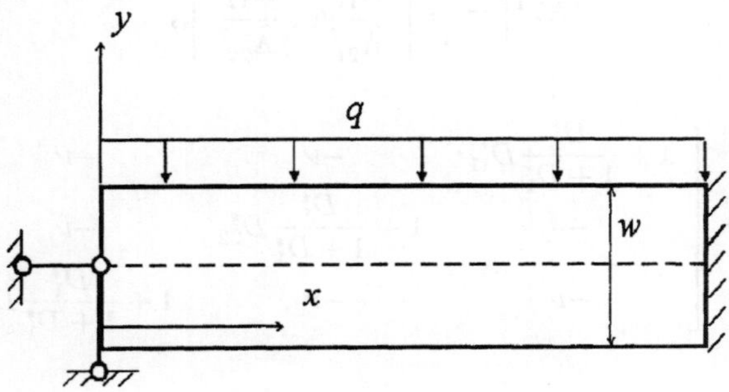

Fig. 10.6. Scheme of structure and load geometry

A local approach to fracture is applied when the modified CDM Litewka model of the orthotropic time-dependent elastic-brittle damage in crystalline metals is used as the constitutive and evolution equations (cf. Sect. 4.2).

The following dimensionless quantities are defined: $\overline{q} = \dfrac{q}{\sigma_{\mathrm{u}} \times 1\,[\mathrm{m}]} = 0.02$, Young's modulus $\overline{E} = E/\sigma_{\mathrm{u}} = 416.7$, Poisson's ratio $\nu = 0.3$, $\overline{C} = C\sigma_{\mathrm{u}} \times 1\,[\mathrm{s}] = 6.81 \times 10^6$, where $\sigma_{\mathrm{u}} = 288$ MPa. The material data corresponds to the carbon steel AISI at a temperature of 811 K (cf. Litewka, 1989).

The constitutive relationships rewritten in the matrix representation, referred to the global frame (1, 2, 3) cf. (Sect. 4.4.2), are of the following form

$$\{\boldsymbol{\varepsilon}\} = \left[\widetilde{\boldsymbol{\Lambda}}_L^{-1}\right]\{\overline{\boldsymbol{\sigma}}\} \tag{10.17}$$

where $\{\boldsymbol{\varepsilon}\}$ and $\{\overline{\boldsymbol{\sigma}}\}$ are the strain and the dimensionless stress vectors

$$\{\boldsymbol{\varepsilon}\} = \begin{Bmatrix} \varepsilon_{11} \\ \varepsilon_{22} \\ \varepsilon_{33} \\ \gamma_{23} \\ \gamma_{31} \\ \gamma_{12} \end{Bmatrix}, \qquad \{\overline{\boldsymbol{\sigma}}\} = \begin{Bmatrix} \overline{\sigma}_{11} \\ \overline{\sigma}_{22} \\ \overline{\sigma}_{33} \\ \overline{\sigma}_{23} \\ \overline{\sigma}_{31} \\ \overline{\sigma}_{12} \end{Bmatrix} \tag{10.18}$$

and $[\overline{\widetilde{\boldsymbol{\Lambda}}_L^{-1}}(\mathbf{D}^*)]$ is the elastic compliance matrix of damaged material (4.87), in general expressed in terms of six components D^*_{11}, D^*_{22}, D^*_{33}, $D^*_{12} = D^*_{21}$, $D^*_{23} = D^*_{32}$, $D^*_{31} = D^*_{13}$ of the modified damage tensor $\mathbf{D}^*$ and its first eigenvalue D^*_{I}:

$$\left[\overline{\widetilde{\boldsymbol{\Lambda}}_L^{-1}}\right] = \frac{1}{\overline{\overline{E}}} \begin{bmatrix} \overline{\widetilde{\boldsymbol{\Lambda}}_{11}^{-1}} & \overline{\widetilde{\boldsymbol{\Lambda}}_{12}^{-1}} \\ \overline{\widetilde{\boldsymbol{\Lambda}}_{21}^{-1}}^{\mathrm{T}} & \overline{\widetilde{\boldsymbol{\Lambda}}_{22}^{-1}} \end{bmatrix}, \tag{10.19}$$

$$\left[\overline{\widetilde{\boldsymbol{\Lambda}}_{11}^{-1}}\right] = \begin{bmatrix} 1+\dfrac{D^*_{\mathrm{I}}}{1+D^*_{\mathrm{I}}}D^*_{11} & -\nu & -\nu \\ -\nu & 1+\dfrac{D^*_{\mathrm{I}}}{1+D^*_{\mathrm{I}}}D^*_{22} & -\nu \\ -\nu & -\nu & 1+\dfrac{D^*_{\mathrm{I}}}{1+D^*_{\mathrm{I}}}D^*_{33} \end{bmatrix}, \tag{10.20}$$

$$\left[\overline{\widetilde{\boldsymbol{\Lambda}}_{22}^{-1}}\right] = \begin{bmatrix} 2+2\nu+\dfrac{D^*_{\mathrm{I}}}{1+D^*_{\mathrm{I}}}(D^*_{22}+D^*_{33}) & \dfrac{D^*_{\mathrm{I}}}{1+D^*_{\mathrm{I}}}D^*_{12} & \dfrac{D^*_{\mathrm{I}}}{1+D^*_{\mathrm{I}}}D^*_{13} \\ \dfrac{D^*_{\mathrm{I}}}{1+D^*_{\mathrm{I}}}D^*_{12} & 2+2\nu+\dfrac{D^*_{\mathrm{I}}}{1+D^*_{\mathrm{I}}}(D^*_{11}+D^*_{33}) & \dfrac{D^*_{\mathrm{I}}}{1+D^*_{\mathrm{I}}}D^*_{23} \\ \dfrac{D^*_{\mathrm{I}}}{1+D^*_{\mathrm{I}}}D^*_{13} & \dfrac{D^*_{\mathrm{I}}}{1+D^*_{\mathrm{I}}}D^*_{23} & 2+2\nu+\dfrac{D^*_{\mathrm{I}}}{1+D^*_{\mathrm{I}}}(D^*_{11}+D^*_{22}) \end{bmatrix}, \tag{10.21}$$

$$\left[\overline{\widetilde{\boldsymbol{\Lambda}}_{12}^{-1}}\right] = \begin{bmatrix} 0 & \dfrac{D^*_{\mathrm{I}}}{1+D^*_{\mathrm{I}}}D^*_{13} & \dfrac{D^*_{\mathrm{I}}}{1+D^*_{\mathrm{I}}}D^*_{12} \\ \dfrac{D^*_{\mathrm{I}}}{1+D^*_{\mathrm{I}}}D^*_{23} & 0 & \dfrac{D^*_{\mathrm{I}}}{1+D^*_{\mathrm{I}}}D^*_{12} \\ \dfrac{D^*_{\mathrm{I}}}{1+D^*_{\mathrm{I}}}D^*_{23} & \dfrac{D^*_{\mathrm{I}}}{1+D^*_{\mathrm{I}}}D^*_{13} & 0 \end{bmatrix}. \tag{10.22}$$

Symbols $\overline{\sigma}_{ij}$ denote dimensionless components of the stress vector, $\overline{\sigma}_{ij} = \sigma_{ij}/\sigma_{\mathrm{u}}$, and $\overline{E}$ is the dimensionless Young's modulus $\overline{E} = E/\sigma_{\mathrm{u}}$, $\overline{\widetilde{\mathbf{\Lambda}}_{11}^{-1}}$, $\overline{\widetilde{\mathbf{\Lambda}}_{22}^{-1}}$, $\overline{\widetilde{\mathbf{\Lambda}}_{12}^{-1}}$, and $\overline{\widetilde{\mathbf{\Lambda}}_{12}^{-1}}^{\mathrm{T}}$ denote submatrices of the 6×6 matrix $\widetilde{\mathbf{\Lambda}}^{-1}$.

For numerical implementation the number of physical equations (10.17) has to be reduced by the number of non zero stress components. Therefore, in the case of a plain state of stress, (10.17) take the form

$$\left\{ \begin{array}{c} \varepsilon_{11} \\ \varepsilon_{22} \\ \gamma_{12} \end{array} \right\} = \frac{1}{\overline{E}} \left[\begin{array}{ccc} 1 + D_1 D_{11}^* & -\nu & D_1 D_{12}^* \\ -\nu & 1 + D_1 D_{22}^* & D_1 D_{12}^* \\ D_1 D_{12}^* & D_1 D_{12}^* & 2 + 2\nu + D_1 \left(D_{11}^* + D_{22}^* \right) \end{array} \right] \left\{ \begin{array}{c} \overline{\sigma}_{11} \\ \overline{\sigma}_{22} \\ \overline{\sigma}_{12} \end{array} \right\}. \tag{10.23}$$

By use of the Zaremba–Jaumann objective derivative $\overset{\nabla}{\mathbf{D}}$

$${}^{t}\overset{\nabla}{D}_{IJ} = {}^{t}\dot{D}_{IJ} - {}^{t}D_{IK}\,{}^{t}S_{KJ} - {}^{t}D_{JL}\,{}^{t}S_{LI}, \tag{10.24}$$

where ${}^{t}\dot{D}_{IJ}$ are components of the time-derivative of the damage tensor evaluated at time t and ${}^{t}S_{IJ}$ are components of the spin tensor

$$[\mathbf{S}] = \left[\begin{array}{ccc} 0 & \dot{\alpha}_1 & -\dot{\alpha}_2 \\ -\dot{\alpha}_1 & 0 & \dot{\alpha}_3 \\ \dot{\alpha}_2 & -\dot{\alpha}_3 & 0 \end{array} \right], \tag{10.25}$$

the components of the damage tensor objective derivative are given here as follows:

$$\begin{aligned} \overset{\nabla}{D}_{11} &= \dot{D}_{11} + 2\dot{\alpha} D_{12}, \\ \overset{\nabla}{D}_{22} &= \dot{D}_{22} - 2\dot{\alpha} D_{12}, \\ \overset{\nabla}{D}_{12} &= \dot{D}_{21} = \dot{\alpha}\left(D_{22} - D_{11} \right), \end{aligned} \tag{10.26}$$

whereas nonobjective derivatives $\dot{D}_{11}$ and $\dot{D}_{22}$ are given by the damage growth rule (4.48) and (4.49)

$$\begin{aligned} \dot{D}_{11} &= \mathsf{K} \left\langle \overline{\sigma}_1 \right\rangle, \\ \dot{D}_{22} &= \mathsf{K} \left\langle \overline{\sigma}_2 \right\rangle, \end{aligned} \tag{10.27}$$

where

$$\mathsf{K} = \frac{\overline{C}}{4\overline{E}^2}\left\{1 - 4\nu\left(\frac{\overline{\sigma}_2}{\overline{\sigma}_1}\right) + 2(1+2\nu)\left(\frac{\overline{\sigma}_2}{\overline{\sigma}_1}\right)^2 - 4\nu\left(\frac{\overline{\sigma}_2}{\overline{\sigma}_1}\right)^3 + \left(\frac{\overline{\sigma}_2}{\overline{\sigma}_1}\right)^4\right.$$
$$\left. +2D_1\left[1 - 2\nu\left(\frac{\overline{\sigma}_2}{\overline{\sigma}_1}\right) + \left(\frac{\overline{\sigma}_2}{\overline{\sigma}_1}\right)^2\right]\left[D_{11}^* + \left(\frac{\overline{\sigma}_2}{\overline{\sigma}_1}\right)^2 D_{22}^*\right]\right\}\overline{\sigma}_1^4. \tag{10.28}$$

The failure criterion (4.43) takes the form

$$C_1\left[1 + \left(\frac{\overline{\sigma}_{22}}{\overline{\sigma}_{11}}\right)\right]^2 + \frac{2}{3}C_2\left[1 + \left(\frac{\overline{\sigma}_{22}}{\overline{\sigma}_{11}}\right)^2 - \left(\frac{\overline{\sigma}_{22}}{\overline{\sigma}_{11}}\right)\right]$$
$$+C_3\left[D_1^* + \left(\frac{\overline{\sigma}_{22}}{\overline{\sigma}_{11}}\right)^2 D_{22}^*\right] - \frac{\sigma_{\mathrm{u}}^4}{\overline{\sigma}_{11}} = 0 \tag{10.29}$$

with constants C_1, C_2, C_3 obtained from (4.51).

10.2.2 FEM mesh generation and results

The constitutive model (cf. Sect. 10.2.1) is implemented in FEM ABAQUS code. The geometry is discretized by fully integrated 2D first-order isoparametric elements CPS4 in a 118×40 mesh, used in conjunction with IRS21A rigid surface elements. In the case of CPS4 elements the so-called selectively reduced integration technique is used which prevents mesh locking. This means that the actual volume change at the Gauss points is replaced by the average volume change of elements. Interface elements are sequentially included into the mesh after the failed CPS4 elements have been removed or the kinematic boundary conditions have been released. Such a procedure is employed because it is anticipated that the structure may again come into contact with the wall after failed elements have been removed from the mesh. Further considerations will often be limited to the area near the wall, as shown in Fig. 10.7, because the initial stress distribution in the domain $\mathfrak{D}$ indicates the damage zone being limited to the close neighborhood of the fixed edge.

An initially heterogeneous elastic stress state results in a nonuniform material softening due to the damage growth. The distribution of the damage tensor component D_{11} at time $t_{I^-} = t_-^{40}$, preceding the macrocrack initiation between the element 40 and the wall, is shown in Fig. 10.8.

Damage is localized in the narrow zone where the initial tensile and shear stress concentration was observed. However, due to the stress redistribution in damaged elements prior to the crack initiation at time $t_{I^-} = t_-^{40}$, before in the first node (node 41, Fig. 10.7) the boundary conditions are released to form the crack of the length of an element, the maximum stress

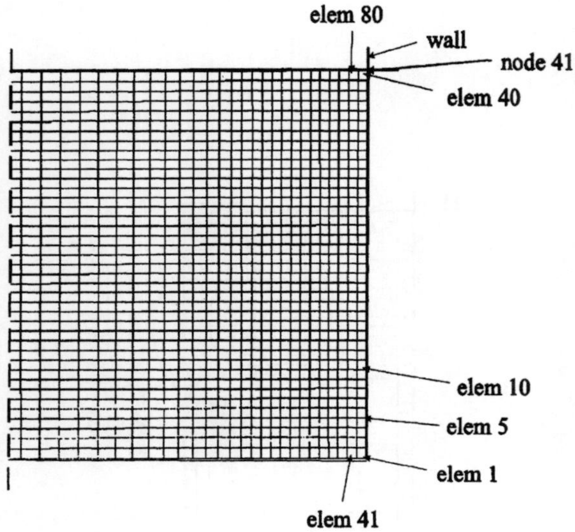

Fig. 10.7. Element numbering in the region near the wall

concentration moves to the elements at a certain distance from the crack tip. The distribution of stress components $\overline{\sigma}_{11} = \sigma_{11}/\sigma_{\rm u}$ and $\overline{\sigma}_{22} = \sigma_{22}/\sigma_{\rm u}$ that corresponds to the damage state shown in Fig. 10.8a is presented in Figs. 10.8b and 10.8c. At time $t_{I+} = t_{+}^{40}$ the cracking process starts from the right-top element (element 40). When the kinematic boundary conditions in node 41 have already been released, further stress redistribution is observed. The tensile stress in the direction normal to the just formed macrocrack is fully released, but element 40 is still carrying the shear stress, which is manifested in shear type mesh deformation, shown in Fig. 10.9.

The evolution of the maximum principal value of the damage tensor D_1 in chosen elements along the wall ($x_1 = 5.0, 0 < x_2 < 1$) is sketched in Fig. 10.10.

The general observation may be summarized in what follows. Due to the stress redistribution from the element which is most exposed to the damage growth, a gradual damage rate drop prior to failure occurs. This is mostly noticed in the first three elements that constitute the crack (elements 40, 39, 38) where the shrinkage of the failure surface is significant (cf. Fig. 4.2). Further, due to the avalanche of the crack length growth, the damage level in the zone neighboring the crack tip is not high enough to significantly change the failure surface. In other words, the damage localization near the crack tip decreases when the crack length increases. So, in contrast to the formulation used by Liu, Murakami, and Kanagawa (1994), there is no need to additionally regularize damage field via a nonlocal damage variable (cf. Sect. 5.2.2). The decrease of the critical damage tensor eigenvalues ob-

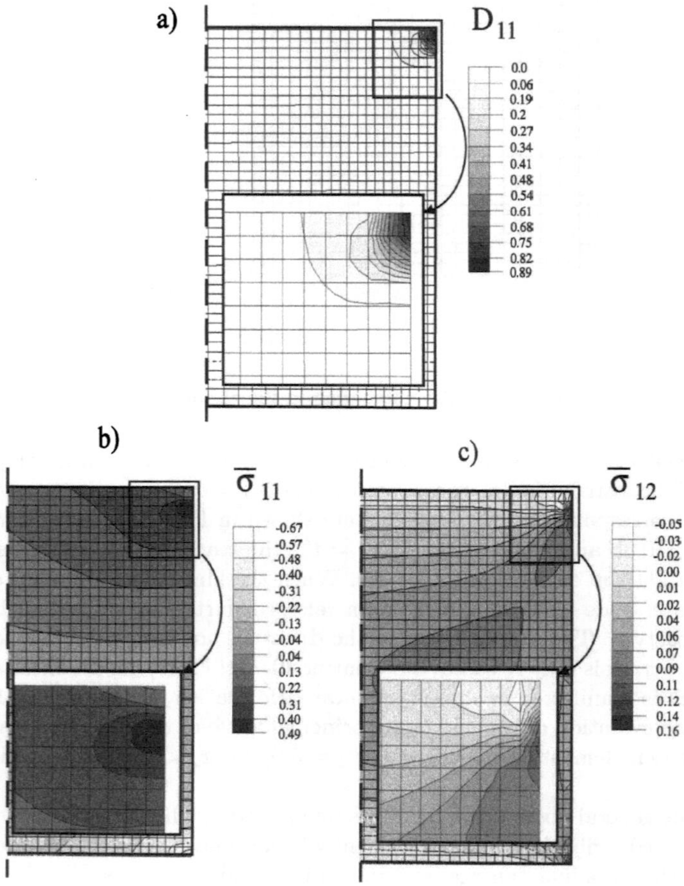

Fig. 10.8. Distribution of a) damage tensor component D_{11}, b) normal stress, c) shear stress, at time t_{I^-}

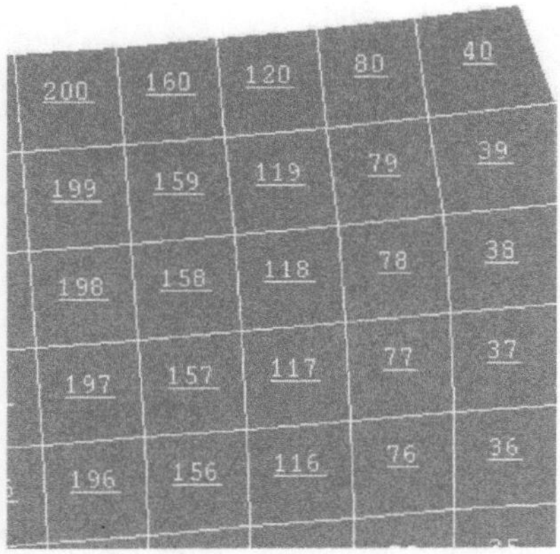

Fig. 10.9. Mesh deformation at time t_{I+}

served in subsequent elements that undergo failure indicates two different types of element failure. First, macrocracks are accompanied by a significant strength reduction; second, in the next elements the failure criterion close to the Huber–Mises–Hencky equivalent stress is satisfied.

The history plot for the stress tensor eigenvalue $\overline{\sigma}_1 = \sigma_1/\sigma_u$ is shown in Fig. 10.10b. At the instant $t_{I+} = t_+^{40}$, when the kinematic boundary conditions in node 41 have been released, the need to confirm to the boundary problem equations results in a discontinuous increase in stress values in neighboring elements. Subsequent stages of the macrocrack development in the deformed mesh are shown in Fig. 10.11.

The stress distribution at time $t_-^{16.56}$ (Fig.10.11c) preceding the crack branching is presented in Fig. 10.12c. After releasing the boundary conditions in node 18 the current stress vector in elements 16 and 56 exceed the actual failure surface, which causes simultaneous failure in both elements. Therefore, the crack deviation from the primary direction is modeled by fully removing elements 16 and 56 from the FEM mesh. The stress state at time $t_+^{16.56}$, just after the mesh modification in the area of the macrocrack tip, is shown in (Fig. 10.12a, b). Eventually, the structure is fully failed when two cracks developing from the top and the bottom of the clamped plate side, of tension and the compression type, meet together to make the structure unserviceable, Fig. 10.11f.

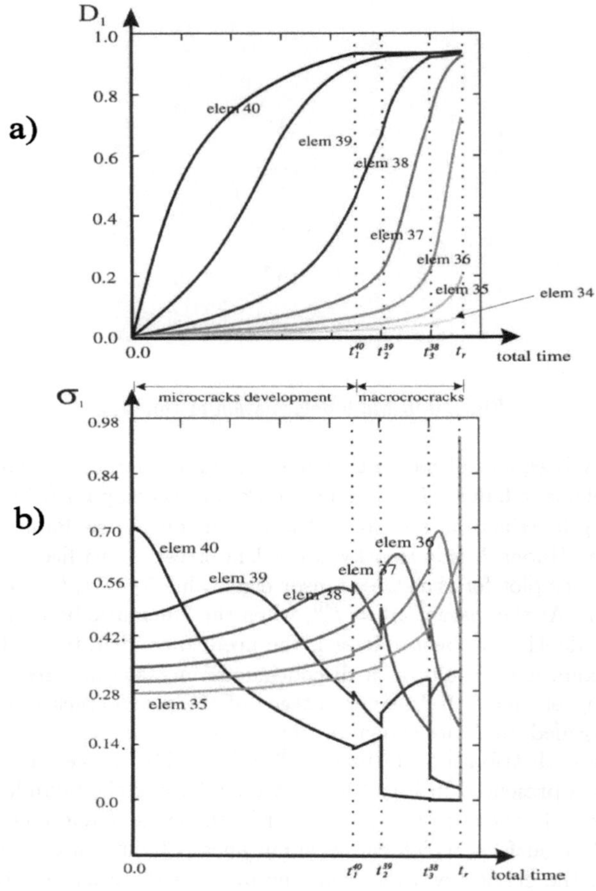

Fig. 10.10. Evolution of a) maximum principal damage value, b) maximum principal stress value in elements along the wall

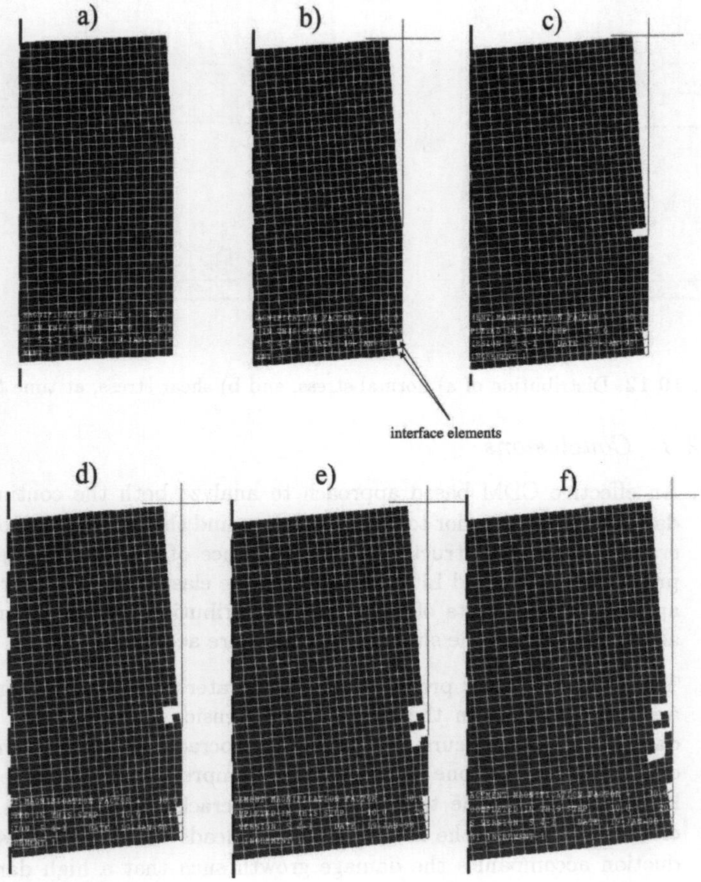

Fig. 10.11. Subsequent stages of macrocrack development

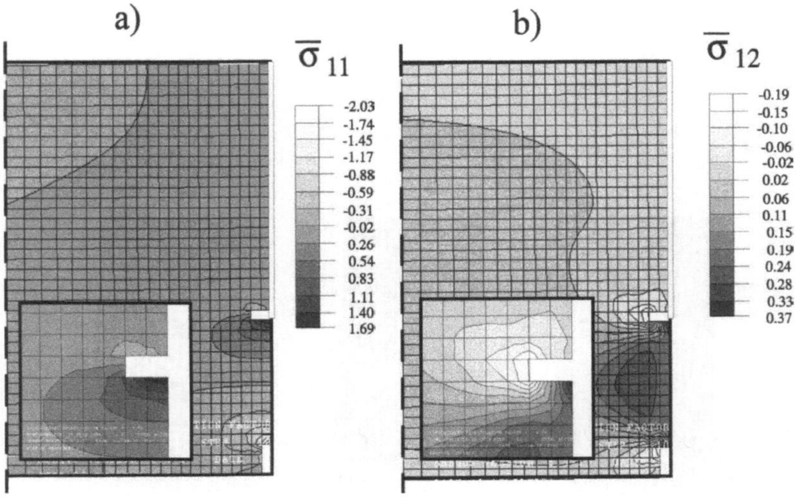

Fig. 10.12. Distribution of a) normal stress, and b) shear stress, at time $t_+^{16,56}$

10.2.3 Conclusions

i. An effective CDM based approach to analyze both the continuum damage evolution prior to crack initiation and the propagation of the crack through the structure in the presence of the damage field is proposed. A modified Litewka model of the elastic-brittle material is applied, where effects of the stress redistribution following damage accumulation and the shear deformation are accounted for.

ii. Two models of crack propagation in the material exposed to damage are distinguished. In the region where tension predominates, high damage advance occurs before the macrocrack is formed. On the other hand, in the zone of predominant compression, no damage evolution(or a little due to shear) prior to crack opening occurs. On crack initiation (in the first element that leads to failure), stress reduction accompanies the damage growth such that a high damage level is reached when the actual stress vector meets the actual failure surface. Next, when crack penetration through the volume is analyzed, the stress increase in the subsequent element is observed due to releasing the stress level in the previous element. Hence, a lower damage advance in the considered element is needed to enable the increasing stress vector to meet the actual failure surface. In other words, two competing phenomena, stress increase due to stress redistribution from the failed zone and failure surface modification due to the damage accumulation, result in damage field regularization near the crack tip when the crack length increases.

iii. The local approach to fracture analysis used here is based on two procedures that describe the crack propagation: a) changing the kinematic boundary conditions on the element face neighboring the crack, or b) fully removing the element, both controlled by an appropriate failure criterion. The description applied shows the phenomenon of the damage localization drop near the crack tip with crack length growth. This behavior is mostly due to the rapid stress increase resulting from the effective structure width reduction in a region where the crack development is expected (near the crack front), such that the continuum damage advance is not too high. In the compressive zones, no damage evolution occurs so the failure criterion is met on the initial failure surface and the corresponding element instantaneously leads to failure.

iv. The crack branching mechanism (or change of its primary direction) can also be detected when the shear-type failure mode in the element neighboring the main crack precedes the tension-type failure mode on the crack primary direction.

v. The structure is totally failed when two main cracks, a tensile-type (from the top) and a compressive-type (from the bottom), meet each other and the effective plate width drops to zero (structure fragmentation).

vi. In contrast to the local approach to creep fracture used by Murakami, Kawai, and Rong (1988), the elastic-brittle damage model developed here seems more promising. The main advantage is the better numerical stability observed when the local damage field near the crack tip is limited by the critical damage level drop with the crack length growth. Additionally, the stress concentration in this zone is also limited by the size of the actual failure surface. Hence, neither additional damage regularization nor other stress limitation methods are required, as discussed in the convergence tests where different mesh patterns with a decreasing elements size are used (cf. Fig. 5.6).

vii. Due to the kinematically controlled crack growth mechanism, the primary crack width is not affected by the element size. However, the secondary crack growth mode, when the element is fully removed from the mesh if the failure criterion in the element is reached, is more mesh-dependent, so that further testing is required.

Part III

Optimal design of structures with respect to creep rupture

11

Formulation of optimal design under creep-damage conditions

11.1 Structural optimization under damage conditions

11.1.1 Optimal design of structures made of inelastic time-dependent materials

When elastic structures are designed for either minimum weight or maximum load under a strength constraint, structures of uniform strength, also called fully stressed designs, are optimal in most cases. In general, however, the condition of uniform strength is neither a necessary nor a sufficient condition of optimality. The exceptions, when structures of uniform strength are nonoptimal, are mainly connected either with the static indeterminacy of a structure or with geometric changes being taken into account. On the other hand, the condition of uniform strength of structures may not be a sufficient optimality condition if it does not result in a unique solution. Hence, following Gallagher (1973), the fully stressed design method (FSDM) is, in general, a first step towards the exact optimal design when more rigorous optimization approaches are used.

When optimization of inelastic structures made of time-dependent solids that suffer from material damage, brittle or ductile, is formulated, the minimum weight (volume) or the maximum load remains the typical design objective, similarly to the corresponding elastic problem. Essential changes are observed in the state and evolution equations as well as the constraints, since a new independent time variable plays an important role. The effect of nonlinear constitutive equations on the optimal shape of structures was discussed by Gajewski (1975). The optimization constraints under conditions of creep damage, ductile or brittle, elastic-brittle damage, thermo-elasto-(visco)plastic damage, etc., may be imposed not only on the strength (rupture or failure), stiffness, and stability, as in the elastic case, but also on a limited stress relaxation, a limited residual displacement, or a lifetime prediction of the first macrocrack initiation ($t_I = t_{\mathrm{R}}$) or the complete failure ($t_{II} = t_{\mathrm{F}}$). Hence, since in an optimization problem the design objectives and constraints may be interchanged, the following global optimization problems, originally proposed for optimal design under creep conditions (cf. Życzkowski, 1988, 1991), may be formulated for optimal design of structures made of damaged time-dependent materials (cf. Table 11.1)

i. minimization of weight under prescribed loadings and lifetime,

ii. minimization of loadings under prescribed weight and lifetime,

iii. maximization of lifetime (t_I or t_{II}) under prescribed weight and loadings.

Table 11.1. Classification of typical problems of optimal design with respect to creep failure (global criteria)

Formulation	Optimality criteria	Constraints
i	$Q \to \min$	$P = \text{const}, \quad t_{I,II} = \text{const}$
ii	$P \to \max$	$Q = \text{const}, \quad t_{I,II} = \text{const}$
iii	$t_{I,II} \to \max$	$P = \text{const}, \quad Q = \text{const}$

The first two problems are, in most cases, inconvenient for practical applications since the lifetime of a structure (t_I or t_{II}) is usually not given in an explicit form but results from the additional constraints imposed on damage variable(s) $\mathcal{D}\,(D, D_\nu, \mathbf{D})$, the magnitudes of which change with time when the appropriate damage evolution law is legislated, e.g., (2.26), (2.35), (2.42), (2.44), (2.46), (2.48), (2.65), (2.71), (2.74), (2.97), if isotropic damage D is assumed, or (4.43), (4.44), (4.46), (4.62), etc., if more general anisotropic damage $\mathbf{D}$ is adopted. Time of first macrocrack initiation t_I is defined here in such a way that the damage variable D (isotropic damage) or the dominant damage component $\sup\{D_{ij}\}$ (anisotropic damage) reaches the critical value D_{crit}. When the constraints are imposed on a ductile creep rupture in Hoff's sense, the condition of vanishing transverse dimensions at a structure cross-section defines the lifetime t^{H}_{DR}. In this case, the geometry changes due to finite strains must be taken into account since infinite strains, at least in one cross-section, constitute the purely ductile failure mechanism of a structure. Representative optimization problems, when constraints are imposed on brittle, ductile or mixed rupture, creep stiffness or creep compliance, creep buckling and dynamic response, were discussed by Życzkowski (1991, 1996).

11.1.2 Optimality criteria for structures made of time-dependent materials

A. Uniform creep strength (UCS)

Structures optimal with respect to brittle rupture, $t_{\text{R}} \to \max$, may often be found among the class of structures of uniform creep strength (UCS) (cf. Życzkowski, 1991). Structures of uniform creep strength with respect to brittle rupture are defined as ones in which macrocracks initiate simultaneously either in every material point $\mathbf{x} \in V$ or along certain characteristic lines or surfaces. Hence, when the simple, scalar Kachanov–Hayhurst isotropic damage growth rule is used (2.35) and the integration is performed from the damage initiation $D(t_0) = 0$ up to formation of the first macrocrack $D(t_I) = D_{\mathrm{cr}}$, the condition of uniform isotropic damage strength (UIDS) takes the following representation:

$$\dot{D} = C\left\langle \frac{\chi[\boldsymbol{\sigma}(\mathbf{x},t)]}{1-D} \right\rangle^r, \ \chi = a\sigma_1 + 3b\sigma_{\mathrm{H}} + c\sigma_{\mathrm{eq}},$$
$$1-(1-D_{\mathrm{cr}})^{r+1} = C(r+1)\int_{t_0}^{t_I} \{\chi[\boldsymbol{\sigma}(\mathbf{x},t)]\}^r \,\mathrm{d}t \tag{11.1}$$

which must be satisfied at $\forall \mathbf{x} \in V$ or at least on a certain surface. For orthotropic damage (Sect. 4.1) the condition of uniform orthotropic damage strength (UODS) can be written as

$$\dot{D}_\nu = C_\nu \left\langle \frac{\sigma_\nu(\mathbf{x},t)}{1-D_\nu} \right\rangle^{r_\nu},$$
$$\sup_{(1,2,3)} \left[\frac{D_\nu(\mathbf{x},t_I)}{D_{\nu\mathrm{cr}}}\right] \equiv 1, \quad \forall \mathbf{x} \in V. \tag{11.2}$$

In a more general case of damage anisotropy the isotropic scalar function of stress and damage tensors $\boldsymbol{\sigma}$ and $\mathbf{D}$ may be postulated as the failure criterion (Sec.4.2.2) at the point $\mathbf{x}$

$$F\left[\boldsymbol{\sigma}(\mathbf{x},t_I), \mathbf{D}(\mathbf{x},t_I)\right] = 0. \tag{11.3}$$

If, for instance, Litewka's model is applied (Sect. 4.2.3) the condition of uniform anisotropic damage strength (UADS) may be furnished as follows:

$$\dot{\mathbf{D}} = C\left\{\Phi^{\mathrm{e}}\left[\boldsymbol{\sigma}(\mathbf{x},t), \mathbf{D}^*(\mathbf{x},t)\right]\right\}^2 \boldsymbol{\sigma}^*,$$
$$F(\boldsymbol{\sigma}, \mathbf{D}^*) = C_1 \mathrm{Tr}^2 \boldsymbol{\sigma}(\mathbf{x},t_I) + C_2 \mathrm{Tr}\left[\boldsymbol{\sigma}'(\mathbf{x},t_I)\right]^2 \tag{11.4}$$
$$+ C_3 \mathrm{Tr}\left[\boldsymbol{\sigma}^2(\mathbf{x},t_I) : \mathbf{D}^*(\mathbf{x},t_I)\right] - \sigma_{\mathrm{u}}^2 = 0, \quad \forall \mathbf{x} \in V,$$

where $\Phi^{\mathrm{e}}[\boldsymbol{\sigma}, \mathbf{D}^*]$ denotes the elastic energy affected by damage (4.39), $\mathbf{D}$ and $\mathbf{D}^*$ denote the second-rank damage tensors, classical (3.3) and modified (3.17), whereas $\boldsymbol{\sigma}^*$ is a modified stress tensor (Sect. 4.2.3).

B. Uniform ductile strength (UDS)

When optimal design with respect to ductile rupture is sought, a geometrically nonlinear finite strain approach is necessary which makes both a formulation and a solution of the optimization problem much more complex.

It was investigated for the first time and developed by Szuwalski (1989, 1991a, 1991b, 1995a, 1995b). Following these papers another classification of structures that are "optimal" in various senses, when geometric changes are significant, may be quoted:

i. Structures of uniform elastic strength in a broader sense (UESb), where the initial equivalent stress $\sigma_{eq}(t_0)$ is proportional at each material point of the structure to the critical stress for the material:

$$\sigma_{eq}(\mathbf{x}, t_0) = c\sigma_{cr}(\mathbf{x}), \quad \forall \mathbf{x} \in V. \tag{11.5}$$

ii. Structures of uniform elastic strength in a narrower sense (UESn), where the initial principal stress components are equal throughout the whole structure:

$$\sigma_\nu(\mathbf{x}, t_0) = \text{const}(\mathbf{x}) \quad \forall \mathbf{x} \in V. \tag{11.6}$$

iii. Structures of uniform creep strength with respect to pure ductile rupture time (UCDS) t_{DR}, where transverse dimensions drop simultaneously to zero at all cross-sections of the structure:

$$h(\mathbf{x}, t_{DR}) = 0, \quad t \to t_{DR}, \quad \forall \mathbf{x} \in V. \tag{11.7}$$

iv. Structures of uniform deformability (UD), where principal strain components are equal in all cross-sections, but vary with time:

$$\varepsilon_\nu(\mathbf{x}, t) = \mathsf{f}(t), \quad t_0 < t < t_{DR}, \quad \forall \mathbf{x} \in V. \tag{11.8}$$

11.1.3 Constraints

The optimality criteria (Sect. 11.1.2) require the appropriate constraints, some of which are listed below

A. Inequality constraints

i. Strength constraints

$$\tilde{\sigma}_{eq}(\boldsymbol{\sigma}, \mathcal{D}) \leq \sigma_{cr}/j, \tag{11.9}$$

e.g.,

$$\tilde{\sigma}_{eq}^{HMH} = \left[\frac{3}{2} \frac{s_{ij} s_{ij}}{(1 - D)^2} \right]^{1/2} \leq \sigma_{cr}/j, \tag{11.10}$$

where σ_{cr} denotes the critical effective equivalent stress for the material and j is the safety factor.

ii. Initial stability constraints (elastic stability condition)

$$n_0 < n_{\mathrm{E}}, \tag{11.11}$$

where n_{E} denotes the basic Eulerian force (if the possibility of creep buckling is not included in the analysis).

iii. Geometric constraints for thickness of the structure h

$$h_{\min} < h(\mathbf{x}) < h_{\max} \tag{11.12}$$

and the prestressing eccentricity e

$$e_{\max} \leq h/2. \tag{11.13}$$

B. Equality constraints (for axisymmetric structures)

i. Condition of constant volume (weight) of a uniform cross-section

$$V = 2\pi \int_0^R h(r) r \mathrm{d}r = \mathrm{const} \tag{11.14}$$

or a two-point sandwich cross-section

$$V = 2\pi \int_0^R \left[\alpha \left(h_{\mathrm{s}} - g_{\mathrm{s}}\right) + 2\beta g_{\mathrm{s}}\right] r \mathrm{d}r = \mathrm{const} \tag{11.15}$$

or

$$\delta V = 2\pi \int_0^R \left[\alpha \left(\delta h_{\mathrm{s}} - \delta g_{\mathrm{s}}\right) + 2\beta \delta g_{\mathrm{s}}\right] r \mathrm{d}r = 0, \tag{11.16}$$

where $h\,(r)$, $h_{\mathrm{s}}\,(r)$, and $g_{\mathrm{s}}\,(r)$ denote thickness of the uniform cross-section, the sandwich cross-section, and the sandwich working layer, respectively, whereas α and β are arbitrary weight factors for the core and layers materials (Fig. 11.1).

ii. Condition of constant lifetime for macrocrack initiation

$$t_I = t_{\mathrm{R}} = \mathrm{const} \tag{11.17}$$

or complete failure (fracturing)

$$t_{II} = t_{\mathrm{F}} = \mathrm{const}, \tag{11.18}$$

where $t_{II} - t_I$ is a safety regime for the structure considered, which is reduced to zero for fully damaged design.

iii. Condition of constant surface loadings (prestressing force excluded)

$$q(\mathbf{x}, t) = q(\mathbf{x}). \tag{11.19}$$

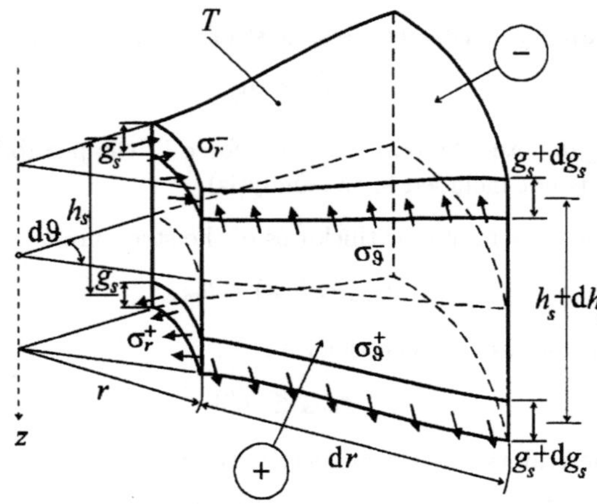

Fig. 11.1. Uniform and substitutive sandwich cross-section

11.1.4 Decision variables

When problems of optimization are formulated for prestressed structures under damage or damage/fracture conditions vectors of control variables involve not only the thickness of a structure $h(\mathbf{x})$ or $h_s(\mathbf{x})$ and $g_s(\mathbf{x})$ for a uniform or sandwich cross-section, respectively, but also parameters of prestressing n_0 or Δ_0 in case of in-plane membrane-type prestressing (a force or a membrane distortion), and m_0 or φ_0 in case of bending-type prestressing (a bending moment or an initial bending distortion). Hence, the corresponding vectors of decision variables are

$$\{\mathbf{c}_{\mathrm{m}}^{\mathrm{u}}\} = \{n_0 \quad \text{or} \quad \Delta_0, h(\mathbf{x})\} \quad \text{or} \quad \{\mathbf{c}_{\mathrm{m}}^{\mathrm{s}}\} = \{n_0 \quad \text{or} \quad \Delta_0, h_s(\mathbf{x}), g_s(\mathbf{x})\}$$

and

$$\{\mathbf{c}_{\mathrm{b}}^{\mathrm{u}}\} = \{m_0 \quad \text{or} \quad \varphi_0, h(\mathbf{x})\} \quad \text{or} \quad \{\mathbf{c}_{\mathrm{b}}^{\mathrm{s}}\} = \{m_0 \quad \text{or} \quad \varphi_0, h_s(\mathbf{x}), g_s(\mathbf{x})\}$$

in case of a uniform cross-section or a sandwich cross-section, respectively. It is important to precisely distinguish the behavior of prestressing, which varies with time, from other loadings which are constant and may appear as the equality constraints. The nature of the prestressing, which is considered as an excitation imposed on the structure, also requires explanation. Generally, internal and the external excitations can be distinguished. The prestressing fibers in reinforced concrete are an example for the first case, whereas a cylindrical shell prestressed by an external circumferential cable illustrates the second. In both these cases, the excitations may have the nature of forces or distortions. Typical examples of excitations, the radial prestressing force n_0 or the displacement type Δ_0, and the radial prestressing moment m_0 or the angle of support φ_0, for membrane and bending states, respectively, are illustrated in Fig. 11.2.

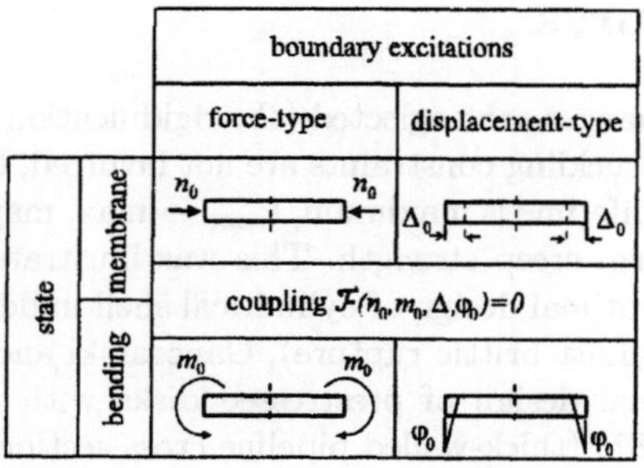

Fig. 11.2. Boundary excitations in axisymmetric plates

Apart from the order of the theory, which may or may not include the coupling between the membrane and bending effects (cf. Sect. 9.2), both membrane and bending states may additionally be coupled to the boundary conditions. Generally, such a coupling can be described by a function $\mathcal{F}$ which depends on the excitation parameters:

$$\mathcal{F}(n_0, m_0, \Delta_0, \varphi_0) = 0, \tag{11.20}$$

where n_0 is the initial prestressing force, Δ_0 the initial distorsion (membrane), m_0 the initial prestressing moment, and φ_0 the initial distorsion (curvature).

From a practical point of view, only a few particular representations of the function $\mathcal{F}$ make sense. These are as follows:

i. Uncoupling, when the function $\mathcal{F}$ depends on the only one of the arguments,

$$\begin{array}{lcl} \mathcal{F}(n_0) = 0 & \text{or} & \mathcal{F}(m_0) = 0 \quad \text{or} \\ \mathcal{F}(\Delta) = 0 & \text{or} & \mathcal{F}(\varphi_0) = 0, \end{array} \tag{11.21}$$

ii. Unilateral coupling of the membrane and bending states, when the function $\mathcal{F}$ may be solved with respect to one of its arguments,

$$m_0 = f(n_0) \quad \text{or} \quad \varphi_0 = f(\Delta_0), \tag{11.22}$$

iii. Bilateral coupling, when the function $\mathcal{F}$ implicitly depends on more than one argument (e.g., plate–shell interaction),

$$\mathcal{F}(\Delta_0, \varphi_0) = 0 \quad \text{or} \quad \mathcal{F}(n_0, m_0). \tag{11.23}$$

11.2 Inelastic structures of uniform strength in various senses versus optimal structures

When geometric changes are neglected (the rigidification principle is used) and creep-damage buckling constraints are not involved, the optimal structure, for which the lifetime is maximum, $t_{\mathbf{max}} \to \max$, may be found among structures of uniform creep strength. This was illustrated by Życzkowski and Rysz (1986) (optimal design of cylindrical shell under combined bending with torsion against brittle rupture), Ganczarski and Skrzypek (1989, 1991, 1992) (optimal design of prestressed disks with respect to brittle rupture), Rysz (1987) (thick-walled pipeline cross-section of uniform creep strength against pressure, axial force, and torsion), and Skrzypek and Egner (1993, 1994) (optimal design and optimal prestressing of disks with respect to creep-brittle rupture).

With geometric changes taken into account, a structure of uniform creep strength is, in general, nonoptimal. Further optimization may be performed by imposing appropriate shape corrections to maximize the lifetime of the structure being optimized. Shapes of flexible beams of uniform creep strength were sough by Życzkowski and Świsterski (1980) by a finite deflections approach. Nonoptimality of the uniform creep strength design was checked by Świsterski et al. (1983), where an eccentrically compressed I–column was optimized against brittle rupture time when finite deflections were admitted. In this case, an essential increase of the lifetime prediction $t_I \to t^{\text{opt}}$ by about 90% when compared to the uniform creep strength t_I^{ucs} was reached when a further parametric optimization procedure was used. The relevant problem was studied by Wróblewski (1989), who checked the nonoptimality of an eccentrically compressed column of uniform creep strength with respect to its lifetime when three rupture mechanisms, the brittle, the ductile, and the brittle-ductile, were applied for lifetime predictions.

With respect to ductile rupture, the structures of uniform deformability (UD) belong to the class of structures of uniform ductile creep strength (UDCS). However, in general, neither structures of uniform ductile creep strength nor structures of uniform deformability are optimal with respect to ductile rupture time t_{DR}. Only in the case when a structure of uniform ductile creep strength is statically determinate it is simultaneously the optimal structure in the sense of maximum ductile rupture lifetime $t_{\text{DR}} = t^{\text{opt}}$. If the above condition does not hold, the conclusion is not true, although the additional shape corrections may be imposed to improve the UDCS solution (cf. Szuwalski, 1989, 1991a, 1991b), as schematically sketched in Fig. 11.3.

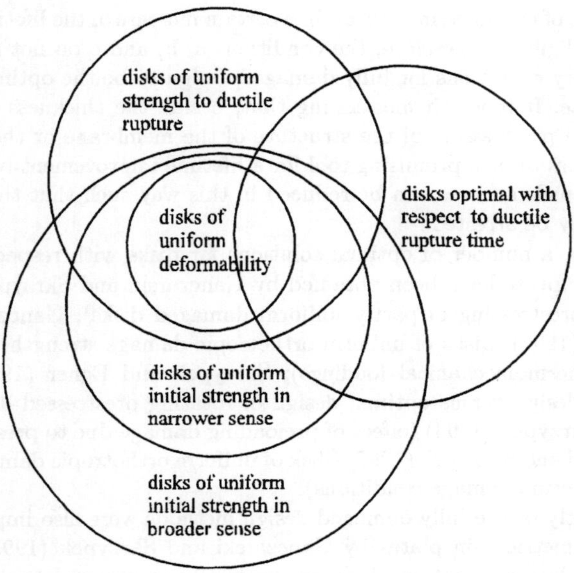

Fig. 11.3. Disks of uniform ductile creep strength in various senses (after Szuwalski, 1993)

Note, however, that a pure ductile failure mechanism in Hoff's sense is strongly limited in practical observations and should rather be enriched with the additional damage evolution by material degradation in a CDM way.

The fully damaged design method is essentially relevant to the fully stressed design method as it was used in elasticity. Roughly speaking, this method leads to exact solutions which are optimal with respect to lifetime $t_I = t^{\text{opt}}$ when the following conditions hold:

a. the structure is statically determinate,

b. a stationary single loading is applied, and

c. geometric changes are neglected.

If the above conditions are violated, the fully damaged design turns out to be only an approximate optimal solution. An exact one may be obtained when more rigorous optimization methods are used. However, Skrzypek and Egner (1993) proved that a disk of fully uniform creep damage strength under steady loadings (non-prestressed) is also the optimal one in the sense of its lifetime. On the other hand, a disk of partly uniform creep damage strength with active lower geometric constraint under unsteady loading conditions (due to the prestressing) is not optimal and, hence, additional corrections of thickness may result in a certain increase of the lifetime (practically negligible). Therefore, the conditions a, b, and c do not appear to be necessary conditions for fully damaged design to be the optimal design in this case. It is worth mentioning that, beside the thickness optimization, initial prestressing of the structure of the membrane or the bending type appears to be a promising tool for a lifetime improvement because, in general, tensile stresses can be reduced in this way such that the damage growth may be arrested.

Recently, a number of optimal solutions for disks with respect to brittle creep rupture have been obtained by Ganczarski and Skrzypek (1989) (optimal prestressing of partly uniform damaged disks); Ganczarski and Skrzypek (1991) (disks of uniform orthotropic damage strength under uncoupled thermomechanical loadings); Skrzypek and Egner (1993) (fully damaged design versus optimal design of rotating prestressed disks); Egner and Skrzypek (1994) (effect of preloading damage due to prestressing); Ganczarski and Skrzypek (1997) (disk of uniform orthotropic damage under coupled thermo-damage conditions).

The partly or the fully damaged design methods were also implemented on axisymmetric thin plates by Ganczarski and Skrzypek (1993) (creep-damaged plate of constant thickness optimally prestressed by the elastic cylindrical shell); Ganczarski and Skrzypek (1994) (initially prestressed sandwich plates with full orthotropic damage at rupture). Optimal prestressing of Reissner's axisymmetric plates with respect to brittle rupture time was also examined by Ganczarski, Freindl, and Skrzypek (1997).

12

Optimal design of axisymmetric disks

12.1 State equations for rotationally symmetric deformation of annular disks of variable thickness

12.1.1 State equations of disks of variable thickness under plane stress conditions

An annular disk of variable thickness $h(r)$ and radii a and b, clamped at the inner edge, is subjected to steady rotation about the axis of symmetry with an angular velocity ω and uniform radial tension along the periphery. Plane stress state $\sigma_z = 0$ and creep incompressibility $\varepsilon_{\mathrm{m}}^{\mathrm{c}} = 0$ are assumed when the transient creep problem is solved in velocities by the use of cylindrical coordinate system r, ϑ, z. Hence, for a rotationally symmetric deformation, the equilibrium equation takes the following form:

$$\frac{1}{h}\frac{\mathrm{d}}{\mathrm{d}r}(h\sigma_r) + \frac{\sigma_r - \sigma_\vartheta}{r} + \rho\omega^2 r = 0. \tag{12.1}$$

Moreover, when an additive decomposition of strains into elastic and creep components is used, the linear geometric equations may be written:

$$\begin{aligned} \varepsilon_r &= \frac{\mathrm{d}u}{\mathrm{d}r} = \varepsilon_r^{\mathrm{e}} + \varepsilon_r^{\mathrm{c}} = \frac{\sigma_r - \nu\sigma_\vartheta}{E} + \varepsilon_r^{\mathrm{c}}, \\ \varepsilon_\vartheta &= \frac{u}{r} = \varepsilon_\vartheta^{\mathrm{e}} + \varepsilon_\vartheta^{\mathrm{c}} = \frac{\sigma_\vartheta - \nu\sigma_r}{E} + \varepsilon_\vartheta^{\mathrm{c}}. \end{aligned} \tag{12.2}$$

Elimination of σ_r and σ_ϑ from (12.1) and (12.2) yields the fundamental equation in terms of the radial displacement u:

$$\begin{aligned} &\frac{\mathrm{d}^2u}{\mathrm{d}r^2} + \left(\frac{1}{h}\frac{\mathrm{d}h}{\mathrm{d}r} + \frac{1}{r}\right)\frac{\mathrm{d}u}{\mathrm{d}r} + \left(\frac{\nu}{h}\frac{\mathrm{d}h}{\mathrm{d}r} - \frac{1}{r}\right)\frac{u}{r} \\ &= \frac{f}{r} + \frac{\mathrm{d}g}{\mathrm{d}r} + \frac{1}{h}\frac{\mathrm{d}h}{\mathrm{d}r}g - kr, \end{aligned} \tag{12.3}$$

where

$$f = (1-\nu)(\varepsilon_r^{\mathrm{c}} - \varepsilon_\vartheta^{\mathrm{c}}), \quad g = \varepsilon_r^{\mathrm{c}} + \nu\varepsilon_\vartheta^{\mathrm{c}}, \quad k = \frac{1-\nu^2}{E}\rho\omega^2. \tag{12.4}$$

Creep strain rates are governed by the Mises-type flow rule associated with the time hardening hypothesis and the Kachanov–Sdobyrev damage growth rule (partly coupled approach):

$$\mathrm{d}\varepsilon_r^{\mathrm{c}} = \frac{\sigma_{\mathrm{eq}}^{m-1}}{(1-D)^m}\left(\sigma_r - \frac{\sigma_\vartheta}{2}\right)\dot{\mathsf{f}}(t)\mathrm{d}t,$$
$$\mathrm{d}\varepsilon_\vartheta^{\mathrm{c}} = \frac{\sigma_{\mathrm{eq}}^{m-1}}{(1-D)^m}\left(\sigma_\vartheta - \frac{\sigma_r}{2}\right)\dot{\mathsf{f}}(t)\mathrm{d}t, \quad \mathrm{d}\varepsilon_z^{\mathrm{c}} = -\left(\mathrm{d}\varepsilon_r^{\mathrm{c}} + \mathrm{d}\varepsilon_\vartheta^{\mathrm{c}}\right), \tag{12.5}$$

$$\mathrm{d}D = C\left\langle\frac{\chi(\boldsymbol{\sigma})}{1-D}\right\rangle^r \mathrm{d}t, \qquad \chi(\boldsymbol{\sigma}) = \delta\sigma_1 + (1-\delta)\sigma_{\mathrm{eq}}. \tag{12.6}$$

Applying the following dimensionless quantities
$\varepsilon_0 = \dfrac{\sigma_0}{E}$, $U = \dfrac{u}{a\varepsilon_0}$, $R = \dfrac{r}{a}$, $F = \dfrac{f}{\varepsilon_0}$, $G = \dfrac{g}{\varepsilon_0}$, $K = \dfrac{ka^2}{\varepsilon_0}$, $S_r = \dfrac{\sigma_r}{\sigma_0}$, $S_\vartheta = \dfrac{\sigma_\vartheta}{\sigma_0}$, $S_{\mathrm{eq}} = \dfrac{\sigma_{\mathrm{eq}}}{\sigma_0}$, $E_r^{\mathrm{c}} = \dfrac{\varepsilon_r^{\mathrm{c}}}{\varepsilon_0}$, $E_\vartheta^{\mathrm{c}} = \dfrac{\varepsilon_\vartheta^{\mathrm{c}}}{\varepsilon_0}$, $T = tE\sigma_0^{m-1}\dot{\mathsf{f}}(t)$, $H = \dfrac{h}{a}$, $P = \dfrac{p}{\sigma_0}$, $R_0 = \dfrac{b}{a} = 5$
the dimensionless form of the governing equations for a disk of variable thickness is obtained

$$\begin{aligned} &R^2\frac{\mathrm{d}^2U}{\mathrm{d}R^2} + \left(\frac{R^2}{H}\frac{\mathrm{d}H}{\mathrm{d}R} + R\right)\frac{\mathrm{d}U}{\mathrm{d}R} \\ &\quad + \left(\frac{\nu R}{H}\frac{\mathrm{d}H}{\mathrm{d}R} - 1\right)U = -KR^3, \qquad (t=0), \\ &R^2\frac{\mathrm{d}^2\dot{U}}{\mathrm{d}R^2} + \left(\frac{R^2}{H}\frac{\mathrm{d}H}{\mathrm{d}R} + R\right)\frac{\mathrm{d}\dot{U}}{\mathrm{d}R} \\ &\quad + \left(\frac{\nu R}{H}\frac{\mathrm{d}H}{\mathrm{d}R} - 1\right)\dot{U} = \dot{F}R + R^2\frac{\mathrm{d}\dot{G}}{\mathrm{d}R} + \frac{1}{H}\frac{\mathrm{d}H}{\mathrm{d}R}\dot{G}, \qquad (t>0), \end{aligned} \tag{12.7}$$

$$\dot{E}_r^{\mathrm{c}} = \frac{S_{\mathrm{eq}}^{m-1}}{(1-D)^m}\left(S_r - \frac{S_\vartheta}{2}\right), \qquad \dot{E}_\vartheta^{\mathrm{c}} = \frac{S_{\mathrm{eq}}^{m-1}}{(1-D)^m}\left(S_\vartheta - \frac{S_r}{2}\right), \tag{12.8}$$

$$\mathrm{d}D = C\sigma_0^n t_I\left\langle\frac{\chi(\boldsymbol{\sigma})}{1-D}\right\rangle^r \mathrm{d}\bar{t}, \tag{12.9}$$

where

$$\bar{t} = \frac{t}{t_I}, \quad \frac{1}{t_I(t)} = E\sigma_0^{m-1}\dot{\mathsf{f}}(t), \quad \dot{F} = (1-\nu)\left(\dot{E}_r^{\mathrm{c}} - \dot{E}_\vartheta^{\mathrm{c}}\right), \quad \dot{G} = \dot{E}_r^{\mathrm{c}} + \nu\dot{E}_\vartheta^{\mathrm{c}}. \tag{12.10}$$

12.1.2 Boundary value problems

Example A

A clamped annular disk of variable thickness $H(R)$ is subjected to steady rotation about the axis with angular velocity ω and the radial tension P_b applied along the periphery (cf. Fig. 12.1).

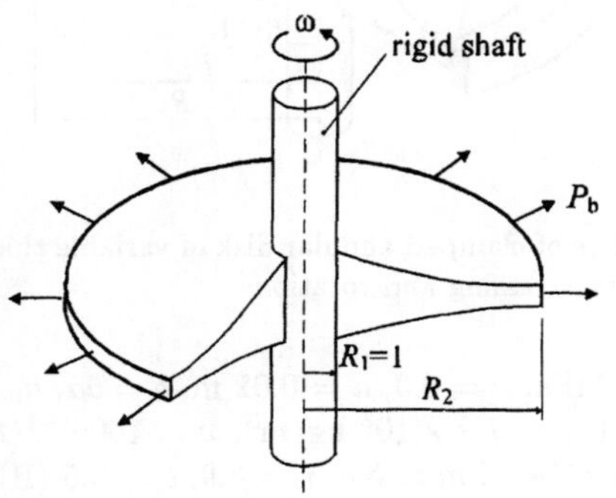

Fig. 12.1. Schematics of clamped annular disk of variable thickness subjected to steady rotation and radial tension

Boundary conditions for the disk of the radii $R_1 = 1$, R_2 are:

$$\begin{aligned} &U(1) = 0, \quad H(R_2)S_r(R_2) = H_0 P_b \quad (\bar{t} = 0), \\ &\dot{U}(1) = 0, \quad \dot{S}_r(R_2) = 0 \quad (\bar{t} > 0). \end{aligned} \tag{12.11}$$

Example B

A clamped annular disk of variable thickness $H(R)$ is subjected to creep-damage under initial prestressing Q and steady rotation ω (cf. Fig. 12.2).

Boundary and continuity conditions for the disk of radii $R_1 = 1$, R_2 (creep) and the prestressing ring of the radii R_2, R_0 (elastic) are:

$$\left.\begin{aligned} &U(1) = 0, \quad &&H(R_2)S_r(R_2) = -H_0 Q \\ &S_r^{\text{ring}}(R_0) = 0 \quad &&S_r^{\text{ring}}(R_2) = -Q, \end{aligned}\right\} \quad (\bar{t} = 0)\,,$$

$$\left.\begin{aligned} &\dot{U}(1) = 0, \quad &&H(R_0)\dot{S}_r(R_2)\mathrm{d}\bar{t} = H_0 \mathrm{d}S_r^{\text{ring}}(R_2) \\ &S_r^{\text{ring}}(R_0) = 0, \quad &&\dot{U}(R_2)\mathrm{d}\bar{t} = \mathrm{d}U^{\text{ring}}(R_2) \end{aligned}\right\} \quad (\bar{t} > 0)\,. \tag{12.12}$$

Data

The calculations are done for the following data:

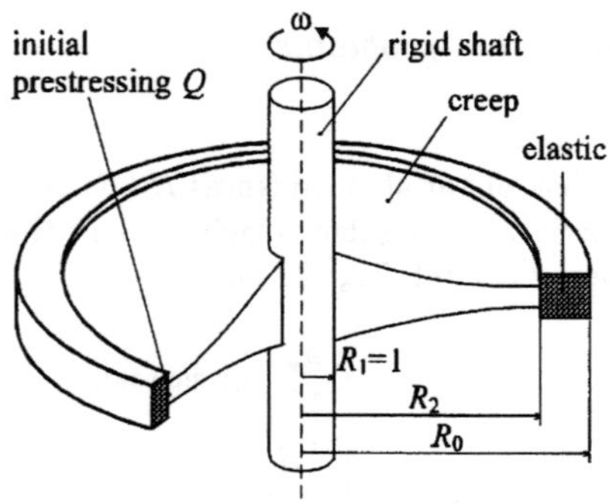

Fig. 12.2. Schematics of clamped annular disk of variable thickness subjected to creep under initial prestressing and rotation

$E = 1.77 \times 10^5$ MPa, $\nu = 0.3$, $a = 0.02$ m, $b = 5a$, $h_0 = 0.004$ m, $\sigma_0 =$ 118 MPa, $P_b = 0.1$, $\rho = 7.9 \times 10^3$ kg/m^3, $\omega = 100$ s^{-1}(A) or 240 s^{-1}(B), $C = 2.13 \times 10^{-42}$Pa^{-r}s^{-1}, $m = 5.6$, $r = 3.9$, $\delta = 0.5$ (B) or 1.0 (A).

12.1.3 Numerical solution by FDM

In order to solve a transient creep-damage problem for the disk of a prescribed thickness $H(R)$ we divide the initial domain $R_1 \leq R \leq R_2$ into a finite number of intervals $N-1$, not necessarily equal, by inserting the ordered set of points R_j, $j = 1, \ldots, N$, where $R_1 = 1$, and $R_N = R_2$. Also we separate a current dimensionless time $\bar{t}$ into discrete intervals delimited by $\bar{t}^k$ with $\bar{t}^0 = 0$ representing the initial condition for creep (the elastic solution). For each disk portion, a piecewise linear approximation of thickness

$$H_I(R) = A_I + B_I R, \qquad I = 1, \ldots, N-1, \tag{12.13}$$

is assumed, where subscript I denotes a number of the interval. Then, at each time step $\bar{t}^k$, a standard finite difference method (FDM) is used on(12.7) in order to find the initial nodal displacements U_j $(\bar{t}^k = 0)$ or velocities $\dot{U}_j$ $(\bar{t}^k > 0)$. Moreover, corresponding stress and strain rates $\dot{S}_j$, $\dot{E}_j$, and the rates of damage function $\dot{D}_j$ are computed. The Runge–Kutta II (RKII) method is applied next to find current values of stress components as well as the damage function. Hence, when geometry of the disk is prescribed the initial-boundary creep-damage problem is solved with the elastic solution considered as the initial condition for creep.

I. Elastic problem $(\bar{t}^k = 0)$

For a disk of constant thickness $H(R) = H_0$, the analytical solution of the reduced fundamental equations (12.7)–(12.9) provides:

$$U^{\mathrm{e}} = -\frac{A}{2}R + \frac{B}{R} - \frac{K}{8}R^3,$$

$$S_r = \frac{1}{1-\nu^2}\left[-(1+\nu)\frac{A}{2} - \frac{(1-\nu)}{R^2}B - (3+\nu)\frac{K}{8}R^2\right],$$

$$S_\vartheta = \frac{1}{1-\nu^2}\left[-(1+\nu)\frac{A}{2} + \frac{(1-\nu)}{R^2}B - (1+3\nu)\frac{K}{8}R^2\right], \tag{12.14}$$

$$E_r^{\mathrm{e}} = -\frac{A}{2} - \frac{B}{R^2} - \frac{3K}{8}R^2, \quad E_\vartheta^{\mathrm{e}} = -\frac{A}{2} + \frac{B}{R^2} - \frac{K}{8}R^2.$$

For a disk of arbitrary thickness $H(R)$, the FDM is used to solve (12.7). Then, in view of Hooke's law, the initial stress and strain components are found:

$$S_r = \frac{1}{1-\nu^2}\left(\frac{\mathrm{d}U}{\mathrm{d}R} + \nu\frac{U}{R}\right), \quad S_\vartheta = \frac{1}{1-\nu^2}\left(\frac{U}{R} + \nu\frac{\mathrm{d}U}{\mathrm{d}R}\right), \tag{12.15}$$

$$E_r^{\mathrm{e}} = S_r - \nu S_\vartheta, \qquad E_\vartheta^{\mathrm{e}} = S_\vartheta - \nu S_r.$$

II. Creep problem ($\bar{t}^k > 0$)

For the next time-step of the process, $\bar{t}^1 = \bar{t}^0 + \Delta\bar{t}$, and for an arbitrary thickness distribution $H(R)$, the creep strain rates $\dot{E}_r^{\mathrm{c}}$, $\dot{E}_\vartheta^{\mathrm{c}}$ are determined from (12.8). The FDM solution of the second fundamental equation (12.7) furnishes then the nodal velocities $\dot{U}$, whereas stress rates $\dot{S}_r$, $\dot{S}_\vartheta$ are obtained as

$$\dot{S}_r = \frac{1}{1-\nu^2}\left[\left(\frac{\mathrm{d}\dot{U}}{\mathrm{d}R} + \nu\frac{\dot{U}}{R}\right) - \dot{G}\right],$$

$$\dot{S}_\vartheta = \frac{1}{1-\nu^2}\left[\left(\frac{\dot{U}}{R} + \nu\frac{\mathrm{d}\dot{U}}{\mathrm{d}R}\right) - \left(\dot{E}_\vartheta^{\mathrm{c}} + \nu\dot{E}_r^{\mathrm{c}}\right)\right]. \tag{12.16}$$

For each subsequent step of the process, $\bar{t}^{k+1} = \bar{t}^k + \Delta\bar{t}$, the current magnitudes of the stress components S_r, S_ϑ are found on the basis of the Runge–Kutta RKII method, whereas the corresponding damage function $D(R)$ is determined from (12.9). Substitution of these values for S_r, S_ϑ and D in (12.8) again sets up the new creep strain rates at $\bar{t}^{k+1}$ and the procedure can be continued for as long as desired.

12.2 Two-step optimization approach

For the prescribed loading parameters, the optimal distribution of disk thickness $H(R)$ and the initial prestressing Q which maximize the time of failure initiation $\bar{t}_I$ (first macrocracks) under the condition of constant volume and the additional geometric constraints are sought:

$$\begin{aligned} &\bar{t}_I[H(R); Q] = \max; \\ &\omega, P_{a,b} = \text{const}, \qquad V = \text{const}, \qquad H_{\text{inf}} \leq H(R) \leq H_{\text{sup}}. \end{aligned} \tag{12.17}$$

As the first step of optimal design, the shape of a disk of uniform creep strength (UCS) is determined, $H_{\text{ucs}}(R)$. In general, we begin the iteration loop with the disk of constant thickness $H(R) = H_0$. The damage distribution $D_0(R)$ that corresponds to the zero-order lifetime estimation $\bar{t}_{R_0}$ (constant thickness) is obtained when the coupled creep-damage problem is solved. Next, the corrections of the disk thickness according to the piecewise linear approximation (12.13) are imposed with the constant volume condition applied. The nodal correction of disk thickness is assumed to be proportional to the power function of the residual value of the nodal continuity function $\psi_j = 1 - D_j$ at rupture time $\bar{t}_{R_0}$. Hence, the thickness correction rule, the constant volume condition (for the corrections), and the continuity of thickness at nodes yield:

$$\begin{aligned} &H_j^k(R_j) - H_j^{k-1}(R_j) = \mathcal{P}\left(\bar{\psi} - \psi_j\right)^{\alpha}, \quad j = 1, \ldots, N, \\ &\sum_{j=1}^{N-1} \int_{R_j}^{R_{j+1}} \left[H_j^k(R) - H_j^{k-1}(R)\right] R\mathrm{d}R = 0, \\ &H_{I-1}^k(R_j) = H_I^k(R_j), \quad j = 2, \ldots, N-1. \end{aligned} \tag{12.18}$$

In the above equations, subscript j stands for the node number, superscript k for the time-step number, and subscript I the number of the spatial interval. Equations (12.18) provide $2N - 1$ conditions for the same number of unknowns: $2(N-1)$ coefficients of linear approximation A_I, B_I $(I = 1, ..., N-1)$ in (12.13) and the reference value $\bar{\psi}$. The step factor $\mathcal{P}$ should be chosen experimentally and the exponent α is adjusted numerically for best convergence. When the initial shape H is improved, we solve the creep problem for the new disk geometry to obtain the corrected distribution of the damage function D_1 at the lifetime $\bar{t}_{R_1}$. The whole procedure is repeated until an error norm for "uniform" damage is satisfied, e.g., $|D_{\min} - D_{\max}| < \varepsilon$. This means that the shape of uniform creep strength with respect to brittle rupture is found (cf. Fig. 12.3).

The first optimization level described above may be insufficient in view of the discussion concerning optimality or nonoptimality of the solution of uniform creep strength $H_{\text{ucs}}(R)$ with respect to the lifetime (cf. Sect. 11.1.5). Therefore, a second optimization step has to be performed in order

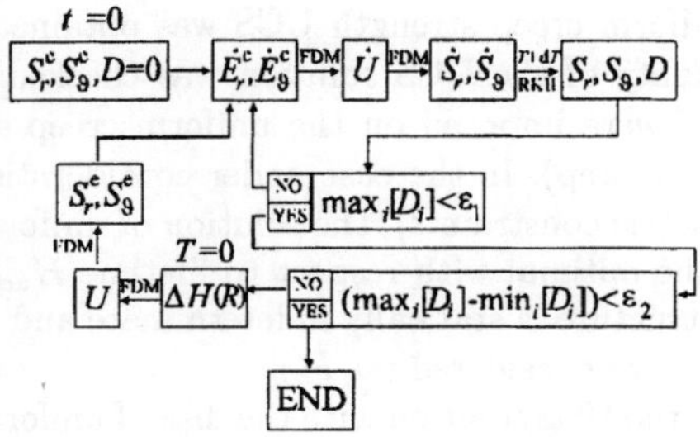

Fig. 12.3. Numerical algorithm for first step to optimal design of disks of uniform damage strength (after Skrzypek and Egner, 1993)

to answer the question whether the UCS shape obtained ensures the maximum lifetime or not. When appropriate corrections to the UCS shape of uniform creep strength are imposed, an improvement of the lifetime may be expected even in the case when the geometry changes are disregarded. In fact, with active zones of the geometric inequality constraints allowed for, the condition of uniform creep strength holds only in the remaining (passive) zones of the disk. Hence, a possible improvement of the disk lifetime may be achieved by corrections of the shape and the length of zones of uniform creep strength in order to maximize the disk lifetime. The following parabolic form of the correction terms of $H_{\text{ucs}}(R)$ is proposed:

$$\begin{aligned} \Delta H(R) &= a_1 R^2 + a_2 R + a_3, \\ H_{\text{opt}}(R) &= H_{\text{ucs}}(R) + \Delta H(R) \end{aligned} \tag{12.19}$$

for which the condition of constant volume,

$$\int_{R_1}^{R_2} \left(a_1 R^2 + a_2 R + a_3\right) R \mathrm{d}R = 0, \tag{12.20}$$

holds; hence, only two parameters remain free to be optimized.

12.3 Example A: Clamped annular disk of uniform damage strength versus uniform elastic strength

Let us consider an annular disk with inner and outer dimensionless radii R_1 and R_2, clamped at the inner edge and free to move at the outer one. The disk is subject to steady rotation about the symmetry axis with angular velocity ω and to uniform tension P_b applied along the periphery (cf. Fig. 12.1). When the optimization procedure described in Sect. 12.2 was used, a disk of uniform creep strength UCS was obtained (first optimization step). Optimality of the UCS solution was checked when the shape corrections (12.19) were imposed on the uniform creep strength solution (second optimization step). In the case under consideration (no prestressing force, no geometric constraints), the solution of uniform creep strength was found to be the optimal with respect to lifetime $H_{\mathrm{ucs}}(R) \equiv H_{\mathrm{opt}}(R)$ even though the structure is statically indeterminate and two independent loading parameters are considered (ω, P_b).

A comparison of the UCS solution with the disk of uniform elastic strength UES is shown in Fig. 12.4. A constant volume condition and the same loading parameters were assumed in both solutions. The proposed design method results in a significant improvement of the disk lifetime when compared to the disk of constant thickness, Table 12.1.

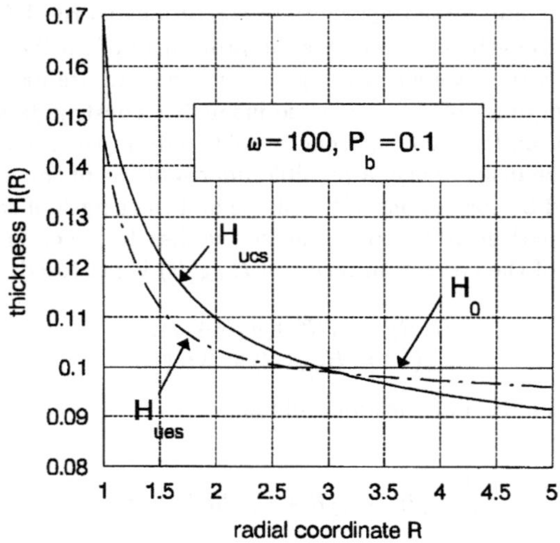

Fig. 12.4. Rotating disk of uniform creep strength versus disk of uniform elastic strength (after Skrzypek and Egner, 1993)

Table 12.1. Lifetime improvement for clamped disks of UES versus UCS (after Skrzypek and Egner, 1993)

Temp.	Lifetime (first macrocracks)		
	Constant thickness H_0	Uniform elastic strength H_{ues}	Uniform creep strength H_{ucs}
773 K	$\bar{t}_I^0 = 74.2$	$\bar{t}_I^{\mathrm{ues}} = 2.9\bar{t}_I^0$	$\bar{t}_I^{\mathrm{ucs}} = 3.9\bar{t}_I^0$
873 K	$\bar{t}_I^0 = 79.0$	$\bar{t}_I^{\mathrm{ues}} = 2.0\bar{t}_I^0$	$\bar{t}_I^{\mathrm{ucs}} = 2.9\bar{t}_I^0$

12.4 Example B: Effect of initial prestressing on the lifetime of disk of uniform creep strength

12.4.1 Prestressed disk of uniform creep strength (UCS)

A rotating disk clamped at the inner edge and prestressed by the elastic ring at the outer edge is analyzed (cf. Fig. 12.2). Both the initial prestressing Q and the distribution of thickness $H(R)$ are subjected to optimization. The effect of initial prestressing as an additional decision variable results in nonuniqueness of the solutions of uniform creep strength. Hence, from among the disks of uniform creep strength, further optimization may be performed with respect to initial prestressing force Q in order to find the optimal shape against brittle rupture and the corresponding prestressing force for which the lifetime is maximized: $\bar{t}_I(H_{ucs}(R), Q_{opt}) \rightarrow \max$. Here $R_1 = 1$, R_2, and R_0 denote dimensionless radii of the disk subjected to optimization and the prestressing ring, respectively. The Kachanov–Sdobyrev damage growth rule is applied in order to find the disk of uniform creep strength, when $\omega = 240\ \mathrm{s}^{-1}$ and $\delta = 0.5$. An additional constraint is imposed on the thickness $H_{inf} = 0.75H_0$. The shape of uniform creep strength without geometric constraints imposed results in an unacceptable distribution of thickness which approaches zero in the middle zone of the disk and, hence, the convergence of the numerical procedure fails dramatically. Moreover, it is supposed that the minimum thickness of the disk must ensure structural stability against creep buckling for the given magnitude of the prestressing force.

The effect of initial radial prestressing Q on the lifetime of a disk of uniform creep strength with the minimum thickness constraint ($H = 0.75H_0$) imposed, is shown in Fig. 12.5.

The magnitude of the optimal initial dimensionless prestressing force which maximizes the lifetime is equal to $Q_{opt} = 0.034$. Note that in the case of a disk of constant thickness the optimal prestressing force is approximately the same, butr the corresponding lifetime is almost 10 times shorter. The shape of the disk of uniform creep strength UCS with a minimum thickness constraint ($H_{inf} = 0.75H_0$) is compared with the corresponding profile obtained when subsequent iterations of thickness were not constrained, Fig. 12.6.

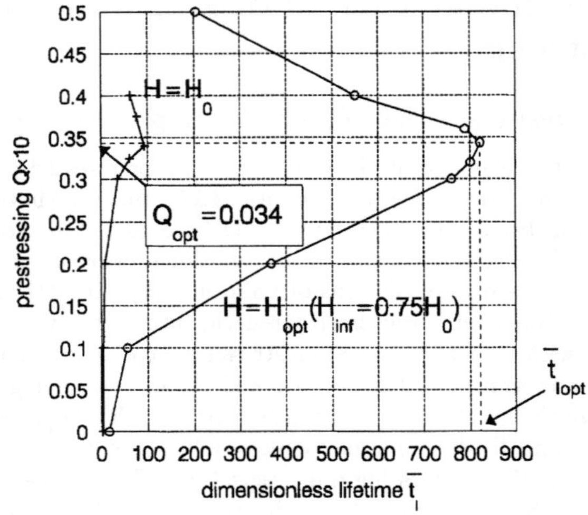

Fig. 12.5. Effect of initial prestressing Q on time to macrocrack initiation $\bar{t}_I$ of a disk of uniform creep strength H_{ucs} versus a disk of constant thickness H_0 (after Skrzypek and Egner, 1993)

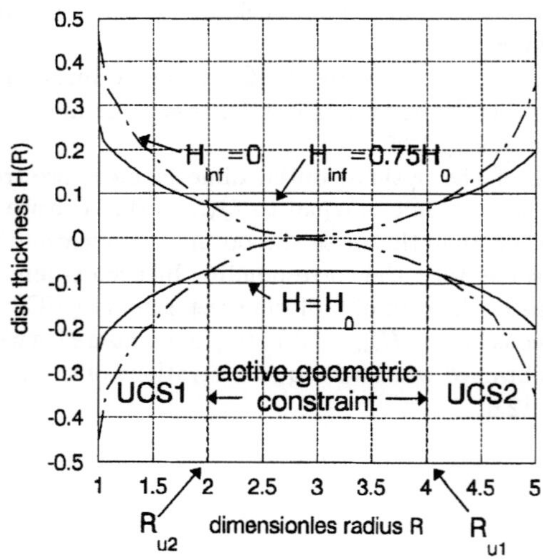

Fig. 12.6. Variation of profiles of optimally prestressed disks of uniform creep strength H_{ucs} with a magnitude of lower geometric constriant H_{inf}

12.4.2 Prestressed disk of uniform creep strength UCS versus the optimal disk

To check the optimality of the solutions of uniform creep strength $H_{ucs}(R)$ the shape corrections (cf. (12.19) and (12.20)) are imposed upon the profile of the disk of uniform creep strength without any volume change.

In the case of prestressed disks when the lower geometric constraint is imposed, correction of both the thickness and the length of zones of uniform creep strength may be subjected to parametric optimization. For instance, in a case when the parabolic shape corrections (12.19) are imposed on each of two zones independently ΔH_1, ΔH_2 with their lengths held constant, only one parameter is free to be optimized in each zone $\varepsilon_1 = \Delta H_1(R_1)$ and $\varepsilon_2 = \Delta H_2(R_2)$ (Fig. 12.7) to yield: $\bar{t}_I(H_{\text{ucs}}) = 293.92$, $\bar{t}_I(\Delta H_1) = 294.22$, $\bar{t}_I(\Delta H_1, \Delta H_2) = \bar{t}_{I_{\text{opt}}} = 294.41$. However, in a more general case when both the thickness and the length are optimized, ΔH_1, ΔH_2 and R_{u1}, R_{u2}, two parameters are free in each zone, $\varepsilon_1 = \Delta H_1(R_1)$, ΔR_{u1} and $\varepsilon_2 = \Delta H_2(R_2)$, ΔR_{u2}. Nevertheless, in the case under consideration the longest lifetime was achieved when the initial lengths of zones of uniform creep strength were unchanged.

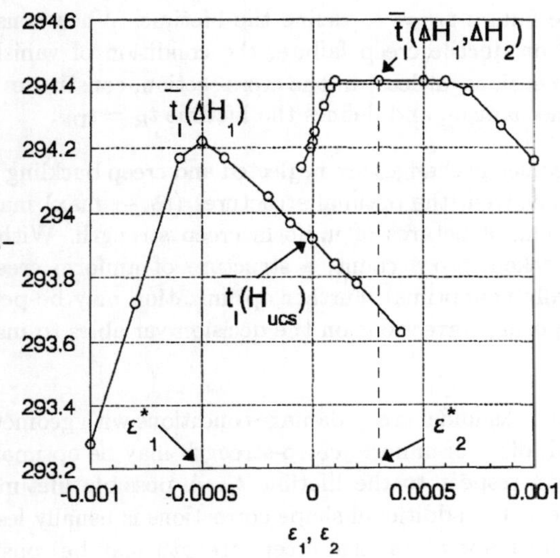

Fig. 12.7. Parametric thickness corrections ΔH_1, ΔH_2 of disk of uniform creep strength H_{ucs} for time to failure improvement $\bar{t}_{I_{\max}}$ (after Skrzypek and Egner, 1993)

12.5 Discussion: Non/optimality of structures of uniform creep strength with respect to lifetime

i. When elastic structures are designed for either minimum weight or maximum load under a strength constraint, the structures of uniform elastic strength, also called the fully stressed designs, are in most cases optimal. In general, when static indeterminacy of structure or geometric changes are taken into account, the condition of uniform strength is neither a necessary nor a sufficient condition of optimality. Hence, the fully stressed design method is, in general, a first step towards the exact optimal solution when more rigorous optimization approaches are used (Gallagher, 1973).

ii. When optimization of structures under creep conditions is formulated, the minimum weight or the maximum load remains a typical design objective, whereas constraints may be imposed not only on the strength (failure), stiffness, and stability as in the elastic case, but also on a limited stress relaxation, a limited residual displacement, or a given lifetime. When constraints are imposed on brittle creep failure, the initiation of first macrocrack $t_I = t_{\mathrm{R}}$ or a complete structure failure $t_{II} = t_{\mathrm{F}}$ define the lifetime. When constraints are imposed on ductile creep failure, the condition of vanishing transverse dimensions, at least in one cross-section, constitutes the ductile failure mechanism, and defines the lifetime $t_{\mathrm{R}} = t_{\mathrm{DF}}$.

iii. When geometric changes are neglected and creep buckling constraints are not involved, the optimal structures ($t_{\mathrm{R}} \rightarrow \max$) may be found from among structures of uniform creep strength. With geometric changes taken into account, a structure of uniform creep strength is generally nonoptimal. Further optimization may be performed by superimposing corrections on the decision variables to maximize the lifetime.

iv. In case of disks under creep-damage conditions with geometric changes neglected, disks of uniform creep-strength may be optimal or nonoptimal with respect to the lifetime t_I. A possible lifetime improvement due to the additional shape corrections is usually less than 1%. Hence, the shape of uniform creep strength may be considered as a sufficiently good approximation for the optimal disk.

v. A disk of uniform creep strength subjected to stationary loadings (not prestrained) was found to be optimal with respect to lifetime. On the other hand, a disk of partially uniform creep strength (zone of active geometric constraint admitted) subjected to nonstationary loadings (due to the initial prestressing) was found to be nonoptimal with respect to lifetime.

vi. When the effect of preloading damage is taken into account (cf. Egner and Skrzypek, 1994) for each prescribed preloading period $\Delta\bar{t}_{\text{pre}}$ the optimum prestressing force may be found. Usually it corresponds to simultaneous initiation of first macrocracks at the inner and the outer fibers of the disk (a switch point where two curves representing different failure mechanisms intersect). However, when the duration of the preloading period is sufficiently long, it may happen that the initial damage during preloading at the inner fiber is rapid enough to reach the maximum net lifetime without the switching effect. In this case the optimal prestressing is determined by the smooth extremum point on the curve $\bar{t}_I^{\text{net}}(Q)$ as shown in Fig. 12.8.

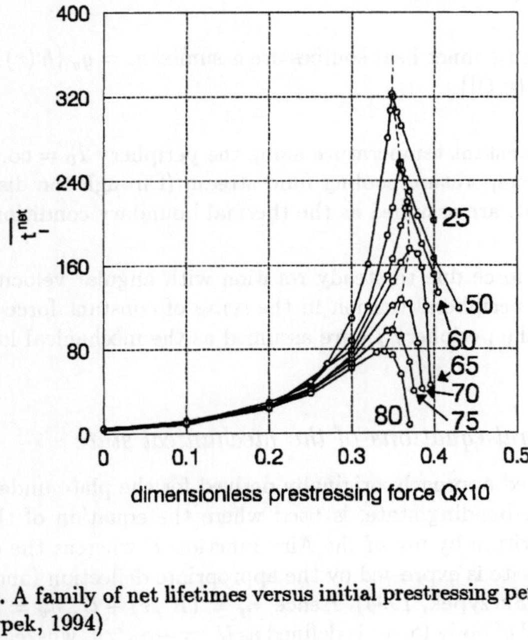

Fig. 12.8. A family of net lifetimes versus initial prestressing period (after Egner and Skrzypek, 1994)

12.6 Example C: Optimal design of rotationally symmetric disks in thermo-damage coupling conditions

12.6.1 Assumptions

i. A thin axisymmetric disk of variable thickness under plane stress conditions is considered (Fig. 12.9).

ii. The geometrically linear theory of small displacement and the additive decomposition of strains are applied: $\boldsymbol{\varepsilon} = \boldsymbol{\varepsilon}^{\mathrm{e}} + \boldsymbol{\varepsilon}^{\mathrm{c}} + \boldsymbol{\varepsilon}^{\mathrm{th}}$.

iii. The fully coupled orthotropic creep-damage approach is used (7.14).

iv. The coupled thermo-damage problem is solved (Model C (7.22)) by the use of the equivalent conductivity concept.

v. A 1D nonstationary temperature field is assumed $T\left[r, D\left(r,t\right)\right]$ (temperature homogenization through the disk thickness) but only quasi-static changes of temperature are allowed ($\dot{T} = 0$).

vi. 1D volumetric inner heat sources are assumed $q_v = q_v \left\{h\left(r\right), \mathrm{d}h\left(r\right) / \mathrm{d}r, T\left[r, D\left(r,t\right)\right]\right\}$.

vii. Uniform constant temperature along the periphery $T_0 = \mathrm{const}$ and a constant temperature cooling fluid stream (through the disk faces), $T_\infty = \mathrm{const}$, are assumed as the thermal boundary conditions.

viii. The body force due to steady rotation with angular velocity ω and a uniform peripheral tension in the sense of constant force per unit length of the periphery p_0 are assumed as the mechanical loadings.

12.6.2 General equations of the mechanical state

The general mixed approach, originally derived for the plate under a combined membrane-bending state, is used where the equation of the membrane state is written by use of the Airy function F whereas the equation of the bending state is expressed by the appropriate deflection function (cf. Ganczarski and Skrzypek, 1994). Hence, $n_r = (F^{'}/r) + U$, $n_\vartheta = F^{''} + U$, where a potential of body forces is defined as $U^{'} = -\varrho\omega^2 r h$, whereas symbol prime stands for the derivative with respect to r. Finally, the fundamental mechanical state equations are furnished:

$$
\left.\begin{array}{l}
\mathcal{F}[F] + (1-\nu)\mathcal{B}(r)\nabla^2\left[\dfrac{U}{\mathcal{B}(r)}\right] + (1-\nu^2)\mathcal{B}(r)\alpha\nabla^2 T = 0 \quad (t=0)\,,\\[2ex]
\left.\begin{array}{l}
\mathcal{F}[\dot{F}] + (1-\nu^2)\mathcal{B}(r)\alpha\nabla^2\dot{T} + \mathcal{B}(r)\nabla^2\left[\dfrac{\dot{n}^{\mathrm{c}}_{\vartheta} - \nu\dot{n}^{\mathrm{c}}_{r}}{\mathcal{B}(r)}\right]\\[2ex]
+\dfrac{1+\nu}{r}\mathcal{B}(r)\dfrac{\mathrm{d}}{\mathrm{d}r}\left[\dfrac{\dot{n}^{\mathrm{c}}_{\vartheta} - \dot{n}^{\mathrm{c}}_{r}}{\mathcal{B}(r)}\right] = 0
\end{array}\right\} \quad (t>0)\,,
\end{array}\right.
\tag{12.21}
$$

where the differential operator $\mathcal{F}[...]$ as well as the auxiliary operators ∇^2, ∇^4, independent of circumferential coordinate, take the form (cf. Sect. 9.2.2 with $k=0$):

$$
\begin{aligned}
\mathcal{F}[...] &= \nabla^4 + \mathcal{B}(r)\frac{\mathrm{d}}{\mathrm{d}r}\left[\frac{1}{\mathcal{B}(r)}\right]\left(2\frac{\mathrm{d}^3...}{\mathrm{d}r^3} + \frac{2-\nu}{r}\frac{\mathrm{d}^2...}{\mathrm{d}r^2} - \frac{1}{r^2}\frac{\mathrm{d}...}{\mathrm{d}r}\right)\\
&\quad +\mathcal{B}(r)\frac{\mathrm{d}^2}{\mathrm{d}r^2}\left[\frac{1}{\mathcal{B}(r)}\right]\left(\frac{\mathrm{d}^2...}{\mathrm{d}r^2} - \frac{\nu}{r}\frac{\mathrm{d}...}{\mathrm{d}r}\right),\\
\nabla^2... &= \frac{\mathrm{d}^2...}{\mathrm{d}r^2} + \frac{1}{r}\frac{\mathrm{d}...}{\mathrm{d}r},\\
\nabla^4... &= \frac{\mathrm{d}^4...}{\mathrm{d}r^4} + \frac{2}{r}\frac{\mathrm{d}^3...}{\mathrm{d}r^3} - \frac{1}{r^2}\frac{\mathrm{d}^2...}{\mathrm{d}r^2} + \frac{1}{r^3}\frac{\mathrm{d}...}{\mathrm{d}r}.
\end{aligned}
\tag{12.22}
$$

The inelastic membrane forces expressed in terms of inelastic strains and the membrane stiffness are defined as follows:

$$
n^{\mathrm{c}}_{r/\vartheta} = \mathcal{B}(r)(\varepsilon^{\mathrm{c}}_{r/\vartheta} + \nu\varepsilon^{\mathrm{c}}_{\vartheta/r}), \qquad \mathcal{B}(r) = \frac{E(r)h(r)}{1-\nu^2}. \tag{12.23}
$$

12.6.3 Constitutive equations for coupled creep-damage problem

Due to nonproportional loadings when the general orthotropic damage rule is applied, the creep process becomes orthotropic as well (damage induced creep orthotropy). Hence, the fully coupled creep damage approach is required, where effective stress components are used instead of simple stress components and time hardening hypothesis governs the creep strain-rate intensity

$$
\dot{\varepsilon}^{\mathrm{c}}_{kl} = \frac{3}{2}\frac{\dot{\varepsilon}^{\mathrm{c}}_{eq}}{\widetilde{\sigma}_{\mathrm{eq}}}\widetilde{s}_{kl}, \qquad \dot{\varepsilon}^{\mathrm{c}}_{\mathrm{eq}} = (\widetilde{\sigma}_{\mathrm{eq}})^{m(T)}\dot{\mathsf{f}}(t),
$$

$$
\widetilde{s}_{r/\vartheta} = \frac{2}{3}\left(\frac{\sigma_{r/\vartheta}}{1-D_{r/\vartheta}} - \frac{\sigma_{\vartheta/r}}{2(1-D_{\vartheta/r})}\right), \qquad k,l = r,\vartheta
$$

$$\widetilde{s}_{r/\vartheta} = \sqrt{\left(\frac{\sigma_r}{1-D_r}\right)^2 + \left(\frac{\sigma_\vartheta}{1-D_\vartheta}\right)^2 - \frac{\sigma_r \sigma_\vartheta}{(1-D_r)(1-D_\vartheta)}} \tag{12.24}$$

$$\dot{D}_\nu = C_\nu(T)\left\langle \frac{\sigma_\nu}{1-D_\nu}\right\rangle^{r_\nu(T)}, \nu = r, \vartheta \tag{12.25}$$

12.6.4 Formulation of coupled thermo-mechanical boundary problems

The mechanical state fulfills (12.21) and the following mechanical boundary conditions:

$$\begin{array}{lll} n_r(0) = n_\vartheta(0), & n_r(R) = p_0 h_0 & (t=0)\,, \\ \dot{n}_r(0) = \dot{n}_\vartheta(0), & \dot{n}_r(R) = 0 & (t>0)\,. \end{array} \tag{12.26}$$

The equation of heat transfer of Model C (7.22), requires the inner heat source intensity to be explicitly defined:

$$\dot{q}_v \stackrel{\text{def}}{=} -\frac{\dot{Q}_v}{\mathrm{d}V} = -\frac{\dot{Q}_v}{r\mathrm{d}\vartheta h \mathrm{d}r}, \tag{12.27}$$

where the surface element and its slope are:

$$\mathrm{d}A = \frac{r\mathrm{d}\vartheta\mathrm{d}r}{\cos\Theta}, \qquad \cos\Theta = \frac{1}{\sqrt{1+\tan^2\Theta}} = \frac{1}{\sqrt{1+(\mathrm{d}h/\mathrm{d}r)^2}}. \tag{12.28}$$

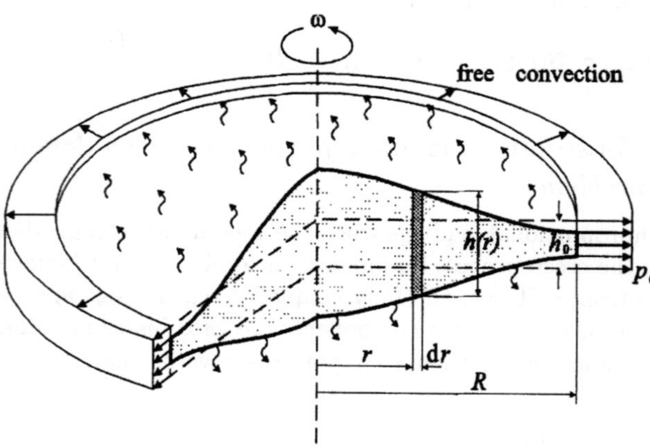

Fig. 12.9. Rotating disk of variable thickness (versus constant thickness disk of the same volume) stretched at periphery and cooled through faces (after Ganczarski and Skrzypek, 1997)

To express the overall effect of convection through both disk faces, the classical Newton law of cooling is applied (cf. Holman, 1990):

$$\dot{Q}_v = 2\beta \mathrm{d}A(T - T_\infty), \tag{12.29}$$

where T_∞ is the temperature of the cooling fluid, hence:

$$\dot{q}_v = -2\beta\frac{\sqrt{1+(\mathrm{d}h/\mathrm{d}r)^2}}{h}(T - T_\infty) \tag{12.30}$$

The appropriate thermal boundary conditions are:

$$\begin{aligned} \mathrm{d}T/\mathrm{d}r|_{r=0} &= 0, \quad T(R) = T_0 \quad (t=0) \\ \mathrm{d}\dot{T}/\mathrm{d}r\Big|_{r=0} &= 0, \quad \dot{T}(R) = 0 \quad (t>0)\,. \end{aligned} \tag{12.31}$$

12.6.5 Optimization problem

Assuming the orthotropic damage law, the structures of uniform creep strength fulfill the condition:

$$\sup_{\forall r \in V}\{D_\nu(r, t_I)\} \cong 1, \qquad \nu = 1,2,3. \tag{12.32}$$

The distribution of disk thickness $h(r)$ is considered as the decision variable when the geometric inequality constraints on the maximum and minimum thicknesses

$$h_{\max} \geq h(r) \geq h_{\min} \tag{12.33}$$

and the constraint on the maximum local gradient of temperature such that the assumption of small thermal displacements is satisfied,

$$\max\{\mathrm{d}T/\mathrm{d}r\} \leq (\mathrm{d}T/\mathrm{d}r)_{\max}, \tag{12.34}$$

are checked during the optimization procedure. Additionally, the condition of constant volume requires:

$$V = 2\pi\int_0^R h(r) r \mathrm{d}r = \mathrm{const} \quad \text{or} \quad \delta V = 2\pi\int_0^R \delta h(r) r \mathrm{d}r = 0. \tag{12.35}$$

The optimization procedure, based on iterative corrections of the decision variable $h\,(r)$ is suggested. When the optimization with respect to uniform creep strength under constant volume of a structure is performed, the nodal increments of the decision variable Δh_j are chosen proportionally to the level of the nodal dominant component of the damage tensor (cf. Ganczarski and Skrzypek, 1994):

$$\Delta h_j = \mathcal{P}\Delta D_j - \Delta h_m, \qquad \Delta D_j = \sup_{(r,\vartheta)} \left\{ D_{r/\vartheta} \right\}_j, \qquad j = 1, ..., N \quad (12.36)$$

where the reference correction Δh_m must satisfy the constant volume condition:

$$\Delta h_m = \frac{\sum_j \mathcal{P}\Delta D_j r_j}{\sum_j r_j}, \tag{12.37}$$

whereas the step factor $\mathcal{P}$ should be chosen experimentally (cf. Sect. 12.2). The process of damage equalization is continued until the following condition is fulfilled:

$$\sup \left\{ D_{r/\vartheta} \right\}_j \leq \text{EPS} \cong 1 \qquad \forall j. \tag{12.38}$$

The suggested procedure is essentially relevant to the concept of the full damage design method. This method leads to exact solutions (optimal with respect to maximal lifetime) when the structure is statically determinate, single loadings are applied, and geometric changes are neglected (cf. Sect. 11.5). If the above constraints are exceeded, the uniform creep strength solution may turn out to be nonoptimal. An exact solution may be obtained when more advanced optimization approaches are used, for which the UCS solution may be regarded as a first approximation.

12.6.6 Numerical algorithm for coupled thermo-creep-damage problem

A modification of the numerical procedure described in Sect. 7.3.4, that accounts for a variable thickness $h(r)$ and the substitutive conductivity concept (Model C) (7.22), is used. Hence, when the FDM is applied to (7.22) with the inner heat source intensity (12.30) we arrive at:

$$\left[\frac{1}{(\Delta r)^2} - \left(\frac{-\lambda_{j-1}^{eq} + \lambda_{j+1}^{eq}}{2\lambda_j^{eq}\Delta r} + \frac{-h_{j-1} + h_{j+1}}{2h_j\Delta r} + \frac{1}{r} \right) \frac{1}{2\Delta r} \right] T_{j-1}$$

$$- \left[\frac{2}{(\Delta r)^2} + \frac{2\beta}{\lambda_j^{eq}} \frac{\sqrt{1 + \left(\frac{-h_{j-1}+h_{j+1}}{2\Delta r} \right)^2}}{h_j} \right] T_j$$

$$+ \left[\frac{1}{(\Delta r)^2} + \left(\frac{-\lambda_{j-1}^{eq} + \lambda_{j+1}^{eq}}{2\lambda_j^{eq}\Delta r} + \frac{-h_{j-1} + h_{j+1}}{2h_j\Delta r} + \frac{1}{r} \right) \frac{1}{2\Delta r} \right] T_{j+1}$$

$$= -\frac{2\beta}{\lambda_j^{\rm eq}} \frac{\sqrt{1+\left(\frac{-h_{j-1}+h_{j+1}}{2\Delta r}\right)^2}}{h_j} T_\infty, \tag{12.39}$$

$$\lambda_j^{\rm eq} = \lambda_0 (1 - D_j) + \sigma\epsilon_0 4 D_j T_j^3 \Delta r.$$

In order to specify the differential operators entering the (12.21), the following FD representation of (12.21)–(12.22) is defined:

$$\nabla^4 F \cong \left[\frac{1}{(\Delta r)^4} - \frac{4r - 3\Delta r}{4r(r-\Delta r)(\Delta r)^3}\right] F_{j-2} + \left[-\frac{4}{(\Delta r)^4} + \frac{2}{r(\Delta r)^3}\right] F_{j-1}$$

$$+ \left[\frac{6}{(\Delta r)^4} + \frac{1}{2(r-\Delta r)} \frac{1}{(r+\Delta r)(\Delta r)^2}\right] F_j$$

$$+ \left[-\frac{4}{(\Delta r)^4} - \frac{2}{r(\Delta r)^3}\right] F_{i+1} + \left[\frac{1}{(\Delta r)^4} + \frac{4r + 3\Delta r}{4r(r+\Delta r)(\Delta r)^3}\right] F_{j+2},$$

$$\mathcal{B}\frac{\mathrm{d}}{\mathrm{d}r}\left(\frac{1}{\mathcal{B}}\right)\left(2\frac{\mathrm{d}^3 F}{\mathrm{d}r^3} + \frac{2-\nu}{r}\frac{\mathrm{d}^2 F}{\mathrm{d}r^2} - \frac{1}{r^2}\frac{\mathrm{d}F}{\mathrm{d}r}\right) \cong \frac{-h_{j-1}+h_{j+1}}{2h_j \Delta r}$$

$$\times \left[\frac{F_{j-2}}{(\Delta r)^3} + \left(\frac{2}{(\Delta r)^3} + \frac{2-\nu}{r(\Delta r)^2} + \frac{1}{2r^2\Delta r}\right) F_{j-1} - 2\frac{2-\nu}{r(\Delta r)^2} F_j\right.$$

$$\left. + \left(-\frac{2}{(\Delta r)^3} + \frac{2-\nu}{r(\Delta r)^2} - \frac{1}{2r^2\Delta r}\right) F_{j+1} + \frac{F_{j+2}}{(\Delta r)^3}\right],$$

$$\mathcal{B}\frac{\mathrm{d}^2}{\mathrm{d}r^2}\left(\frac{1}{\mathcal{B}}\right)\left(\frac{\mathrm{d}^2 F}{\mathrm{d}x^2} - \frac{\nu}{r}\frac{\mathrm{d}F}{\mathrm{d}r}\right) \cong \left[\frac{(-h_{j-1}+h_{j+1})^2}{2h_j^2(\Delta r)^2} - \frac{h_{j-1} - 2h_j + h_{j+1}}{h_j(\Delta r)^2}\right]$$

$$\times \left[\left(\frac{1}{(\Delta r)^2} + \frac{\nu}{2r\Delta r}\right) F_{j-1} - \frac{2}{(\Delta r)^2} F_j + \left(\frac{1}{(\Delta r)^2} - \frac{\nu}{2r\Delta r}\right) F_{j+1}\right],$$

$$(1-\nu)\mathcal{B}\nabla^2\left(\frac{U}{\mathcal{B}}\right) \cong (1-\nu) h_j \left\{\left[\frac{1}{(\Delta r)^2} - \frac{1}{2r\Delta r}\right]\frac{U_{j-1}}{h_{j-1}} - \frac{2}{(\Delta r)^2}\frac{U_j}{h_j}\right.$$

$$\left. + \left[\frac{1}{(\Delta r)^2} + \frac{1}{2r\Delta r}\right]\frac{U_{j+1}}{h_{j+1}}\right\},$$

$$(1-\nu^2)\mathcal{B}\alpha\nabla^2 T \cong E\alpha h_j \left\{\left[\frac{1}{(\Delta r)^2} - \frac{1}{2r\Delta r}\right] T_{j-1} - \frac{2}{(\Delta r)^2} T_j\right.$$

$$+\left[\frac{1}{(\Delta r)^2}+\frac{1}{2r\Delta r}\right]T_{j+1}\bigg\},$$

$$\mathcal{B}\nabla^2\left(\frac{n_\vartheta^c-\nu n_r^c}{\mathcal{B}}\right)\cong h_j\left\{\left[\frac{1}{(\Delta r)^2}-\frac{1}{2r\Delta r}\right]\frac{n_{\vartheta_{j-1}}^c-\nu n_{r_{j-1}}^c}{h_{j-1}}\right.$$

$$\left.-\frac{2}{(\Delta r)^2}\frac{n_{\vartheta_j}^c-\nu n_{rj}^c}{h_j}+\left[\frac{1}{(\Delta r)^2}+\frac{1}{2r\Delta r}\right]\frac{n_{\vartheta_{j+1}}^c-\nu n_{r_{j+1}}^c}{h_{j+1}}\right\},$$

$$\frac{1+\nu}{r}\mathcal{B}\frac{\mathrm{d}}{\mathrm{d}r}\left(\frac{n_\vartheta^c-n_r^c}{\mathcal{B}}\right)\cong\frac{1+\nu}{2r\Delta r}h_j\left[-\frac{n_{\vartheta_{j-1}}^c-\nu n_{r_{j-1}}^c}{h_{j-1}}+\frac{n_{\vartheta_{j+1}}^c-\nu n_{r_{j+1}}^c}{h_{j+1}}\right].\tag{12.40}$$

The numerical procedure begins when the elastic solutions of the thermal and the coupled mechanical problems are known. Assuming an initially constant structure thickness $[h]_j\equiv h_0$ and initial components of the damage tensor $[D_{r/\vartheta}]_j\equiv 0$, the elastic solution is obtained in the following way. Applying the stage algorithm (Fig. 12.10), the equation of heat transfer, which is linear for the elastic problem, is solved to yield the initial distribution of temperature $[T^e]_j$. Then, equations of the mechanical state are solved, providing the distribution of the Airy function $[F^e]_j$ and the vector of elastic state $[T^e, n^e_{r/\vartheta}, \sigma^e_{r/\vartheta}]_j$. Next, the program enters the creep loop which requires the vector of effective stress intensity, and components of the damage tensor and strain rates $[\widetilde{\sigma}_{eq}, \dot{D}_{r/\vartheta}, \dot{\varepsilon}^c_{r/\vartheta}]_j$ are computed. The thermal problem (12.39) is nonlinear, hence, by substituting the previous solution for temperature $[T^*]_j$ to the equivalent coefficient of thermal conductivity λ^{eq} the solution of (12.39) provides the updated temperature distribution $[T]_j$, which is considered next as an approximate solution for λ^{eq}. The procedure is repeated until the calculated functions $[T]_j$ differs from $[T^*]_j$ with a given accuracy. As a consequence, when rates of change of both the temperature $[\dot{T}]_j$ and the inelastic forces are known $[\dot{n}^c_{r/\vartheta}]_j$, rates of the Airy function $[\dot{F}]_j$ are found by solution of (12.21) and, finally, the vector of state is determined $[\dot{T}, \dot{n}_{r/\vartheta}, \dot{\sigma}_{r/\vartheta}]_j$. In the next time step, applying the Runge–Kutta II method for the thermal state and the mechanical state, the 'new' vector of state is computed, and the program jumps to the beginning of the creep loop.

The numerical procedure is repeated until the highest value of the damage tensor reaches the critical level, when the program quits the loop via the conditional statement.

12.6.7 Results

All numerical examples presented in this chapter deal with disks made of ASTM–321 stainless steel with the following mechanical and thermal properties (cf. Holman, 1990): E = 170 GPa, $\sigma_{0.2}$ = 120 MPa, ν = 0.3,

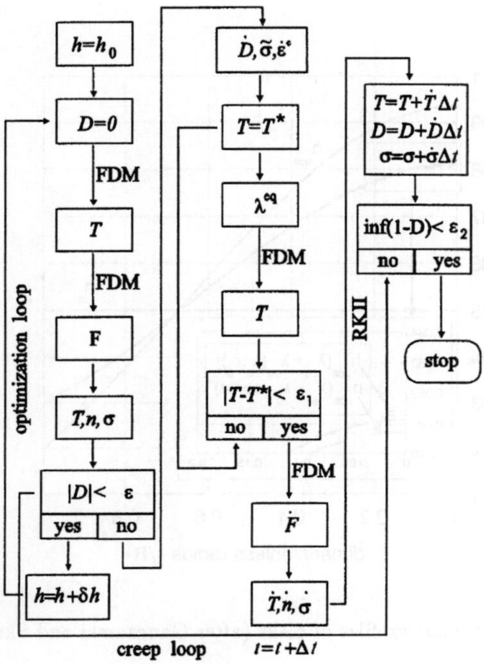

Fig. 12.10. Numerical algorithm for coupled thermo–mechanical problem (after Ganczarski and Skrzypek, 1997)

$\varrho = 7850$ kg/m^3, $\alpha = 1.85 \times 10^{-5}$ K^{-1}, $\lambda_0 = 20$ Wm^{-1}K^{-1}, $\beta = 15$ Wm^{-2}K^{-1}, $R = 1.0$ m, $h_0 = 0.05$ m, $p_0 = 0.1 \times \sigma_{0.2}$, $\omega = 100$ s^{-1}, $T_0 = 798$ K (525°C), $T_\infty = 773$ K (500°C), $\sigma = 5.669 \times 10^{-8}$ Wm^{-2}K^{-4}, $\epsilon_0 = 0.5$. The temperature dependent material functions for creep rupture are presented in Table 12.2, where $\sigma^5_{c_B}$ denotes the stress necessary to cause creep rupture after 10^5 hr.

Table 12.2. Temperature dependent material functions

T (K)	T (°C)	m	r	$\sigma^5_{c_B}$ (MPa)	C (Pa^{-r}s^{-1})
773	500	5.6	3.9	210	1.98×10^{-42}
873	600	4.5	3.1	100	1.07×10^{-34}
923	650	4.0	2.8	60	1.21×10^{-31}

The disk of uniform creep strength, the disk of constant thickness, and the disk of uniform elastic strength (in the sense of the Galileo hypothesis) are compared in Fig. 12.11.

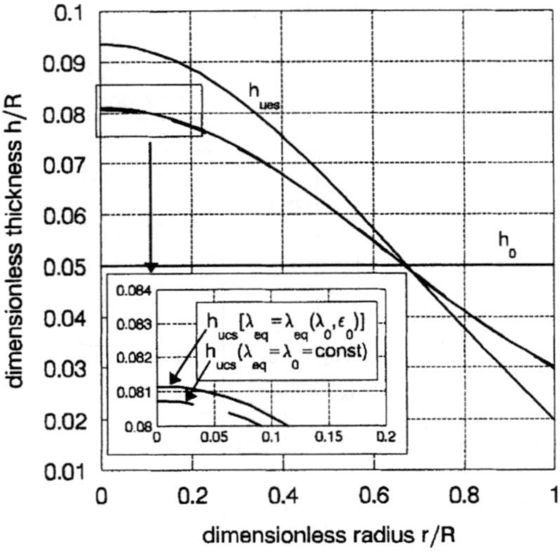

Fig. 12.11. Optimal profiles of disks (after Ganczarski and Skrzypek, 1997)

In the case of a disk of uniform creep strength where the thermo-damage coupling is disregarded $h_{ucs}(\lambda_{eq} = \lambda_0)$, and a disk of uniform creep strength where the thermo-damage coupling is taken into account $h_{ucs}\,[\lambda_{eq} = \lambda_{eq}\,(\lambda_0, \epsilon_0)]$, differences between optimal profiles are negligible (window in Fig. 12.11). However, essential differences in lifetimes are observed (Table 12.3).

Two rupture mechanisms accompany the process of disk design. In the case of the disk of constant thickness $h_0\,[\lambda_{eq} = \lambda_{eq}\,(\lambda_0, \epsilon_0)]$, the distribution of the continuity components at the instant of rupture is presented in Fig. 12.12. The damage accumulation with respect to circumferential component D_ϑ concentrates here near the centre.

The situation becomes much more complex when the uniform creep strength solution is obtained in a disk suffering from orthotropic damage. The uniform creep strength is understood here in the sense of equalizing the dominant damage components (D_r or D_ϑ) along the radius of a disk. In a numerical sense, it means that at failure the dominant damage ($\sup_\nu D_\nu$) or continuity ($\inf_\nu \Psi_\nu$) components reach the critical level $D_{\nu_{crit}}$ or $\Psi_{\nu_{crit}}$, respectively, with a given accuracy. For the purpose of this example we assume $D_{\nu_{crit}} = 0.78 \pm 0.05$, $\Psi_{\nu_{crit}} = 0.22 \pm 0.05$, respectively. The failure mechanism consists of two zones, the first of which refers to a bi-directional system of microcracks (D_r and D_ϑ) around the central point of a disk ($r/R \leq 0.3$), whereas the second refers to the radial direction (D_r) in the remaining portion of the disk. In other words, the fully failed zone

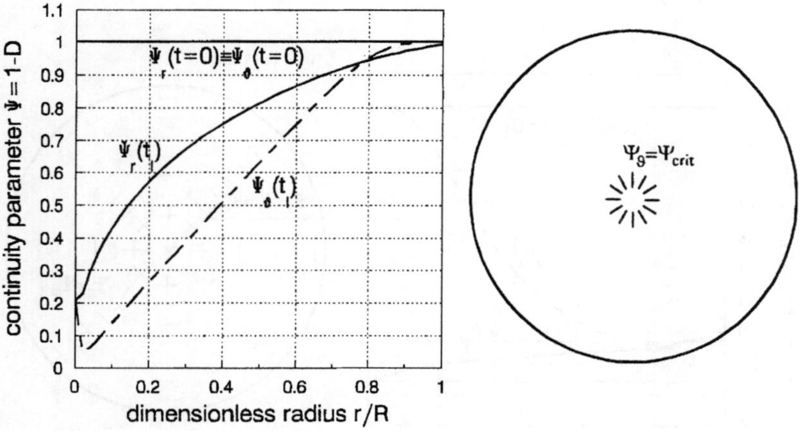

Fig. 12.12. Distribution of continuity tensor components in disk of constant thickness (after Ganczarski and Skrzypek, 1997)

(the central part) and the partly failed one (the outer part) constitute the failure mechanism of a disk (Fig. 12.13) (cf. Malinin and Rżysko, 1981).

The corresponding distribution of temperature versus initial temperature in a disk of constant thickness are shown in Fig. 12.14. During the creep-damage process the local temperature decreases, moderating damage accumulation, what leads to a longer lifetime compared to the case when the thermo-damage coupling is disregarded $h_{\mathrm{ucs}}(\lambda_{\mathrm{eq}} = \lambda_0)$.

Lifetimes of all previously discussed cases are compared in Table 12.3.

Table 12.3. Comparison of lifetime for optimally designed disks (Model C (7.22))

	constant thickness $h(r) = h_0$; thermo-damage coupling (7.22)	
	no $\lambda_{\mathrm{eq}} = \lambda_0$	yes $\lambda_{\mathrm{eq}} = \lambda_{\mathrm{eq}}(\lambda_0, \epsilon_0)$
lifetime	$t_{n_0}^{(\lambda_{\mathrm{eq}}=\lambda_0)} = t_{\mathrm{ref}}$	$t_{n_0}^{(\lambda_{\mathrm{eq}}=\lambda_{\mathrm{eq}}(\lambda_0,\epsilon_0))} = 1.01 t_{\mathrm{ref}}$
	uniform elastic strength $h_{\mathrm{ues}}(r)$; thermo-damage coupling (7.22)	
	no $\lambda_{\mathrm{eq}} = \lambda_0$	yes $\lambda_{\mathrm{eq}} = \lambda_{\mathrm{eq}}(\lambda_0, \epsilon_0)$
lifetime		$t_{\mathrm{ues}}^{(\lambda_{\mathrm{eq}}=\lambda_{\mathrm{eq}}(\lambda_0,\epsilon_0))} = 1.03 t_{\mathrm{ref}}$
	uniform creep strength $h_{\mathrm{ucs}}(x)$; thermo-damage coupling (7.22)	
	no $\lambda_{\mathrm{eq}} = \lambda_0$	yes $\lambda_{\mathrm{eq}} = \lambda_{\mathrm{eq}}(\lambda_0, \epsilon_0)$
lifetime	$t_{\mathrm{ucs}}^{(\lambda_{\mathrm{eq}}=\lambda_0)} = 4.43 t_{\mathrm{ref}}$	$t_{\mathrm{ucs}}^{(\lambda_{\mathrm{eq}}=\lambda_{\mathrm{eq}}(\lambda_0,\epsilon_0))} = 4.70 t_{\mathrm{ref}}$

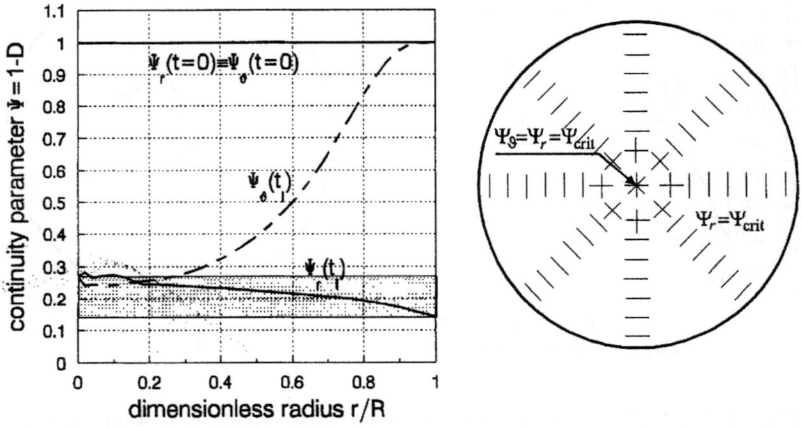

Fig. 12.13. Distribution of continuity tensor components in disk of uniform creep strength (after Ganczarski and Skrzypek, 1997)

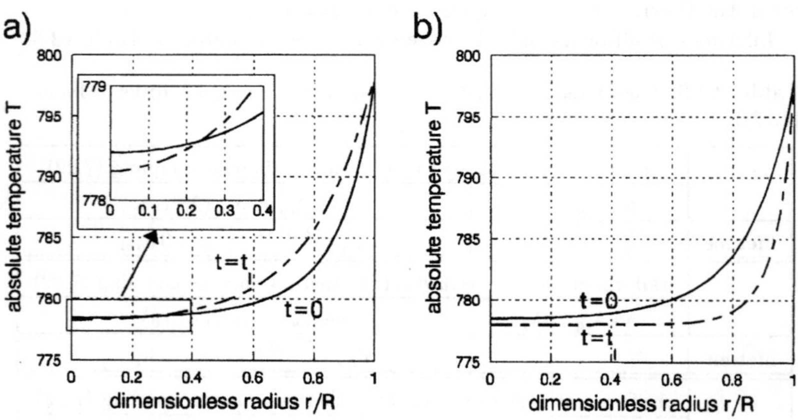

Fig. 12.14. Distribution of temperature a) in disk of constant thickness, b) in disk of uniform creep strength (after Ganczarski and Skrzypek, 1997)

13

Optimal design of thin axisymmetric plates

13.1 Effect of membrane prestressing on the optimal design of sandwich plates with respect to orthotropic creep damage

13.1.1 Initial, boundary and continuity conditions for plates of variable thickness in-plane prestressed by the elastic ring

Two boundary problems are considered:

Example A

A simply supported plate is prestressed by an elastic ring imposed on the plate with initial fit δ, which produces an initial radial force n_0 (Fig. 13.1). $\delta \overset{\text{def}}{=} R_{\text{ring}} - R_{\text{plate}}$ denotes the difference of initial radii of the ring and the plate but some changes of the prestressing force n_0 result from the creep–damage process in the plate:

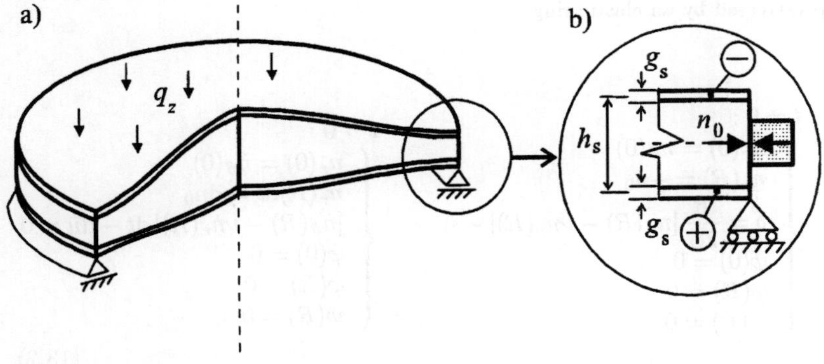

Fig. 13.1. Layout of a simply supported sandwich plate of variable core depth, in-plane prestressed by a ring

$$t=0: \quad \begin{cases} n_r(0) = n_\vartheta(0) \\ n_r(R) = -n_0 \\ \delta = \dfrac{R}{2G}\left[n_\vartheta(R) - \nu n_r(R)\right] - \bar{u} \\ \varphi(0) = 0 \\ m_r(R) = 0 \\ w(R) = 0 \end{cases} \qquad t>0: \quad \begin{cases} \dot{n}_r(0) = \dot{n}_\vartheta(0) \\ \dot{n}_r(R)\mathrm{d}t = \mathrm{d}n_0 \\ \left[\dot{n}_\vartheta(R) - \nu\dot{n}_r(R)\right]\mathrm{d}t - \mathrm{d}\bar{u} = 0 \\ \dot{\varphi}(0) = 0 \\ \dot{m}_r(R) = 0 \\ \dot{w}(R) = 0. \end{cases} \tag{13.1}$$

Example B

A clamped plate is prestressed by an elastic ring imposed on the plate with initial fit δ, which produces an initial radial force n_0 (Fig. 13.2):

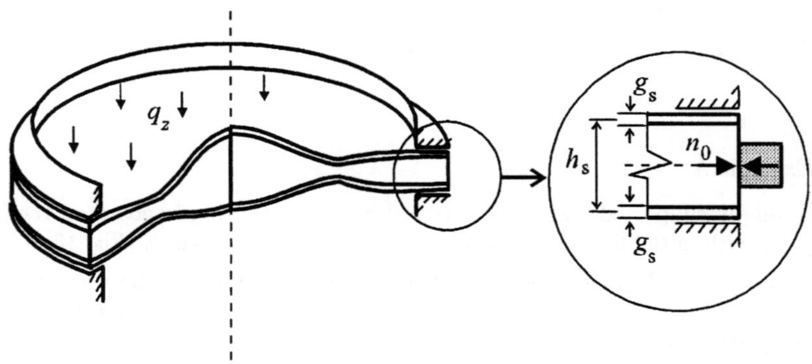

Fig. 13.2. Layout of a clamped sandwich plate of variable core depth in-plane prestressed by an elastic ring

$$t=0: \quad \begin{cases} n_r(0) = n_\vartheta(0) \\ n_r(R) = -n_0 \\ \delta = \dfrac{R}{2G}\left[n_\vartheta(R) - \nu n_r(R)\right] - \bar{u} \\ \varphi(0) = 0 \\ \varphi(R) = 0 \\ w(R) = 0 \end{cases} \qquad t>0: \quad \begin{cases} \dot{n}_r(0) = \dot{n}_\vartheta(0) \\ \dot{n}_r(R)\mathrm{d}t = \mathrm{d}n_0 \\ \left[\dot{n}_\vartheta(R) - \nu\dot{n}_r(R)\right]\mathrm{d}t - \mathrm{d}\bar{u} = 0 \\ \dot{\varphi}(0) = 0 \\ \dot{\varphi}(R) = 0 \\ \dot{w}(R) = 0. \end{cases} \tag{13.2}$$

In the above formulas, $\varphi = -\dfrac{\mathrm{d}w}{\mathrm{d}r}$ is the angular deflection of the plate, n_0 is the peripheral prestressing force, and $\bar{u}$ the peripheral radial displacement.

In both cases under consideration, the prestressing problem may be classified as the mixed-type force-distortion boundary excitation problem (Fig. 11.2) since neither force nor boundary displacement are explicitly given but result from the interaction between the plate and the elastic ring. The stiffnesses of both elements eventually affect the response of the structure to the initial prestressing imposed.

13.1.2 Optimization methods

According to the optimality criteria presented in Chap. 11, three numerical procedures of optimization are suggested, all based on iterative correction of the vector of decision variables.

1. When the first procedure for optimization with respect to uniform creep strength under constant loadings and constant volume of a structure is used, increments of decision variables are proportionally chosen to the levels of continuity function (cf. Ganczarski, 1992):

$$\begin{aligned} \Delta g_{s_j} &= \mathcal{P}_1 \Delta\psi_j - \Delta g_m, \quad \Delta h_{s_j} = \mathcal{P}_2 \Delta\psi_j - \Delta h_m, \\ \Delta\psi_j &= 1 - \inf(\psi^{\pm}_{r/\vartheta})_j, \end{aligned} \tag{13.3}$$

where the average corrections $\Delta g_m, \Delta h_m$ must satisfy the constant volume condition

$$\Delta g_m = \frac{\sum_j \mathcal{P}_1 \Delta\psi_j r_j}{\sum_j r_j}, \quad \Delta h_m = \frac{\sum_j \mathcal{P}_2 \Delta\psi_j r_j}{\sum_j r_j}, \tag{13.4}$$

whereas the step factors $\mathcal{P}_1$, $\mathcal{P}_2$ should be chosen experimentally. When the most general approach is used, thicknesses of the working layers g_s and the core h_s may be changed independently; however, in this example proportional changes are assumed, when $\zeta = g_s/h_s = 1/5$, and $\mathcal{P}_1 = \mathcal{P}_2 = \mathcal{P}$ is held, such that the working layer to core depth ratio of the section is fixed and, in consequence, only one independent decision variable ζ remains. The process of damage equalization is continued until the following condition is fulfilled:

$$\inf(\psi^{\pm}_{r/\vartheta})_j \leq \mathrm{EPS1} \cong 0 \quad \forall j. \tag{13.5}$$

2. When the procedure for optimization with respect to uniform creep strength under constant loading and prescribed lifetime t_I is applied, a modification of the strategy discussed above is proposed. At each optimization step k the volume is subsequently decreased according to the modified shape corrections

$$\Delta g_{s_j} = \zeta \Delta h_{s_j} = -\mathcal{P}\left[\max(\Delta\psi_j) - \Delta\psi_j\right] \tag{13.6}$$

when $t_{I_k} \geq t_I$, else the shape corrections are governed by (13.3) under the constant volume until the condition $t_{I_k} = t_I$ is fulfilled.

3. The numerical procedure applied in the case of optimization with respect to maximum lifetime t_I differs slightly from the approaches presented above as far as the global nature of the objective function is concerned. It starts from a known vector of decision variables, for instance assuming a shape of constant thickness and parameters of prestressing equal to zero, and at the end of the creep process the shape corrections are imposed under constant volume according to the rule (13.3) as long as the global condition $t_{I_k} > t_{I_{k-1}}$ holds and the stability and geometric constraints are satisfied. Then the procedure is stopped, since further thickness corrections (13.3) result in diminished lifetime.

13.1.3 Results: Plates of a variable core depth of uniform creep strength or/and optimal with respect to lifetime under constant volume

In case of a simply supported plate of variable core depth (h_s = var, g_s = const) the terms associated with derivatives of membrane stiffness in (9.19) and (9.20) are omitted. Thus, the following mixed optimization problems are formulated:

1. The distribution of core depth $h_s(r)$ and the parameter of initial prestressing n_0 are sought, $\mathbf{c}_{\text{ucs}} = \{n_0, h_s(r)\}_{\text{ucs}}$, under the constant volume constraint, such that the uniform creep strength UCS is achieved.
2. In case when the above criterion can not be fulfilled, the vector of decision variables $\mathbf{c}_{t_{I\max}} = \{n_0, h_s(r)\}_{t_{I\max}}$ which maximizes the lifetime under the elastic stability, constant volume, and lower geometric constraints, is sought.

Example A: Simply supported plate

Starting from the solution of UCS obtained for a non-prestressed plate ($n_0 = 0$) it is seen that almost whole the bottom working layer suffers damage with respect to the radial component of damage function $D_r^+ \cong 1$, except for a narrow zone where the geometric constraint is active, point A in Fig. 13.3 and shape A in Fig. 13.4. Therefore, the plate is classified as a structure of partly uniform creep strength. The improvement of the plate lifetime compared to the plate of constant thickness, versus the initial prestressing force n_0, is presented in Fig. 13.3. It is easy to notice that increasing the initial prestressing force initially causes the lifetime for the UCS solution to grow. However, the lifetime reaches the maximum at point B (global optimum), where the force equals $n_0/\sigma_0 R = -2.5 \times 10^{-3}$, then it drops with a further prestressing increase. Subsequent trials of the prestressing increase ($n_0/\sigma_0 R < -3.2 \times 10^{-3}$) lead to the optimization range where the criterion of maximum lifetime becomes predominant. Optimal profiles of the plate corresponding to selected points A, B, C, D and E from Fig. 13.3 are shown in Fig. 13.4. When the initial prestressing increases, the zones of constant thickness become deeper and broader. Finally, at point E, the zone of constant thickness is extended over the whole plate and further thickness optimization becomes impossible. Comparison of the lifetime improvements for the simply supported plates, when the core thickness $h_s(r)$ and/or prestressing force n_0 as well as the proportionally changed core to working layers thickness ratio h_s/g_s are considered as control variables for the optimization problem, is summarized in Table 13.1.

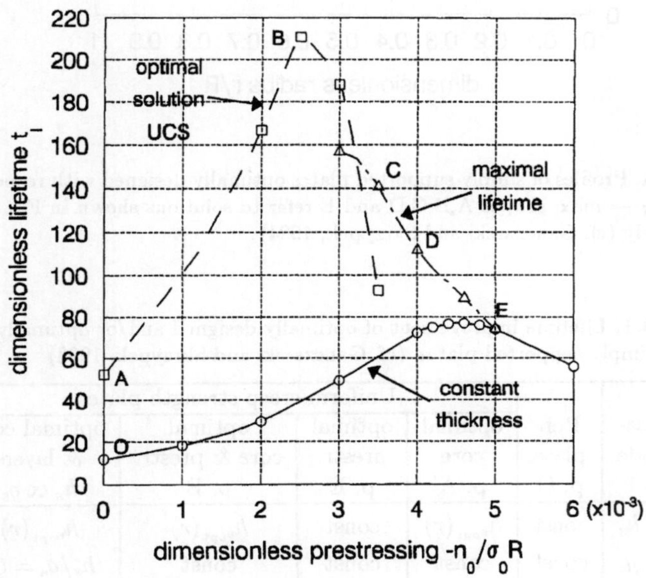

Fig. 13.3. Effect of initial prestressing $n_0/\sigma_0 R$ on lifetime of simply supported plates optimally designed with respect to UCS or $t_I \rightarrow \max$ (after Ganczarski and Skrzypek, 1994)

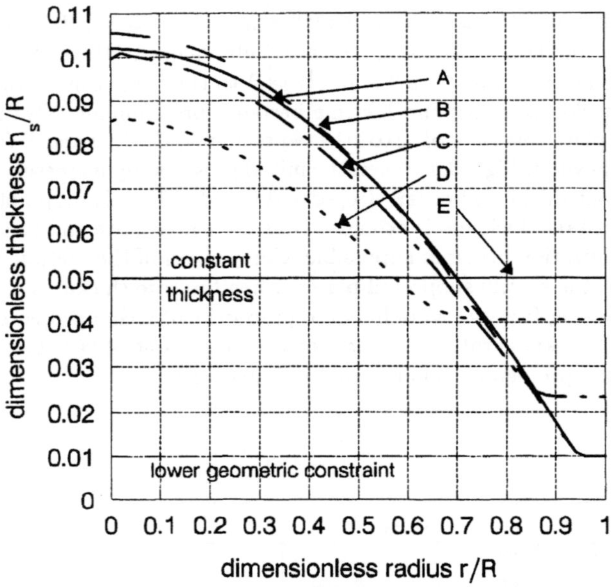

Fig. 13.4. Profiles of simply supported plates optimally designed with respect to UCS or $t_I \to \max$; shapes A,B,C,D and E refer to solutions shown in Fig. 13.3, respectively (cf. Ganczarski and Skrzypek, 1994)

Table 13.1. Lifetime improvement of optimally designed and/or optimally prestressed simply supported plates. (cf. Ganczarski and Skrzypek, 1993)

Optimization mode	Ref. plate p. O	Uniform creep strength plates: optimal core p. A	optimal prestr. p. E	optimal core & prestr. p. B	optimal core & layers $h_s \propto g_s$
core h_s	const	$h_{s_{opt}}(r)$	const	$h_{s_{opt}}(r)$	$h_{s_{opt}}(r)$
layer g_s	const	const	const	const	$h_s/g_s = \zeta_0$
prestr. n_0	0	0	$n_{0_{opt}}$	$n_{0_{opt}}$	0
Lifetime	t_I^{ss}	$3.94\ t_I^{ss}$	$5.72\ t_I^{ss}$	$15.84\ t_I^{ss}$	$6.19\ t_I^{ss}$

Example B: Clamped plate

In the case of a clamped plate, two damage zones with respect to the radial component of the damage function D_r, one in the top D_r^- and other in the bottom D_r^+ working layers, are produced. The bottom $D_r^+ \cong 1$ and the top $D_r^- \cong 1$ damage zones are separated by a narrow zone where the geometric constraint is active. The lifetimes of optimal plates compared to plates of constant thickness are presented in Fig. 13.5 for the case when both the thickness and the initial prestressing are subject to optimization. The solution is quantitatively similar to that obtained for the simply supported plate, as far as a shift of the maximum lifetime towards lower magnitudes of initial forces, when compared to the plate of constant thickness, is observed. However, the range of initial prestressing where the structure ought to be optimized with respect to maximal lifetime is broader, and the global optimum of the lifetime is found for $n_{0_{\mathrm{opt}}}/\sigma_0 R = -3.0 \times 10^{-3}$. The corresponding optimal profiles, approaching the shape of constant thickness when the magnitude of prestressing increases, are presented in Fig. 13.6.

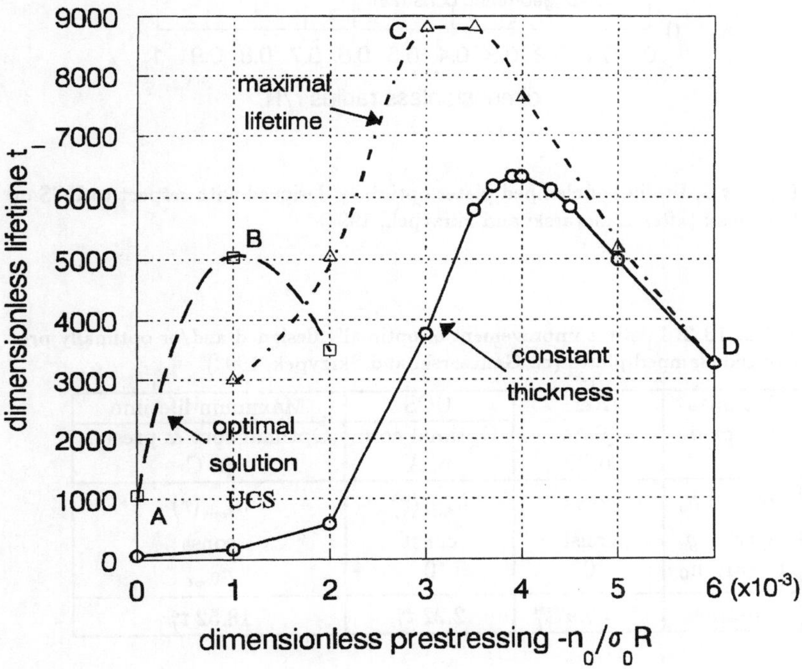

Fig. 13.5. Influence of initial prestressing force $n_0/\sigma_0 R$ on lifetime of clamped plates designed with respect to UCS or $t_I \rightarrow \max$ (after Ganczarski and Skrzypek, 1994)

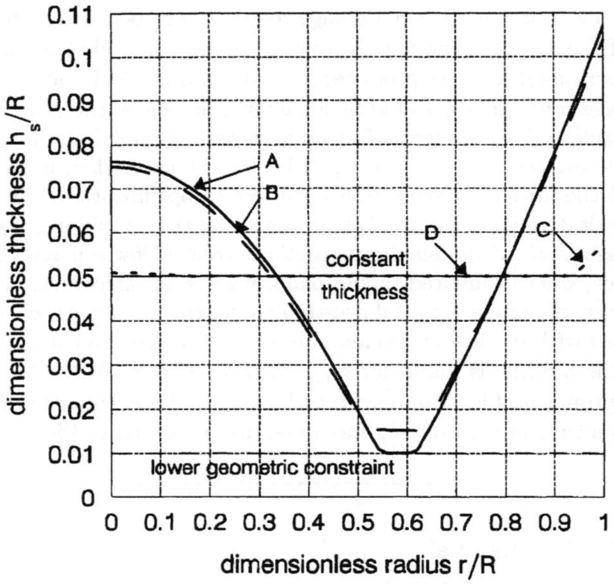

Fig. 13.6. Profiles of clamped plates optimally designed with respect to UCS or $t_I \to$ max (after Ganczarski and Skrzypek, 1994)

Table 13.2. Lifetime improvement of optimally designed and/or optimally prestressed clamped plates (cf. Ganczarski and Skrzypek, 1993)

Optimiza-tion mode	Ref. plate p. O	UCS	Maximum lifetime
		Optimal core p. A	Optimal core & prestr. p. C
core h_s	const	$h_{s_{opt}}(r)$	$h_{s_{opt}}(r)$
layer g_s	const	const	const
prestr. n_0	0	0	$n_{0_{opt}}$
Lifetime	$t_I^c = 5.97 t_I^{ss}$	$2.22\ t_I^c$	$18.52\ t_I^c$

Example C: Uniform creep strength plates of variable core depth under constant lifetime

Design with respect to uniform creep strength when the lifetime is prescribed, despite numerical difficulties associated with the shooting method required, is important for practical applications. It allows one to reduce the volume of a structure which is often the objective of optimal design. This example deals with the optimization of a non-prestressed ($n_0 = 0$) simply supported plate as described by equations (9.19) and (9.20). The optimization problem, based on the local criterion, is formulated as follows: the distribution of the core depth $h_s(r)$ under prescribed lifetime $t_I = \text{const}$ and geometric constraint is sought for, such that uniform creep strength is achieved. The optimal shape of a partly uniform creep strength, with a peripheral zone of active geometric constraint, is shown in Fig. 13.7. The volume of the optimal plate has been reduced to 45% when compared to a plate of constant thickness.

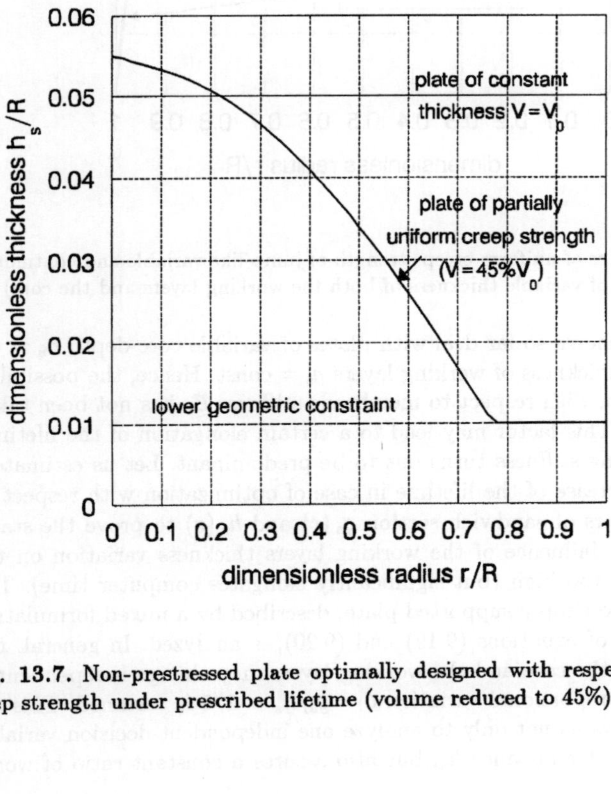

Fig. 13.7. Non-prestressed plate optimally designed with respect to uniform creep strength under prescribed lifetime (volume reduced to 45%)

Example D: Plates of uniform creep strength of variable thicknesses of both the core and working layers

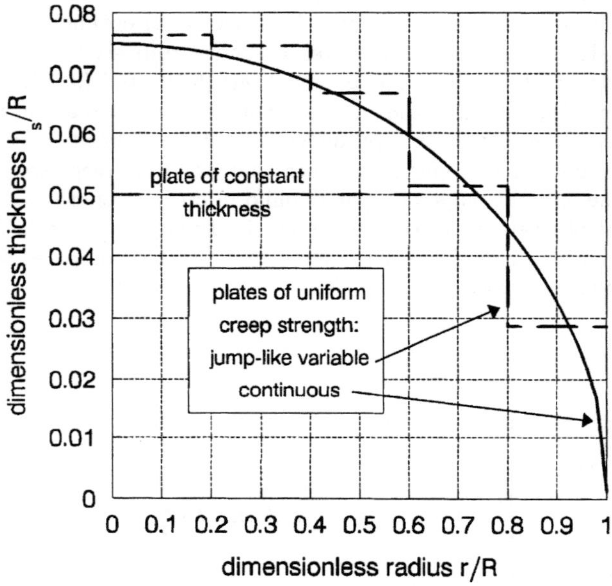

Fig. 13.8. Plates of uniform creep strength of jump-like variable and continuous thickness (case of variable thickness of both the working layers and the core)

All solutions shown so far deal with plates of variable core depth $h_s = \text{var}$ but constant thickness of working layers $g_s = \text{const}$. Hence, the possibility of optimization with respect to membrane stiffness $\mathcal{B}_s$ has not been taken into account. This factor may lead to a certain elongation of the lifetime, but the bending stiffness turns out to be predominant. Let us estimate a percentage increase of the lifetime in case of optimization with respect to both parameters of sandwich section $g_s(r)$ and $h_s(r)$ to prove the statement that the influence of the working layers thickness variation on the lifetime is not too high (but significantly elongates computer time). The non-prestressed simply supported plate, described by a mixed formulation of the system of equations (9.19) and (9.20), is analyzed. In general, the thicknesses of the core and the working layers may change independently, but here the proportional variation $\zeta = g_s/h_s = 1/5$ is assumed. This assumption allows us not only to analyze one independent decision variable instead of two, for instance h_s, but also assures a constant ratio of working layers thickness to core depth. On the other hand, this assumption causes the lower geometric constraint to be passive because the thickness of working layers is changed proportionally to the thickness of the core and, hence, working layers will never be in contact. Nevertheless, optimization of the core depth distribution $h_s(r)$ and simultaneously $g_s = \zeta \times h_s$ leads to structures of uniform creep strength only when the constraint of constant volume is applied. Additional troubles, when the problem formulated above is solved, are connected with a singularity of the radial stress in the supported section of the plate. To overcome this problem, the de l'Hôspital principle may successfully be used because, approaching zero thickness $h_s \longrightarrow 0$, an infinite radial curvature $\kappa_r \longrightarrow \infty$ is reached. The above inconvenience has been avoided by assuming an arbitrary minimum thickness constraint $h_s \geq h_{s_0}/50$, where h_{s_0} is the uniform plate core depth, and by the de l'Hôspital principle associated with the backwards computed finite differences. The optimal shape of uniform creep strength, versus the profile of a jump-like variable thickness, is shown in Fig. 13.8 (cf. Ganczarski, 1992 also, Ganczarski and Skrzypek, 1994).

13.2 Discussion: Sandwich-plates optimization with respect to creep rupture via thickness and prestressing design

i. It is shown that in case of sandwich plates a combined thickness and initial prestressing optimization leads to a significant improvement of the time of initiation of brittle failure, when compared to plates of constant thickness, $t^{ss}_{I_{opt}} = 15.84 t^{ss}_{I_{const}}$ or $t^{c}_{I_{opt}} = 18.52 t^{c}_{I_{const}}$, in case of a simply supported or clamped plate, respectively.

ii. A two-step optimization procedure, first the prestressing optimization of the plate of constant thickness, and next additional thickness improvements with the aim of either uniform creep strength or maximum lifetime, is proposed. It has been shown that optimal prestressing for uniform thickness plates becomes nonoptimal for the optimal design plates. Hence, the thickness corrections should be inspected over a wide range of prestressing forces, rather than around the "optimal" prestressing force for the uniform plate.

iii. Simultaneous plate thickness and in-plane prestressing optimization allows a significantly enhancement of the lifetime for both simply supported and clamped plates, unless the critical magnitude of the initial prestressing that corresponds to the appropriate lifetime of plates of constant thickness $n^{ss}_{0\text{crit}}/\sigma_0 R = -5.0 \times 10^{-3}$ in case of a simply supported plate, or $n^{\mathrm{c}}_{0\mathrm{crit}}/\sigma_0 R = -6.0 \times 10^{-3}$, in case of a clamped plate, is exceeded. Over-prestressing of the plate $|n_0| > |n_{0\mathrm{crit}}|$, e.g., $n_0^{\mathrm{ss}}/\sigma_0 R = -5.0 \times 10^{-3}$ or $n_0^{\mathrm{c}}/\sigma_0 R = -6.0 \times 10^{-3}$, makes further thickness design undesirable since plates of uniform thickness ensure a longer lifetime.

iv. Sandwich plate thickness optimization by varying core depth but holding thickness of the working layers constant (h_{s} = var, g_{s} = const), which affects the variable bending stiffness $\mathcal{D}_{\mathrm{s}}(r)$ = var but not the membrane stiffness $\mathcal{B}_{\mathrm{s}}(r)$ = const, is recommended. Independent optimization with respect to both bending stiffness $\mathcal{D}_{\mathrm{s}}(r)$ = var and membrane stiffness $\mathcal{B}_{\mathrm{s}}(r)$ = var not only doubles the number of decision variables (two independent unknown functions $h_{\mathrm{s}}(r)$ and $g_{\mathrm{s}}(r)$ instead of one) but also may obviously lead to violation of the substitutive sandwich section assumption $g_{\mathrm{s}} << h_{\mathrm{s}}$. Instead, the proportional variation of both thicknesses, $\xi = g_{\mathrm{s}}/h_{\mathrm{s}} = 1/5$, has been examined in the case of a non-prestressed ($n_0 = 0$) simply supported plate to reach a lifetime improvement by the factor 6.19 when compared to uniform thickness, by contrast to the factor 3.94 when the single core depth is considered as a design variable.

v. Design with respect to uniform creep strength when the lifetime is prescribed furnishes the optimal profile that allows a significant reduction of volume (up to 45% in case of a non-prestressed simply supported plate). However, this dual formulation associated with the need to the shooting method in order to ensure the implicit lifetime to failure initiation t_I = const is highly computer time consuming and, hence, cannot be recommended for practical applications.

References

Altenbach, H., Altenbach, J., Schiesse, P. (1990): Konzept der Schädigungsmechanik und ihre Anwendung bei der werkstoffmechanischen Bauteilanalyse, *Techn. Mechanik*, 11, 2, 81–93.

Altenbach, H., Morachkovsky, O., Naumenko, K., Sychov, A. (1997): Geometrically nonlinear bending of thin-walled shells and plates under creep-damage conditions, *Arch. Appl. Mech.*, 67, 339–352.

Altenbach, J., Altenbach, H., Naumenko, K. (1997): Lebendauerabschätzung dünnwandrige Flächentragwerke auf der Grundlage phënomenologischer Materialmodelle für Krichen und Schädigung, *Techn. Mechanik*, 12, 4, pp. 353–364.

Armstrong, P.J., Frederick, C.O. (1966): A mathematical representation of the multiaxial Bauschinger effect, C.E.G.B. Report RD/B/N 731.

Bailey, R.W. (1929): Creep of steel under simple and compound stresses and the use of high initial temperature in steam power plant, *Trans. Tokyo Sectional Meeting of the World Power Conf.* 1089.

Baasar, H., Gross, D. (1998): Damage and strain localization during crack propagation in thin-walled shells, in *Mechanics of Materials with Intrinsic Length Scale*, Proc. 2 Eur. Mechanics of Material Conf., 23–26 Feb. 1998, Magdeburg, 22–26.

Bathe, K.J. (1982): *Finite Element Procedures in Engineering Analysis*, Prentice-Hall.

Ben-Hatira, F., Forster, Ch., Saanouni, K. (1992): Prediction of flow localization and failure in finite elastoplasticity with damage, *Eur. J. of Finite Elements.*

Betten, J. (1983): Damage tensors in continuum mechanics, *J. Méc. Théor. Appl.*, 2, 1, 13–32.

Betten, J. (1986): Application of tensor functions to the formulation of constitutive equations involving damage and initial anisotropy, *Eng. Fracture Mechanics*, 25, 573–584.

Betten, J. (1992): Application of tensor functions in continuum damage mechanics, *Int. J. Damage Mech.*, 1, 47–59.

Białkiewicz, J., Kuna-Ciskał, H. (1996): Shear effect in rupture mechanics of middle-thick plates, *Eng. Fracture Mech.*, 54, 3, 361–370.

Bodnar, A., Chrzanowski, M., Latus, P., Madej, J. (1994): Safety of materials and structures in creep conditions, *Int. J. Pres. Vess. and Piping*, 59, 161–174.

Bodnar, A., Chrzanowski, M., Nowak, K. (1996): Brittle failure lines in creeping plates, *Int. J. Pres. Vess. and Piping*, 66, 253–261.

Bodnar, A., Chrzanowski, M. (1994): Cracking of creeping plates in terms of continuum damage mechanics, *Mech. Teor. Stos.*, 32, 1, 31–42.
Bodnar, A., Chrzanowski, M. (1996): Numerical simulation of cracking in plates, *Arch. Bud. Masz.*, 43, 2–3, 131–139.
Boyle, J.T., Spence, J. (1983): *Stress Analysis for Creep*, Butterworths, London.
Broberg, H. (1974): *Damage Measures in Creep Deformation and Rupture*, Swedish Solid Mechanics Report, 8, 100–104.
Budiansky, B., O'Connel, R.J. (1976): Elastic moduli of cracked solid, *Int. J. Sol. Struct.*, 12, 81–97.
Carslow, H.S, Jaeger, J.C. (1959): *Conduction of Heat in Solids*, Clarendon Press, Oxford.
Chaboche, J.L. (1974): Une Loi Différentielle d'Endommagement de Fatigue avec Cumulation non Linéaire, *Revue Française de Mécanique*, 50–51, English trans. in *Annales de I'IBTP*, HS 39, 1977.
Chaboche, J.L. (1978): *Description Thermodynamique et Phénoménologique de la Viscoplasticité Cyclique avec Endommagement,* Thése Univ. Paris VI et Publication ONERA, No. 1978-3.
Chaboche, J.L. (1979): Le Concept de Contrainte Effective Appliqué à l'Élasticité et á la Viscoplasticité en Présence d'un Endommagement Anisotrope, *Col. EUROMECH 115,* Grenoble 1979, Editions du CNRS, 1982.
Chaboche, J.L. (1981): Continuous damage mechanics – a tool to describe phenomena before crack initiation, *Nucl. Eng. Des.*, 64, 2, 233–247.
Chaboche, J.L. (1982): *Mechanical Behavior of Anisotropic Solids*, ed. Boehler, J.P., Martinus Nijhoff.
Chaboche, J.L., Rousselier, G. (1983): On the plastic and viscoplastic constitutive equations – P.1: Rules developed with internal variable concept, P.2: Application of internal variable concepts to the 316 stainless steel, *J. Pressure Vessel Technol.,* 105, 5, 153–164.
Chaboche, J.L. (1988): Continuum damage mechanics: Part I: General concepts, Part II: Damage growth, crack initiation, and crack growth, *J. Appl. Mech.*, 55, 3, 59–71.
Chaboche, J.L. (1993): Development of continuum damage mechanics for elastic solids sustaining anisotropic and unilateral damage, *Int. J. Damage Mech.*, 2, 311–329.
Chaboche, J.L., Lesne, P.M., Moire, J.F. (1995): Continuum damage mechanics, anisotropy and damage deactivation for brittle materials like concrete and ceramic composites, *Int. J. Damage Mech.*, 4, 5–22.
Charewicz, A., Daniel, I.M. (1985): Fatigue damage mechanisms and residual properties of graphite/epoxy laminates, *Symp. IUTAM on Mechanics of Damage and Fatigue,* Haifa, Israel.
Chen, X.F., Chow, C.L. (1995): On damage strain energy release rate Y, *Int. J. Damage Mech.*, 4, 3, 251–263.
Chow, C.L. Wang, J. (1987a): An anisotropic theory of elasticity for continuum damage mechanics, *Int. J. Fracture*, 33, 3–16.

Chow, C.L. Wang, J. (1987b): An anisotropic theory of continuum damage mechanics for ductile materials, *Eng. Frac. Mech.*, 27, 547–558.

Chow, C.L. Lu, T.J. (1992): An analytical and experimental study of mixed-mode ductile fracture under nonproportional loading, *Int. J. Damage Mech.*, 1, 191–236.

Christensen, R.M., Lo, K.H. (1979): Solution for effective shear properties of three phase sphere and cylinder models, *J. Mech. Phys. Solids*, 27, 315–330.

Chrzanowski, M. (1976): Use of the damage concept in describing creep-fatigue interaction under prescribed stress, *Int. J. Mech. Sci.*, 18, 69–73.

Chrzanowski, M., Madej, J. (1980): Construction of the failure curves based on the damage parameter concept, *Mech. Teor. Stos.*, 4, 18, 587–601 (in Polish).

Chrzanowski, M., Madej, J., Latus, P. (1991): A CDM motivated safety factor for creeping materials and structures, *Trans. SMIRT 11*, Ed. H. Shibata, Tokyo, 1991, L, 355–365.

Coffin, L.F. (1954): A study of the effects of cyclic thermal stresses in a ductile metal, *Trans. of the ASME,* 76, 931.

Coffin, L.F. (1970): The deformation and fracture of a ductile metal under superimposed cyclic and monotonic strain, *Achievment of High Fatigue Resistance in Metals and Alloys,* ASTM STP 467, American Society for Testing and Materials, 53–76.

Contesti, E. (1986): *Endommagement en Fluage: Expérience et Modélisation,* Réunion GRECO Grandes Déformations et Endommagement – GIS Rupture á Chaud, Aussois, France.

Cordebois, J.P., Sidoroff, F. (1979): Damage induced elastic anisotropy, Coll. Euromech 115, Villard de Lans, also in *Mechanical Behavior of Anisotropic Solids*, ed. Boehler, J. P., Martinus Nijhoff, Boston, 1983, 761–774.

Cordebois, J.P., Sidoroff, F. (1982): Endommagement anisotrope en élasticité et plasticité, *J. Méc. Théor. Appl.*, Numero Spécial: 45.

Carroll, M.M., Holt, A.C. (1972): Static and dynamic pore- collapse relations for ductile solids, *J. Appl. Phys.,* 43, 1626–1636.

Curran, D.R., Seaman, L., Shockey, D.A. (1977): Dynamic failure in solids, *Physics Today,* 46–55.

Curran, D.R., Seaman, L., Shockey, D. A. (1987): Dynamic failure of solids, *Physics Reports,* V. 147, 253–388.

Davison, L. Stevens, A.L. (1973): Thermodynamical constitution of spalling elastic bodies, *J. Appl. Phys.*, 44, 2, 668.

Davison, L., Stevens, L.E., Kipp, M.E. (1977): Theory of spall damage accumulation in ductile metals, *J. Mech. Phys. Solids*, 25, 11–28.

Davison, L., Kipp, M.E. (1978): Calculation of spall damage accumulation in ductile metals, *Proc. IUTAM Symp. High Velocity Deformation of Solids*, Tokyo, Japan 1977; Ed. K. Kawata and J. Shiratori, 163–175, Springer, New York.

Delobelle, P. (1985): *Etude en Contraintes Biaxiales des Lois de Comportement d'un Acier Inoxydable du type 17 - 21 SPH - Modélisation et Identification Introduction de l'Endommagement, Cas de L'INCONEL 718,* These de Doctorat d'Etat, Besancon.
Dragon, A. Mróz, Z. (1979): A continuum model for plastic-brittle behaviour of rock and concrete, *Int. J. Eng. Sci.*, 17, 2, 121.
Dragon, A. Chihab, A. (1985): Quantifying of ductile fracture damage evolution by homogenization approach, *SMiRT 8*, Brussels.
Dufailly, J. (1980): *Modelization Mécanique et Identification de l'Endommagement Plastique des Materiaux,* These de Doctorat, l'Université de Paris 6.
Dufailly, J., Lemaitre, J. (1995): Modeling very low cycle fatigue, *Int. J. Damage Mech.*, 4, 153–170.
Dunne, F.P.E., Hayhurst, D.R. (1992a): Continuum damage based constitutive eqaution for copper under high temperature creep and cyclic plasticity, *Proc. R. Soc. Lond.*, A 437, 545–566.
Dunne, F.P.E., Hayhurst, D.R. (1992b): Modelling of combined high–temperature creep and cyclic plasticity in components using Continuum Damage Mechanics, *Proc. R. Soc. Lond.*, A 437, 567–589.
Dunne, F.P.E., Hayhurst, D.R. (1994): Efficient cycle jumping techniques for the modelling of materials and structures under cyclic mechanical and thermal loadings, *Eur. J. Mech., A/Solids*, 13, 5, 639–660.
Dunne, F.P.E., Hayhurst, D.R. (1995): Physically based temperature dependence of elastic-viscoplastic constitutive equations for copper between 20 and 500°C, *Philosophical Mag.*, A., 74, 2, 359–382.
Dyson, B.F. (1993): A new mechanism and constitutive law for creep of precipitation hardened engineering alloy, *NPL Report DAMM* A 102, 1993.
Egner, W., Skrzypek, J. (1994): Effect of pre-loading damage on the net-lifetime of optimally prestressed rotating disks, *Arch. Appl. Mech.*, 64, 447–456.
Fournier, D., Pineau, A. (1977): Low cycle fatigue behavior of INCONEL 718 at 298K and 823K, *Matallurgical Transactions*, A, 8A, 1095.
Fung, Y.C. (1965): *Foundations of Solid Mechanics*, Prentice-Hall.
Gajewski, A. (1975): *Optimal strength design in case of physically nonlinear materials*, Politechnika Krakowska, Zesz. Nauk. Nr 5, Kraków (in Polish).
Gallagher, R.H. (1973): Fully stressed design, in *Optimal Structural Design,* Eds. R.H. Gallagher and O.C. Zienkiewicz, John Wiley, New York, 19–23.
Ganczarski, A., Skrzypek, J. (1989): Optimal prestressing and design of rotating disks against brittle rupture under unsteady creep conditions (in Polish), *Eng. Trans.*, 37, 4, 627–649.
Ganczarski, A. Skrzypek, J. (1991): On optimal design of disks with respect to creep rupture, Proc. of *IUTAM Symp. Creep in Struct.*, Ed. M. Życzkowski, Springer, 571–577.
Ganczarski, A., Skrzypek, J. (1992): Optimal shape of prestressed disks against brittle rupture under unsteady creep conditions, *Struct. Optim.*, 4, 47–54.

Ganczarski, A., Skrzypek, J. (1993a): Axisymmetric plates optimally designed against brittle rupture, in Proc. World Congr. on *Optimal Design of Structural Systems, Structural Optimization 93*, Rio de Janeiro, 1993, Ed. J. Herskovits, Vol. I, 197–204.

Ganczarski, A., Skrzypek, J. (1993b): Brittle-rupture mechanisms of axisymmetric plates subject to creep under surface and thermal loadings, in Proc. of *SMIRT–12*, Stuttgart, 1993, Ed. K. Kussmaul, Vol. L, 263–268.

Ganczarski, A. (1993): *Analysis of brittle damage and optimal design of prestressed axisymmetric structures in creep conditions under combined loadings,* PhD Thesis, Cracow Univ. of Technol., Cracow 1993.

Ganczarski, A. Skrzypek, J. (1994a): Effect of initial prestressing on the optimal design of plates with respect to orthotropic brittle rupture, *Arch. Mech.*, 46, 4, 463–483.

Ganczarski A., Skrzypek, J. (1994b): Influence of shear stress on the orthotropic brittle rupture, *2 Eur. Solid Mechanics Conference,* F1, Genoa.

Ganczarski, A., Skrzypek, J. (1995): Concept of thermo-damage coupling in continuum damage mechanics, *First Int. Symp. Thermal Stresses '95,* Hamamatsu, Japan, Act City, June 5–7, 1995, Eds. R. B. Hetnarski, N. Noda, 83–86.

Ganczarski, A., Freindl, L., Skrzypek, J. (1997): Orthotropic brittle rupture of Reissner's prestressed plates, Proc. 5 Int. Conf. on Computational Plasticity, Barcelona, 17–20 March 1997, *Computational Plasticity, Fundamentals and Applications*, CIMNE, Eds. D. R. Owen, E. Onãte, E. Hinton, 1904–1909.

Ganczarski, A., Skrzypek, J. (1997): Modeling of damage effect on heat transfer in solids, *Sec. Int. Symp. Thermal Stresses'97*, Rochester Inst. of Technology, Rochester, NY, June 8–11, 1997, Eds. R. B. Hetnarski, N. Noda, 213–216.

Ganczarski, A., Skrzypek, J. (1997): Optimal design of rotationally symmetric disks in thermo-damage coupling conditions, *Techn. Mechanik*, 17, 4, 365–378.

Germain, P., Nguyen, Q.S., Suquet, P. (1983): Continuum Thermodynamics, *ASME J. Appl. Mech.,* 50, 1010–1020.

Gittus, J. (1978): *Irradiation Effects in Crystaline Solids,* Applied Science Publ., London.

Grabacki, J. (1989): On continuous description of damage, *Mech. Teor. Stos.*, 27, 2, 271–291.

Grabacki, J. (1991): On some description of damage process, *Eur. J. Mech., A/Solids*, 10, 3, 309–325.

Grabacki, J. (1992): Mechanics of materials with internal structures, *Zesz. Nauk. Politechniki Krakowskiej*, Monogr. 131, Kraków (in Polish).

Grabacki, J.K. (1994): Constitutive equations for some damaged materials, *Eur. J. Mech., A/Solids*, 13, 1, 51–71.

Grady, D.E. (1982): Local inertial effects in dynamic fragmentation, *J. Appl. Phys.*, 53, 322–325.

Gurland, J. (1972): *Acta Metall.*, 20, 5, 735.

Gurson, A.L. (1977): Continuum theory of ductile rupture by void nucleation and growth: Part I – Yield criteria and flow rules for porous ductile media, *Trans. ASME, J. Eng. Mat. Tech.*, 99, 1, 2–13.

Hashin, Z. (1983): Analysis of composite materials, a survey, *J. Appl. Mech.*, 50, 481–505.

Hashin, Z. (1988): The differential scheme and its application to cracked materials, *J. Mech. Phys. Solids*, 36, 719–734.

Hayakawa, K., Murakami, S. (1997): Thermodynamical modeling of elastic-plastic damage and experimental validation of damage potential, *Int. J. Damage Mech.*, 6, October, 333–362.

Hayakawa, K., Murakami, S. (1998): Space of damage conjugate force and damage potential of elastic-plastic-damage materials, in *Damage Mechanics in Engineering Materials*, Ed. G. Z. Voyiadjis, J.-W. W. Ju, J.-L. Chaboche, Elsevier Science, Amsterdam, 27–44.

Hayhurst, D.R. (1972): Creep rupture under multiaxial state of stress, *J. Mech. Phys. Solids,* 20, 6, 381–390.

Hayhurst, D.R., Leckie, F.A. (1973): The effect of creep constitutive and damage relationships upon rupture time of solid circular torsion bar, *J. Mech. Phys. Solids*, 21, 6, 431–446.

Hayhurst, D.R., Dimmer, P.R., Chernuka, M.W. (1975): Estimates of the creep rupture lifetimes of structures using the finite element methods, *J. Mech. Phys. Solids*, 23, 335.

Hayhurst, D.R. (1983): On the role of creep continuum damage in structural mechanics, in *Engineering Approaches to High Temperature Design,* Eds. Wilshire, Owen, Swansea, Pineridge Press.

Hayhurst, D.R., Dimmer, P.R., Morrison, C.J. (1984): Development of continuum damage in the creep rupture of notched bars, *Phil. Trans. R. Soc.* (London) A311, 103.

Holman, J.P. (1990): *Heat Transfer*, McGraw-Hill, 12–13, and 508–509.

Hult, J. (1988): Stiffness and strength of damaged materials, *Z. angew. Math. Mech.*, T31–T39.

Johnson, A.E., Henderson, J., Mathur, V.D. (1956): Combined stress creep fracture of commercial copper at 250°, *The Engineer*, 24, 261–265.

Johnson, A.E., Henderson, J., Khan, B. (1962): *Complex-stress creep, relaxation and fracture of metallic alloys*, HMSO, Edinbourgh.

Johnson, J.N. (1981): Dynamic fracture and spallation in ductile solids, *J. Appl. Phys.*, 53, 2812–2825.

Kachanov, L.M. (1958): Time of the rupture process under creep conditions, *Izv. AN SSR, Otd. Tekh. Nauk*, 8, 26–31.

Kachanov, L.M. (1974): *Foundations of Fracture Mechanics*, Moscow, Nauka (in Russian).

Kachanov, L.M. (1986): *Introduction to Continuum Damage Mechanics*, The Netherlands: Martinus Nijhoff.

Kachanov, M.I. (1972): *Mekh. Tverdogo Tiela*, (in Russian), 2, 54.

Kachanov, M.I. (1992): Effective elastic properties of cracked solids: critical review of some basic concepts, *Appl. Mech. Rev.*, 45, 8, 304–335.

Karihaloo, B.L. and Fu, D. (1989): A damage-based constitutive law for plain concrete in tension, *Eur. J. Mech./Solids*, 8, 373–384.

Kaviany, M. (1995): *Principles of Heat Transfer im Porous Media*, Springer, New York.

Knott, J.F. (1973): *Fundamentels of Fracture Mechanics,* Butterworths.

Kondaurov, W.I. (1988): Continuous damage of nonlinear elastic media (in Russian), *Izv. AN SSSR: Mech. Tverdogo Tela*, 52, 2, 302–310.

Korbel, A., Korbel, K., Pęcherski, R.B. (1998): Catastrophic slip phenomena in crystalline materials, in *Damage Mechanics in Engineering Materials*, Ed. G.Z. Voyiadjis, J.-W.W. Ju, J.-L. Chaboche, Elsevier Science, Amsterdam, 237–255.

Kowalewski, Z.L. (1991a): The influence of deformation history on creep of pure copper, IUTAM Symp. *Creep in Structures*, Ed. M. Życzkowski, 1990, 115–122.

Kowalewski, Z.L. (1991b): Creep behavior of copper under plane stress state, *Int. J. Plasticity*, 7, 387–404.

Kowalewski, Z.L., Hayhurst, D.R., Dyson, B.F. (1994): Mechanisms-based creep constitutive equations for an aluminium alloy, *J. Strain Analysis*, 29, 4, 309–316.

Kowalewski, Z.L., Lin, J., Hayhurst, D.R. (1994): Experimental and theoretical evaluation of a high-accuracy uni-axial creep testpiece with slit extensometers ridges, *Int. J. Mech. Sci.*, 36, 8, 751–769.

Kowalewski, Z.L. (1995): Experimental evaluation of the influence of stress state type on creep characteristics of copper at 523 K, *Arch. Mech.*, 47, 1, 13–26.

Kowalewski, Z.L. (1996a): Creep rupture of copper under complex stress state at elevated temperatures, C494/037, *I. Mech. E.*, 113–122

Kowalewski, Z.L. (1996b): Experimental creep study of metals under multiaxial stress conditions, *Mech. Teor. Stos., Journ. Appl. Mech.*, 2, 34, 405–422.

Kowalewski, Z.L. (1996c): Biaxial creep study of copper on the basis of isochronous creep surafces, *Arch. Mech.*, 48, 1, 89–109.

Krajcinovic, D., Fonseka, G.U. (1981): The continuous damage theory of brittle materials, Part I: General theory, *Trans ASME, J. Appl. Mech.*, 48, 4, 809–815.

Krajcinovic, D., Silva, M.A.G. (1982): Statistical aspect of the continuous damage theory, *Int. J. Solids Structures,* 18, 7, 551–562.

Krajcinovic, D. (1983): Constitutive equations for damaging materials, *Trans. ASME, J. Appl. Mech.*, 50, 2, 355–360.

Krajcinovic, D. (1989): Damage mechanics, *Mech. Mater.*, 8, 117–197.

Krajcinovic, D. (1995): Continuum Damage Mechanics: when and how?, *Int. J. Damage Mechanics*, 4, July, 217–229.

Krajcinovic, D. (1996): *Damage Mechanics*, North Holland Series in Appl. Math. and Mech., Elsevier, Amsterdam.

Kraus, H. (1980): *Creep analysis*, John Wiley & Sons, New York.
Kunin, I.A. (1983): *Elastic media with microstructure*, Three-dimensional models, Springer, Berlin.
Ladeveze, P. (1983): On an anisotropic damage theory in *Failure Criteria of Structural Media*, CNRS Int. Coll. No 351, Villard-de-Lans, ed. Boehler, Balkema, Rotterdam, 1993.
Lacy, T.E., McDowell, D.L., Willice, P.A., Ramesh Talreja (1997): On representation of damage evolution in continuum damage mechanics, *Int. J. Damage Mech.*, 6, January, 62–95.
Ladeveze, P. (1990): A damage approach for composite structures theory and identification, in *Mechanical Identification of Composites*, ed. A. Vautrin and H. Sol, Elsevier, 44–57.
Leckie, F.A., Hayhurst, D.R. (1974): Creep rupture of structures, *Proc. Roy. Soc. London*, A 340, 323–347.
Leckie, F.A., Onat, E.T. (1981): *Physical Non-linearities in Structural Analysis*, ed. Hult, J. and Lemaitre, J., Springer, Berlin.
Lekhnitskii, S.G. (1981): *Theory of Elasticity of an Anisotropic Body*, Engl. Transl., Mir Publishers, Russian edition: Nauka 1977.
Lemaitre, J. (1971): Evaluation of dissipation and damage in metals, *Proc. I.C.M.* Kyoto, Japan, Vol. 1.
Lemaitre, J., Chaboche, J.L. (1975): A non-linear model of creep fatigue damage cumulation and interaction, Proc. IUTAM Symp. *Mechanics of Visco-Elastic Media and Bodies*, ed. J. Hult, Gothenburg, Sweden, Springer, 291–301.
Lemaitre, J., Chaboche, J.L. (1978): Aspect phenomenologique de la rapture per endommagement, *J. de Méchanique applique*, 2, 317–365.
Lemaitre, J. (1984): How to use damage mechanics, *Nuclear Engng. and Design*, 80, 233–245.
Lemaitre, J. (1985): A continuum damage mechanics model for ductile fracture, *ASME J. Engng. Mat. and Technology*, 107, 83–89.
Lemaitre, J., Chaboche, J.L. (1985): *Méchanique des Matériaux Solides*, Dunod Publ., Paris.
Lemaitre, J. (1987): Formulation and identification of damage kinetic constitutive equations, in continuum damage mechanics-theory and application, *CISM Courses and Lectures*, 295, (ed. Krajcinowic, D. and Lemaitre, J.), Springer, Berlin, 37–89.
Lemaitre, J. (1990): Micro–mechanics of crack initiation, *Int. J. of Fracture*, 42, 87–99.
Lemaitre, J. (1992)(1996), *A Course on Damage Mechanics*, 2nd rev. and enlarged edn. 1996 Springer, Berlin.
LeRoy, G., Embury, J.D., Edward, G., Ashby, M.F. (1981): A model of ductile fracture based on the nucleation and growth of voids, *Acta Metallurgica*, 40, 1509–1522.
Lin, J., Dunne, F.P.E., Hayhurst, D.R. (1996): Physically based temperature dependence of elastic-viscoplastic constitutive equations for copper between 20 and 500°C, *Philosophical Mag.*, A., 74, 2, 359–382.

Lin, J., Dunne, F.P.E., Hayhurst, D.R. (1998): Aspects of testpiece design responsible for errors in cyclic plasticity experiments, *Int. J. Damage Mech.* (to be published).

Lis, Z. (1992): Creep damage of solids under non-proportional loading, *Z. angew. Math. Mech.*, 72, 4, T164–T166.

Lis, Z., Litewka, A. (1996): Mathematical model of creep rupture under cyclic loading, *Math. Modelling and Scient. Computing*, 6 (submitted for publication).

Litewka, A. (1985): Effective material constants for orthotropically damaged elastic solid, *Arch. Mech.*, 37, 6, 631–642.

Litewka, A. (1986): On stiffness and strength reduction of solids due to crack development, *Eng. Fract. Mechanics*, 25, 5/6, 637–643.

Litewka, A. (1987): Analytical and experimental study of fracture of damaging solids, in Proc. IUTAM/ICM Symp., *Yielding, Damage, and Failure of Anisotropic Solids*, Villard-de-Lans, 24–28 Aug. 1987, 655–665.

Litewka, A. (1989): Creep rupture of metals under multi-axial state of stress, *Arch. Mech.*, 41, 1, 3–23.

Litewka, A., Hult, J. (1989): One parameter CDM model for creep rupture prediction, *Eur. J. Mech., A/Solids*, 8, 3, 185–200.

Liu, Y., Murakami, S., Kanagawa, Y. (1994): Mesh-dependence and stress singularity in finite element analysis of creep crack growth by Continuum Damage Mechanics, *Eur. J. Mech. A/Solids*, 13, 395–417.

Lubarda, V.A., Krajcinovic, D. (1993): Damage tensors and the crack density distribution, *Int. J. Solids Struct.*, 30, 20, 2859–2877.

Malinin, N.N., Khadjinsky, G.M. (1972): Theory of creep with anisotropic hardening, *Int. J. Mech. Sci.*, 14, 235–246.

Martin, J.B., Leckie, F.A. (1972): *J. Mech. Phys. Solids*, 20, 4, 223.

Manson, S.S. (1954): *Behaviour of materials under conditions of thermal stress*, NASA, Techn. Note 2933.

Manson, S.S. (1979): Some useful concepts for the designer in treating cumulative damage at elevated temperature, *I.C.M. 3*, Cambridge, Vol. 1, 13–45.

Marco, S.M., Starkey, W. L. (1954): A concept of fatigue damage, *Trans. ASME*, 74, 4, 627–632.

Marigo, J.J. (1985): Modelling of brittle and fatigue damage for elastic material by growth of microvoids, *Engng. Fracture Mechanics*, 21, 4, 861–874.

Mazars, J. (1985): Description of micro and macroscale damage of concrete structures, *Symp. on Mechanics of Damage and Fatigue*, Haifa, Israel.

Mori, T., Tanaka, K. (1973): Average stress in matrix and average elastic energy of materials with misfitting inclusions, *Acta Mech.*, 21, 571–574.

Morrow, J.D. (1964): Cyclic plastic strain energy and fatigue of metals, *Symp. A.S.T.M.*, Chicago.

Mou Y.H., Han, R.P.S. (1996): Damage evolution in ductile materials, *Int. J. Damage Mech.*, 5, July, 241–258.

Mróz, Z., Seweryn, A. (1998): Damage evolution rule for multiaxial variable loading, in *Damage Mechanics in Engineering Materials*, Ed. G. Z. Voyiadjis, J.-W.W. Ju, J.-L. Chaboche, Elsevier Science, Amsterdam, 145–162.

Murakami, S., Ohno, N. (1981): A continuum theory of creep and creep damage, in *Creep in Structuers,* ed. Ponter, A. R. S. and Hayhurst, D. R., Springer, Berlin, 442–444.

Murakami, S. (1983): Notion of continuum damage mechanics and its applications to anisotropic creep damage theory, *J. Eng. Mater. Technol.*, 105, 99.

Murakami, S. (1986): *Failure Criteria of Structured Media,* ed. Boehler, A. A. Balkema.

Murakami, S. (1987): *Proc. 2nd Int. Conf. on Constitutive Laws for Engng Materials*, ed. Desai, C. S. and Krempl, E., Elsevier.

Murakami, S. (1988): Mechanical modelling of material damage, *J. Appl. Mech.*, 55, 6, 280–286.

Murakami, S., Kawai, M., Rong, H. (1988): Finite element analysis of creep crack growth by a local approach, *Int. J. Mech. Sci.*, 30, 7, 491–502.

Murakami, S., Mizuno, M., Okamoto, T. (1991): Mechanical modeling of creep, swelling and damage under irradiation for polycrystalline metals, *Nucl. Engng. Design*, 131, 147–155.

Murakami, S., Mizuno, M. (1992): A constitutive equation of creep, swelling and damage under neutron irradiation applicable to multiaxial and variable states of stress., *Int. J. Solids Structures*, 29, 19, 2319–2328.

Murakami, S., Sanomura, Y., Saitoh, K. (1986): Formulation of cross-hardening in creep and its effect on the creep damage process of copper, *J. Eng. Mat. Techn.*, 108, 167–173.

Murakami, S., Liu, Y. (1995): Mesh-dependence in local approach to creep fracture, *Int. J. Damage Mech.*, 4, July, 230–250.

Murakami, S., Kamiya, K. (1997): Constitutive and damage evolution equations of elastic-brittle materials based on irreversible thermodynamics, *Int. J. Solids Struct.*, 39, 4, 473–486.

Murakami, S., Hayakawa, K. and Liu, Y. (1998): Damage evolution and damage surface of elastic-plastic-damage materials under multiaxial loading, *Int. J. Damage Mech.*, 7, April, 103–128.

Murzewski, J. (1957): Une theorie statistique du corps fragile quasihomogené, Proc. *IX–e Congrés International de Méchanique Appliqué*, Université de Bruxelles, 5, 313–320.

Murzewski, J. (1958): The tensor of failure and its application to determination of the strength of welded joints, *Bull. Acad. Polon. Sci.*, 6, 3, 159–163.

Murzewski, J. (1992): Brittle and ductile damage of stochastically homogeneous solids, *Int. J. Damage Mech.*, 1, 276–289.

Najar, J. (1994): Brittle residual strain and continuum damage at variable unaxial loading, *Int. J. Damage Mech.*, 3, July, 260–277.

Naumenko, K. (1996): *Modellierung und Berechnung der Langzeitfestigkeit dünwandiger Flächentragwerke unter Einbeziehung von Werkstoffkriechen und Schädigung*, Otto-von-Guericke Universität Magdeburg, Preprint MBI–96–2, April 1996.

Nemat-Nasser, S., Hori, M. (1993): *Micromechanics: Overall Properties of Heterogeneous Materials*, Amsterdam, North-Holland.

Nemes, J.A., Eftis, J., Randles, P.W. (1990): Viscoplastic constitutive modeling of high strain – rate deformation, material damage and spall fracture, *Trans. ASME*, 57, 6, 282–291.

Needleman, A., Tvergaard, V., Giessen, E. (1995): Evolution of void shape and size in creeping solids, *Int. J. Damage Mech.*, 4, 2, 134–152.

Norton, F.H. (1929): *The Creep of Steel at High Temperature,* New York, McGraw-Hill.

Nowacki, W. (1970): *Theory of Elasticity* (in Polish), PWN Polish Scientific Publ., Warsaw.

O'Donnel, M.P., Hurst, R.C., Taylor, D. (1998): Observations of the micromechanisms affecting the fracture path for thermal fatigue-creep loading of a 315L stainless steel, in *Creep and Fatigue Crack Growth in High Temperature Plant*, Proc. Int. HIDA Conf. 15–17 April 1998, Sacley/Paris, S3–23–1 S3–23–4.

Odqvist, F.K.G., Hult, J. (1962): *Kriechfestigkeit metallischer Werkstoffe*, Springer, Berlin.

Odqvist, F.K.G. (1966): *Mathematical Theory of Creep and Creep Rupture*, Oxford Mathematical Monographs, Clarendon Press, Oxford.

Odqvist, F.K.G., Hult, J. (1971): Some aspects of creep rupture, *Archiv for Fysik*, 19, 379.

Ohashi, Y., Ohno, N., Kawai, M. (1982): Evaluation of creep constitutive equations for type 304 stainless steel under repeated multiaxial loadings, *J. Eng. Materials and Technology*, 104, 6, 159–164.

Othman, A.M., Hayhurst, D.R, Dyson, B.F. (1993): Skeletal point stresses in circumferentially notched tension bars undergoing tertiary creep modelled with physically based constitutive equations, *Proc. R. Soc. London*, 441, 343–358.

Penny, R.K., Sim, R.G. (1969): Time dependent creep of plates and pressure containers, *Int. Conf. Pressure Vessels Design*, Delft, 1969.

Penny, R.K., Marriott, D.L. (1995): *Design for Creep*, Chapmani & Hall, London.

Perzyna, P. (1986): Internal state variable description of dynamic fracture of ductile solids, *Int. J. Solids Structures,* 22, 7, 797–818.

Plumtree, A., Lemaitre, J. (1979): Application of damage concepts to predict creep-fatigue failures, *ASME, Press. Vessels and Piping Conf.,* Montreal.

Qi, W. (1998): Modellierung der Kriechschädigung einkriestalliner Superlegierungen im Hochtemperaturbereich, Fortschritt-Berichte VDI, 18: Mechanik/Bruchmechanik, No 230, Düsseldorf.

Qi, W., Bertram, A. (1997): Anisotropic creep damage modeling of single crystal superalloys, *Techn. Mechanik*, 17, 4, 313–332.
Rabotnov, Ju.N. (1963): Damage from creep (in Russian), *Zhurn. Prikl. Mekh. Tekhn. Phys.*, 2, 113–123.
Rabotnov, Ju.N. (1968): Creep rupture, *Proc. 12 Int. Congr. Appl. Mech.*, Stanford, Calif., 342–349.
Rabotnov, Ju.N. (1969): *Creep Problems in Structural Members,* (Engl. trans. by F. A. Leckie), North-Holland, Amsterdam.
Randy, J.Gu., Cozzarelli, F.A. (1988): The strain-controlled creep damage law and its application to the rupture analysis of thick-walled tubes, *Int. J. Non-linear Mechanics*, 23, 2, 147–165.
Rides, M., Cocks, A.C., Hayhurst, D.R. (1989): The elastic response of damaged materials, *J. Appl. Mech.*, 56, 9, 493–498.
Robinson, E.L. (1952): Effect of temperature variation on the long time rupture strength of steel, *Trans. ASME,* 74, 5, 777–780.
Rousselier, G. (1981): Finite deformation constitutive relations including ductile fracture damage, *IUTAM Symp. on Three-Dimensional Constitutive Relations and Ductile Fracture,* Eds. Dourdan, Nemat-Nasser, North-Holland, 331–355.
Rousselier, G. (1986): Les Modéles de Rupture Ductile et Leurs Possibilités Actuelles dans le Cadre de l'Approche Locale de la Rupture, *Sẽminaire International Approches Locales de la Rupture,* Moret-sur-Loing, France.
Rudnicki, J.W., Rice, J.R. (1975): Conditions for the localization of deformation in pressure-sensitive dilatant materials, *J. Mech. Phys. Solids*, 23, 371–394.
Runesson, K., Ottosen, N.S., Peric, D. (1991): Plane stress and strain discontinuous bifurcations in elastic-plastic materials at plane stress and plane strain, *Int. J. of Plasticity*, 27, 99–121.
Rysz, M. (1987): Optimal design of a thick-walled pipeline cross-section against creep rupture, *Acta Mechanica*, 66, 83–102.
Rysz, M., Życzkowski, M. (1988): Optimal design of a thin walled cross-section subject to bending with torsion against creep rupture, *Int. J. Mech. Sci.,* 30, 2, 127–136.
Saanouni, K., Forster, CH., Ben Hatira, F. (1994): On the anelastic flow with damage, *Int. J. Damage Mech.*, 3, April, 140–169.
Sayers, C., Kachanov, M. (1991): A simple technique for finding effective elastic constants of cracked solids for arbitrary crack orientation statistics, *Int. J. Solids Struct.*, 7(6), 671–680.
Schisse, P. (1994): *Ein Beitrag zur Berechnung des Deformationsverhaltens anisotrop geschädigter Kontinua unter Berücksichtigung der thermoplastischen Kopplung*, Ruhr Universität Bochum, Institut für Mechanik, No 89.
Schmitt, J.H., Jalinier, J.M. (1982): *Acta Metall.,* 30, 9, 1789.
Schreyer, H.L., Zhou, S. (1995): A unified approach for predicting material failure and decohesion, AMD–Vol. 200/MD–Vol. 57, *Plastic and Fracture Instabilities in Materials, ASME,* 11, 203–214.

Sdobyrev, V.P. (1959): Criteria of a long term strength for some heat-resisting alloys under combined stress state, *Izv. AN SSSR, Otdel. Tekhn. Nauk, Mekh. Mashinostr.*, 6, 93–99.
Seaman, L., Curran, D.R., Shockey, D.A. (1976): Plasticity theory for porous solids, *Int. J. Mech. Sci.*, 18, 285–291.
Sidoroff, F. (1981): Description of anisotropic damage application to elasticity, in *IUTAM Coll. on Physical Nonlinearities in Structural Analysis*, Springer, Berlin, 237–244.
Simo, J.C., Ju, J.W. (1987): Strain- and stress-based continuum damage models. I – Formulation, II – Computational aspects, *Int. J. Solids Struct.*, 23, 7, 821–869 and 841–869.
Singh, U.K., Digby, P.J. (1989): A continuum damage models for simulation of the progressive failure of brittle rocks, *Int. J. Solids Struct.*, 25, 6, 647–663.
Skoczeń, B. (1996): Generalization of the Coffin equations with respect to the effect of large mean plastic strain, *Journ. of Eng. Materials and Technol.*, ASME, 387–393.
Skrzypek, J.J. (1993): *Plasticity and Creep, Theory, Examples, and Problems,* Ed. R. B. Hetnarski, Begell House – CRC Press.
Skrzypek, J., Egner, W. (1993): On the optimality of disks of uniform creep strength against brittle rupture, *Eng. Opt.*, 21, 243–264.
Skrzypek, J., Ganczarski, A. (1998a): Application of the orthotropic damage growth rule to variable principal directions, *Int. J. Damage Mech.*, 7, Apr., 180–206.
Skrzypek, J., Ganczarski, A. (1998b): Modeling of damage effect on heat transfer in time-dependent non-homogeneous solids, *J. Thermal Stresses*, 21, 3–4, 205–231.
Skrzypek, J.J., Kuna-Ciskał, H., Ganczarski, A. (1998c): On CDM Modelling of Pre- and Post-Critical Failure Models in the Elastic-Brittle Structures, in Beiträge zur Festschrift zum 60. Geburtstag von Prof.Dr.-Ing. Peter Gummert, Mechanik, Berlin, 203–228.
Stigh, U. (1985): *Material Damage and Constitutive Properties*, Chalmers University of Technology, Göteborg, Sweden.
Suquet, P. (1982); *Plasticité et Homogénéisation*, Thése Doctorat d'Etat, Univ. Paris VI.
Szabo, L., Balla, M. (1989): Comparison of some stress rates, *Int. J. Solids Struct.*, 25, 3, 279–297.
Szuwalski, K. (1989): Optimal design of bars under nonuniform tension with respect to ductile creep rupture, *Mech. Struct. and Machines*, 3, 17, 303–319.
Szuwalski, K. (1991a): Optimal design of structural members with respect to ductile creep rupture (in Polish), *Zeszyt Nauk. Polit.* Krak., Monogr. 114, Kraków.
Szuwalski, K. (1991b): Bars of uniform strength vs. optimal with respect to ductile creep rupture time, in *Proc. IUTAM Symp. Creep in Structures IV,* 1990, Ed. M. Życzkowski, Springer, Berlin, 637–643.

Szuwalski, K. (1995a): Optimal design of disks with respect to ductile creep rupture, *Struct. Optim.*, 10, 54–60.
Szuwalski, K. (1995b): Some general theorems on optimality of disks with respect to ductile creep rupture, *Proc. First World Congr. Struct. and Multidisciplinary Optim. (WCSM–1)*, Goslar 1995.
Świsterski, W., Wróblewski, A., Życzkowski, M. (1983): Geometrically nonlinear eccentrically compressed columns of uniform creep strength vs. optimal columns, *Int J. Non–linear Mech.,* 18, 4, 287–296.
Taher, S.F., Baluch, M.H., Al-Gadhib, A.H. (1994): Towards a canonical elastoplastic damage model, *Eng. Fracture Mechanics,* 48, 2, 151–166.
Tetelman, A.S., McEvily, A.J., Jr. (1970): *Fructure of Structural Materials,* John Wiley.
Tohgo, K., Suzuki, N., Ishii, H. (1996): Influence of debonding damage on a crack tip field in particulate–reinforced ductile–matrix composite, 5, 2, 150–170.
Tomkins, B. (1981): *Creep and Fatigue in High Temperature Alloys*, Ed. Bresses, J., Elsevier Applied Science.
Townley, C.H.A., et. al. (1991): High temperature design data for ferritic pressure vessel steels, *Creep of Steel Working Party (CSWP)*, Inst. of Mechanical Eng., J. Mech. E. London.
Trąpczyński, W.A., Hayhurst, D.R., Leckie, F.A. (1981): Creep rupture of copper and aluminium under non-proportional loading, *J. Mech. Phys. Solids*, 29, 5/6, 353–374.
Tvergaard, V. (1981): Material failure by void coalescence in localized shear bands, *DCAMM Report* No. 221, Techn. Univ. of Denmark, Lyngby, 1–26.
Tvergaard, V. (1988): Numerical study of localization in a void sheet, *DCAMM Report* No. 377, Techn. Univ. of Denmark, Lyngby, 1–19.
Vakulenko, A.A., Kachanov, M. L. (1971): Continuum theory of medium with cracks, *Izv. A.N. SSSR, M.T.T.*, 4, 159–166 (in Russian).
Voyiadjis, G.Z., Kattan, P.I. (1992): A plasticity-damage theory for large deformation of solids, Part I: Theoretical formulation, *Int. J. Eng. Sci.*, 30, 9, 1089–1108.
Voyiadjis, G.Z., Venson, A. R. (1995): Experimental damage investigation of a SiC–Ti aluminum metal matrix composite, *Int. J. Damage Mach.*, 4, Oct. 1995, 338–361.
Voyiadjis, G.Z., Park, T. (1996): Anisotropic damage for the characterization of the onset of macro-crack initiation in metals, *Int. J. Damage Mech.*, 5, 1, 68–92.
Voyiadjis, G.Z., Park, T. (1998): Kinematics of large elastoplastic damage deformation, in *Damage Mechanics in Engineering Materials*, Ed. G.Z. Voyiadjis, J.-W.W. Ju, J.-L. Chaboche, Elsevier Science, Amsterdam, 45–64.
Wang, J. (1992): Low cycle fatigue and cycle dependent creep with continuum damage mechanics, *Int. J. Damage Mech.*, 1, April, 237.
Woo, C.W., Li, D.L. (1993): A universal physically consistent definition of material damage, *Int. J. Solids Structures*, 30, 15, 2097–2108.

Wróblewski, A. (1989): On optimality of eccentrically compressed column under creep condition, *Rozpr. Inż.*, 37, 2, 269–281 (in Polish).

Yokobori, A.T., Shibata, M., Tabuchi, M., Fuji, A. (1998): Comparative study of the estimation of creep crack growth behaviour of TiAl by using precrack and a notch CT specimens, in *Creep and Fatigue Crack Growth in High Temperature Plant*, Proc. Int. HIDA Conf. 15–17 April 1998, Sacley/ Paris, S1–9–1 S1–9–6.

Zang, P., Lee, H. (1993): Creep damage and fracture at high temperature, *Eng. Fracture Mechanics*, 44, 2, 283–288.

Zheng, Q.-S., Betten, J. (1996): On damage effective stress and equivalence hypothesis, *Int. J. Damage Mech.*, 5, 3, 219–240.

Żuchowski, R. (1986): *Analysis of failure processes of metals under conditions of thermal fatigue*, Sci. Pap. Inst. Mat. Sci. and Appl. Mech., Techn. Univ. Wrocław, Monographs 18.

Życzkowski, M., Świsterski, W. (1980): Optimal design of flexible beams with respect to creep rupture time, *Proc. IUTAM Symp. Structural Control*, Waterloo 1979, North-Holland, 795–810.

Życzkowski, M., Rysz, M. (1986): Optimization of cylindrical shells under combined loading against brittle creep rupture, in Proc. IUTAM Symp. *Inelastic Behaviour and Shells,* Rio de Janeiro 1985, Springer, Berlin, 385–401.

Życzkowski, M. (1988): Optimal structural design under creep conditions, *Appl. Mech. Reviews,* 41, 12, 453–461.

Życzkowski, M. (1991): Problems of structural optimization under creep conditions (1), in Proc. *IUTAM Symp. Creep in Structures IV,* 1990, Ed. M. Życzkowski, Springer , Berlin, 519–530.

Życzkowski, M. (1996): Optimal structural design under creep conditions (2), *Appl. Mech. Reviews*, 49, 9, 433–446.

Index

MIX
Papier aus verantwortungsvollen Quellen
Paper from responsible sources
FSC® C105338

If you have any concerns about our products,
you can contact us on
ProductSafety@springernature.com

In case Publisher is established outside the EU,
the EU authorized representative is:
Springer Nature Customer Service Center GmbH
Europaplatz 3, 69115 Heidelberg, Germany

Printed by Libri Plureos GmbH
in Hamburg, Germany